Feynman
and
Computation

Other Books by Anthony J. G. Hey:

Einstein's Mirror (with Patrick Walters)
The Quantum Universe (with Patrick Walters)
Gauge Theories in Particle Physics (with I.J.R. Aitchison)
Feynman Lectures on Computation (edited with Robin W. Allen)

Feynman
and
Computation

Exploring

the Limits of

Computers

Edited by Anthony J.G. Hey

Advanced Book Program

CRC Press
Taylor & Francis Group
Boca Raton London New York

CRC Press is an imprint of the
Taylor & Francis Group, an **informa** business

The Advanced Book Program

Originally published in 1999 as a hardcover by Perseus Books Publishing

Published 2002 by Westview Press

Published 2018 by CRC Press
Taylor & Francis Group
6000 Broken Sound Parkway NW, Suite 300
Boca Raton, FL 33487-2742

CRC Press is an imprint of the Taylor & Francis Group, an informa business

Visit the Taylor & Francis Web site at
http://www.taylorandfrancis.com

and the CRC Press Web site at
http://www.crcpress.com

A Cataloging-in-Publication data record for this book is available
from the Library of Congress.

Cover design by Suzanne Heiser
Set in LaTex by the editor Anthony J. G. Hey

ISBN 13: 978-0-8133-4039-5 (pbk)

Contents

IV: Parallel Computation

V: Fundamentals

Contributors

Dr Charles Bennett
IBM Thomas J Watson Research Center
PO Box 218
Yorktown Heights
New York 10598, USA

Dr Paul Benioff
Physics Division
Argonne National Laboratory, Argonne
IL 60439, USA

Professor Geoffrey Fox
University of Syracuse
111 College Place, Syracuse
New York 13244-4100, USA

Ed Fredkin
Distinguished Service Professor
School of Computer Science
Carnegie Mellon University
5000 Forbes Avenue, Pittsburgh
PA 15213, USA

W. Daniel Hillis
Walt Disney Imagineering
1401 Flower Street
P.O.Box 5020, Glendale
CA 91221, USA

Professor John Hopfield
Department of Molecular Biology
Pinceton University, Princeton
NJ 08544-1014, USA

Dr Richard Hughes
Physics Division
Los Alamos National Laboratory
Los Alamos
NM 87545, USA

Dr Rolf Landauer
IBM Thomas J Watson Research Center
Yorktown Heights
New York 10598, USA

Professor Norman Margolus
Laboratory for Computer Science
Massachusetts Institute of Technology
NE43-428, Cambridge
MA 02139, USA

Professor Carver Mead
California Institute of Technology
Moore Laboratory, Pasadena
CA 91125, USA

Professor Marvin Minsky
Media Labs
Massachussets Institute of Technology
Cambridge
MA 02139, USA

Professor Gerry Sussman
Artificial Intelligence Laboratory
Massachussets Institute of Technology
Cambridge
MA 02139, USA

Professor Tommaso (Tom) Toffoli
ECE Dept
Boston University, Boston
MA 02215, USA

Professor J A Wheeler
Physics Department
Princeton University, Princeton
New Jersey, USA

Professor Wojciech H Zurek
Theoretical Division
Los Alamos National Laboratory
Los Alamos
NM 87545, USA

Feynman and Computation

An Overview

Introduction

When the *Feynman Lectures on Computation* were finally published in September 1996, I promised a complementary volume that would address the 'advanced topics' covered in Feynman's course but excluded from the published version. Over the years that Feynman taught the course, he invited guest lectures from experts like Marvin Minsky, Charles Bennett, John Cocke and several others. Since these lectures covered topics at the research frontier, things have moved on considerably in most areas during the past decade or so. Thus, rather than attempt to transcribe the often incomplete records of these original lectures, it seemed more relevant, appropriate and exciting to invite the original lecturers to contribute updated versions of their lectures. In this way, a much more accurate impression of the intellectual breadth and stimulation of Feynman's course on computation would be achieved. In spite of some change to this philosophy along the way, I am satisfied that this book completes the picture and provides a published record of Feynman's long-standing and deep interest in the fundamentals of computers. The contributions are organised into five sections whose rationale I will briefly describe, although it should be emphasized that there are many delightful and intriguing cross links and interconnections between the different contributions.

Feynman's Course on Computation

The first part is concerned with the evolution of the Feynman computation lectures from the viewpoint of the three colleagues who participated in their construction. The contributions consist of brief reminiscences together with a reprint from each on a topic in which they had shared a mutual interest with Feynman. In 1981/82 and 1982/83, Feynman, John Hopfield and Carver Mead gave an interdisciplinary course at Caltech entitled 'The Physics of Computation.' The different memories that John Hopfield and Carver Mead have of the course make interesting reading.

Feynman was hospitalized with cancer during the first year and Hopfield remembers this year of the course as 'a disaster,' with himself and Mead wandering 'over an immense continent of intellectual terrain without a map.' Mead is more charitable in his remembrances but both agreed that the course left many students mystified. After a second year of the course, in which Feynman was able to play an active role, the three concluded that there was enough material for three courses and that each would go his own way.

The next year, 1983/84, Gerry Sussman was visiting Caltech on sabbatical leave from MIT intending to work on astrophysics. Back at MIT, Sussman supervised Feynman's son, Carl Feynman, as a student in Computer Science, and at Caltech, Feynman had enjoyed Abelson and Sussman's famous 'Yellow Wizard Book' on 'The Structure and Interpretation of Computer Programs.' Feynman therefore invited Sussman to lunch at the Athenaeum, the Caltech Faculty Club, and agreed a characteristic 'deal' with him — Sussman would help Feynman develop his course on the 'Potentialities and Limitations of Computing Machines' in return for Feynman having lunch with him after the lectures. As Sussman says, 'that was one of the best deals I ever made in my life.'

Included with these reminiscences are three reprints which indicate the breadth of Feynman's interests — Hopfield on the collective computational properties of neural networks, Mead on an unconventional approach to electrodynamics without Maxwell, and Sussman and Wisdom on numerical integrations of the orbit of Pluto, carried out on their 'digital orrery'.

Reducing the Size

Part 2 is concerned with the limitations due to size. The section begins with a reprint of Feynman's famous 1959 lecture 'There's Plenty of Room at the Bottom', subtitled 'an invitation to enter a new field of physics'. In this astonishing lecture, given as an after-dinner speech at a meeting of the American Physical Society, Feynman talks about 'the problem of manipulating and controlling things on a small scale', by which he means the 'staggeringly small world that is below'. He goes on to speculate that 'in the year 2000, when they look back at this age, they will wonder why it was not until the year 1960 that anybody began seriously to move in this direction'. In this talk Feynman also offers two prizes of $1000 — one 'to the first guy who makes an operating electric motor... [which] is only 1/64 inch cube', and a second 'to the first guy who can take the information on the page of a book and put it on an area 1/25,000 smaller in linear scale in such a manner that it can be read by an electron microscope.' He paid out on both — the first, less than a year later, to Bill McLellan, a Caltech alumnus, for a miniature motor which satisfied the specifications but which was somewhat of a disappointment to Feynman in that it required no new technical advances. Feynman gave an updated version of his talk in 1983 to the Jet Propulsion Laboratory. He predicted 'that

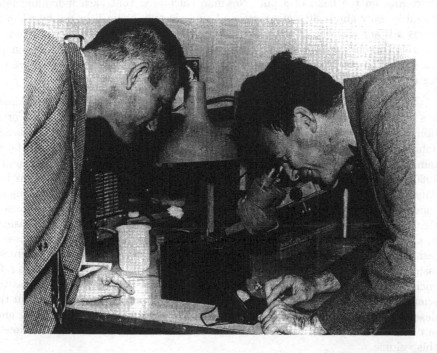

Fig. 1. Feynman examining Bill McLellan's micromotor. The motor was six thousandths of an inch in diameter and could generate one millionth of a horsepower. By the time McLellan brought his motor for Feynman to examine for the prize, there had been many other would-be inventors anxious to show Feynman their versions of a micromotor. Feynman knew at once that McLellan was different — he was the only one to have brought a microscope to view the motor.

with today's technology we can easily... construct motors a fortieth of that size in each dimension, 64,000 times smaller than... McLellan's motor, and we can make thousands of them at a time.'

It was not for another 26 years that he had to pay out on the second prize, this time to a Stanford graduate student named Tom Newman. The scale of Feynman's challenge was equivalent to writing all twenty-four volumes of the *Encyclopedia Brittanica* on the head of a pin: Newman calculated that each individual letter would be only about fifty atoms wide. Using electron-beam lithography when his thesis advisor was out of town, he was eventually able to write the first page of Charles Dickens' *A Tale of Two Cities* at 1/25,000 reduction in scale. Feynman's paper is often credited with starting the field of nanotechnology and there are now regular 'Feynman Nanotechnology Prize' competitions.

The second chapter in this section is contributed by Rolf Landauer, who himself has made major contributions to our understanding of computational and informational limits. Here, Landauer discusses a seminal paper by his late IBM colleague John Swanson, which addressed the question of 'how much memory could be obtained from a given quantity of storage material'. Swanson's paper appeared in 1960, around the same time as Feynman's 'Room at the Bottom' paper. In Landauer's opinion, 'Feynman's paper, with its proposal of small machines making still smaller machines, was that of a supremely gifted visionary and amateur; Swanson's that of a professional in the field.' Landauer also deplores the impact of fashions in science — while acknowledging that Feynman 'was very far from a follower of fashions'. Nevertheless, such was Feynman's influence that he could very often *start* fashions, and an unfortunate side-effect of his somewhat cavalier attitude to referencing relevant prior work — that he himself had not needed to read — was that scientists such as Rolf Landauer and Paul Benioff did not always get the credit they deserved. This was an unintended side-effect of Feynman's way of working and I am sure that Feynman would have approved of their thoughtful contributions to this volume.

The third chapter in this section is a reprint of a paper by Carver Mead, in which he revisits his semiconductor scaling predictions that formed the basis of Moore's Law. In 1968, Gordon Moore had asked Mead 'whether [quantum] tunneling would be a major limitation on how small we could make transistors in an integrated circuit'. As Mead says, this question took him on a detour that lasted nearly 30 years. Contrary to all expectations at the time, Mead found that the technology could be scaled down in such a way that everything got better — circuits would run faster and take less power! Gordon Moore and Intel have been confirming Mead's prediction for the last 30 years.

In the last part of this section, Marvin Minsky updates and reflects upon his 1982 paper on Cellular Vacuum — together with some thoughts about Richard Feynman and some trenchant comments about the importance of the *certainties* of quantum mechanics. Minsky also recalls Feynman's suspicions of continuous functions and how he liked the idea that space-time might in fact be discrete: 'How could there possibly be an infinite amount of information in any finite volume?'

Quantum Limits

Computational limitations due to quantum mechanics is the theme of the next section. There is no better place to begin than with a reprint of Feynman's famous paper in which he first suggested the possibility of a 'quantum computer'. 'Simulating Physics with Computers' was given as a 'keynote speech' at a 1981 conference at MIT on the 'Physics of Computation', organized by Ed Fredkin, Rolf Landauer and Tom Toffoli. At the conference, after claiming not to 'know what a keynote speech is', Feynman proceeded to give a masterful keynote presentation. In his talk, he credited his entire interest in the subject to Ed Fredkin and thanked him for 'wonderful, intense and interminable arguments'. Feynman begins by discussing the question of whether a universal computer can simulate physics *exactly* and then goes on to consider whether a 'classical' computer can efficiently simulate quantum mechanics and its quantum probabilities. Only Feynman could discuss 'hidden variables', the Einstein-Podolsky-Rosen paradox and produce a proof of Bell's Theorem, without mentioning John Bell. In fact, the paper contains no references at all — but it does contain the idea of simulating a quantum system using a new type of non-Turing, 'quantum computer.' It is also interesting to see Feynman confessing that he's 'not sure there's no real problem' with quantum mechanics.

The next three chapters are by three of the leaders of the new research fields of quantum information theory and quantum computing. Paul Benioff, like Rolf Landauer, has probably received insufficient credit for his pioneering contributions to the field. In his paper on 'Quantum Robots', he explores some of the self-referential aspects of quantum mechanics and considers the quantum mechanical description of robots to carry out quantum experiments.

Charles Bennett, famous for his resolution of the problem of Maxwell's Demon and for his demonstration of the feasibility of reversible computation, has made important contributions both to the theory of quantum cryptography and quantum teleportation. In a wonderful advertisement, shown to me gleefully by Rolf Landauer, IBM Marketing Department went overboard on Bennett's work on teleportation. Invoking images of 'Star Trek', the ad proclaimed "An IBM scientist and his colleagues have discovered a way to make an object disintegrate in one place and reappear intact in another" An elderly lady pictured in the ad talking on the telephone to a friend says "Stand by. I'll teleport you some goulash." Her promise may be 'a little premature,' the ad says, but 'IBM is working on it.' Charles Bennett was embarrassed by these claims and was later quoted as saying "In any organization there's a certain tension between the research end and the advertising end. I struggled hard with them over it, but perhaps I didn't struggle hard enough." His

November 15, 1960

Mr. William H. McLellan
Electro-Optical Systems, Inc.
125 North Vinedo Avenue
Pasadena, California

Dear Mr. McLellan:

        I can't get my mind off the fascinating motor you
showed me Saturday. How could it b e made so small?

        Before you showed it to me I told you I hadn't
formally set up that prize I mentioned in my Engineering and
Science article. The reason I delayed was to try to formulate
it to avoid legal arguments (such as showing a pulsing mercury
drop controlled by magnetic fields outside and calling it a
motor), to try to find some organization that would act as judges
for me, to straighten out any tax questions, etc. But I kept
putting it off and never did get around to it.

        But what you showed me was exactly what I had had in
mind when I wrote the article, and you are the first to show me
anything like it. So, I would like to give you the enclosed
prize. You certainly deserve it.

        I am only slightly disappointed that no major new
technique needed to be developed to make the motor. I was sure
I had it small enough that you couldn't do it directly, but you
did. Congratulations!
        Now don't start writing small.
        I don't intend to make good on the other one. Since
writing the article I've gotten married and bought a house!

                                        Sincerely yours,

                                        Richard P. Feynman

RPF:n

Fig. 2. The letter from Feynman to McLellan acompanying the $1,000 cheque. In the letter
Feynman admits to some disappointment that he had not specified the motor small enough
to require some new engineering techniques — but was honest enough to acknowledge that
McLellan had made exactly the sort of motor he had had in mind when he issued the
challenge.

paper in this volume discusses recent developments in quantum information theory including applications of 'quantum entanglement' — a term used by Schrödinger as long ago as 1935 — and possible 'entanglement purification' techniques.

The last chapter in this section is by Richard Hughes, who was at Caltech with Feynman in 1981 when I was there on sabbatical. Hughes now leads a multidisciplinary research team at Los Alamos National Laboratory that has constructed working quantum cryptographic systems. His contribution surveys the fundamentals of quantum algorithms and the prospects for realising quantum computing systems using ion trap technology.

Parallel Computation

Feynman's first involvement with parallel computing probably dates back to his time at Los Alamos during the Manhattan Project. There was a problem with the 'IBM group', who were performing calculations of the energy released for different designs of the plutonium implosion bomb. At this date in 1944, the IBM machines used by the IBM group were not computers but multipliers, adders, sorters and collators. The problem was that the group had only managed to complete three calculations in nine months prior to Feynman taking charge. After he assumed control, there was a complete transformation and the group were able to complete nine calculations in three months, three times as many in a third of the time. How was this done? As Feynman explains in *Surely You're Joking, Mr. Feynman*, his team used parallel processing to allow them to work on two or three problems at the same time. Unfortunately, this spectacular increase in productivity resulted in management assuming that a single job took only two weeks or so — and that a month was plenty of time to do the calculation for the final Trinity test configuration. Feynman and his team then had to do the much more difficult task of figuring out how to parallelise a single problem.

During the 1980's, Feynman became familiar with two pioneering parallel computing systems — the Connection Machine, made by Thinking Machines Corporation in Boston, and the Cosmic Cube, built by Geoffrey Fox and Chuck Seitz at Caltech. Parallel computing was one of the 'advanced topics' discussed in the lecture course and both types of parallel architecture — SIMD or Single Instruction Multiple Data, exemplified by the Connection Machine, and MIMD or Multiple Instruction Multiple Data, exemplified by the Cosmic Cube — were analysed in some detail. Parallel computing was in its infancy in the early 1980's, and in the first chapter of this next section Feynman talks optimistically of the future for parallel computing. This chapter is a reprint of a little-known talk he gave in Japan as the 1985 Nishina Memorial Lecture. In addition to discussing possible energy consumption problems and size limitations of future computers, Feynman is very positive about the role for parallel computing in the future.

By contrast, in the next chapter, Geoffrey Fox reflects on the failure of parallel

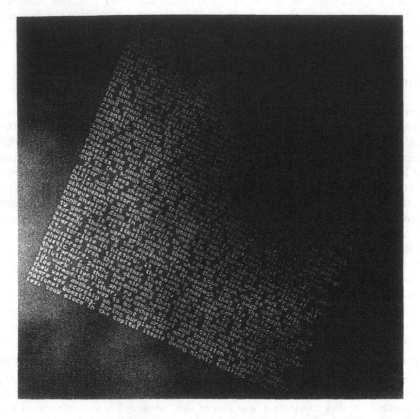

Fig. 3. Tom Newman programmed the electron beam lithography machine to write out the first page of the novel *A Tale of Two Cities* by Charles Dickens. The reduction in size is 25,000 to 1 and each letter is only about 50 atoms wide.

computing and computational physics to become a major focus for growth over the last ten years. In his view, the problem is not that parallel computing cannot be made to work effectively for many types of scientific problems. Instead, the outstanding problem is that the size of the parallel computer market has been insufficient to solve the difficult problem of developing high quality, high-level parallel programming environments that are both easy to use and also offer a straightforward migration path for users with a significant investment in existing sequential software. Feynman's optimistic suggestion that 'programmers will just have to learn how to do it,' while true for 'Grand Challenge' scientific problems, has not yet come true in a commercial sense. Fox offers an alternative vision of the future, encapsulated by the term 'Internetics'. It is in the context of the World Wide Web and commodity hardware and software, that parallel computing has a commercially viable future.

In the third chapter of this section, Feynman's first-hand involvement with parallel computing is chronicled by parallel computer pioneer Danny Hillis. Feynman's son Carl, then an undergraduate at MIT, was helping Hillis with his ambitious thesis project to design a new type of parallel computer powerful enough to solve common sense reasoning problems. Over lunch, one day in the spring of 1983, Hillis told Feynman he was founding a company to build this machine. After saying that this was 'positively the dopiest idea I ever heard', Feynman agreed to work as a consultant for the new company. As Hillis recounts, when Feynman was told the name of the company 'Thinking Machines Corporation' he was delighted. "That's good. Now I don't have to explain to people that I work with a bunch of loonies. I can just tell them the name of the company." What shines through the article by Hillis is Feynman's need to be involved with the details — with the implementation of Hopfield's neural networks, with a clever algorithm for computing a logarithm, and with Quantum Chromo-Dynamics using a parallel-processing version of BASIC he had devised. Feynman's triumph came with the design of the message router that enabled the 64,000 processors of the machine to communicate. Using an analysis based on differential equations, he had come up with a more efficient design than that of the engineers who had used conventional discrete methods in their analysis. Hillis describes how engineering constraints on chip size forced them to set aside their initial distrust of Feynman's solution and use it in the final machine design.

One of the earliest applications on the Connection Machine was John Conway's 'Game of Life', which is an example of a cellular automaton model. Feynman was always interested in the idea that down at the bottom, space and time might actually be discrete. What we observe as continuous physics might be merely the large-scale average of the behaviour of vast numbers of tiny cells. In the final contribution to this section, Norman Margolus, who gave two lectures in Feynman's original course — one on reversible logic and billiard ball computers, and a second on cellular automata — updates these ideas and explores them in detail. Margolus describes how he and Tom Toffoli built several generations of 'SIMD' cellular automata computers but does not include his account of one of Feynman's not-so-epic adventures. Margolus had given Feynman one of their cellular automata computers for his own use and as a result, Feynman had gone out to buy a color monitor, to better display the 'live' simulations. On leaving the store Feynman tripped and bumped his head badly. Although he did not think it serious at the time, he became unable to drive a car safely or even answer basic questions in his famous Physics-X class at Caltech. Apparently, although alarmed by these symptoms, Feynman put them down to his getting old and 'losing it.' It was some time before the doctors understood what the problem was and drilled a hole in his head to relieve the pressure so that Feynman became smart again! Losing your intelligence is not such a funny adventure. I am also grateful to Norman for his suggestion to include Feynman's paper on 'Simulating Physics with Computers' in this volume.

Fundamentals

The last part is entitled 'Fundamentals' and leads off with a reprint of John Archibald Wheeler's paper on 'Information, Physics, Quantum — The Search for Links'. As Rolf Landauer has said, 'Wheeler's impact on quantum computation has been substantial — through his papers, his involvement in meetings, and particularly through his students and associates.' Feynman was an early student of Wheeler, of course, and so was Wojciech Zurek, also a contributor to this volume. In Zurek's view, the paper by Wheeler, first published in 1989, is 'still a great, forward-looking call to arms'. The credo of the paper is summarized by the slogan *It from Bit* — the hypothesis that every item of the physical world, be it particle or field of force, ultimately derives its very existence from apparatus-solicited answers to binary, yes/no questions.

Another influential figure in the computational community is Ed Fredkin, who first met Feynman in 1962. Fredkin and Marvin Minsky were in Pasadena with nothing to do one evening and they 'sort of invited themselves' to Feynman's house. The three discussed many things until the early hours of the morning and, in particular, the problem of whether a computer could perform algebraic manipulations. Fredkin credits the origin of MIT's MACSYMA algebraic computing project to that discussion in Pasadena. In his chapter, Fredkin discusses his time at Caltech as a Fairchild Scholar in 1974, and his preoccupation with reversible dynamics. The deal this time was that Feynman would teach Fredkin quantum mechanics and Fredkin would teach Feynman computer science. Fredkin believes he got the better of the deal: 'It was very hard to teach Feynman something because he didn't want to let anyone teach him anything. What Feynman always wanted was to be told a few hints as to what the problem was and then to figure it out for himself. When you tried to save him time by just telling him what he needed to know, he got angry because you would be depriving him of the satisfaction of discovering it for himself.' Besides learning quantum mechanics, Fredkin's other assignment to himself during this year was to understand the problem of reversible computation. They had a

4 Part II/Wednesday, July 30, 1986

Los Angeles Times
A Times Mirror Newspaper

Small Wonder

In the Middle Ages scholars wondered how many angels would fit on the head of a pin. This puzzle was never satisfactorily answered. But as the result of a recent technological advance at Stanford University we now know that the entire Encyclopaedia Britannica would comfortably fit there.

In 1960 Richard Feynman, the Caltech physicist, offered a $1,000 prize to anyone who could make a printed page 25,000 times smaller while still allowing it to be read. A Stanford graduate student, Tom Newman, has now done it, and Feynman has paid him the grand.

Newman's technique is based on the same technology that is used to imprint electronic circuits on those tiny computer chips that are everywhere. Newman uses several electron beams to trace letters made up of dots that are 60 atoms wide. The resulting text can be read with an electron microscope.

Some technological advances bring instant rewards to humanity, while some have no practical use—at least for the moment. They are just amazing. In the latter category, chalk one up for Tom Newman, with an assist from Richard Feynman.

Fig. 4. A Los Angeles Times article reporting the winner of the second Feynman challenge, more than 26 years after Feynman first announced the prize. Tom Newman rang up Feynman to ask if the challenge was still open and then completed the project when his thesis advisor was away from Stanford.

wonderful year of creative arguments and Fredkin invented Conservative Logic and the 'Fredkin Gate' — which led to the billiard ball computer. During one of their arguments Feynman got so exasperated that he broke off the argument and started to quiz Fredkin about quantum mechanics. After a while he stopped the quiz and said "The trouble with you is not that you don't understand quantum mechanics."

The last two contributions are by two of the major figures in the field. The first is by Tom Toffoli, who helped organise the 1981 MIT conference on the *Physics of Computation* at which Feynman spoke. Toffoli is also credited by Feynman 'for his help with the references' in his *Optics News* paper on quantum computers, reprinted in the *Lectures on Computation*. Toffoli has had a long-standing interest in cellular automata and, using the idea of a fine-grained dynamical substrate underlying the observed dynamics, he speculates on a possible link between 'action integrals' in physics and the concept of 'computation capacity.' On this view, the 'principle of least action', so often used by Feynman, is 'an expression not of Nature's *parsimony* but of Nature's *prodigality*: A system's natural trajectory is the one that will hog most computational resources.'

The final chapter is by Wojciech Zurek whose initial interest in the subject of physics and information was stimulated by John Wheeler. It was also Wheeler who insisted that Zurek maintain a regular dialogue with Feynman when Zurek was appointed as a Tolman Fellow at Caltech in 1981. In his article, Zurek goes beyond the seminal work of Landauer and Bennett in exploring any threat to the second law of thermodynamics posed by an intelligent version of Maxwell's Demon. The intellectual abilities of the demon are assumed to be equivalent to those of a universal Turing machine and the notion of algorithmic information content provides a measure of the storage space required to describe the system.

Feynman Stories

In this eclectic collection of papers on computation and information theory, the reader will find a number of 'new' Feynman stories. Murray Gell-Mann, his long-time colleague at Caltech, always deplored the way Feynman 'surrounded himself with a cloud of myth' and the fact that 'he spent a great deal of time and energy generating anecdotes about himself'. In fact, I think the stories generate themselves. For example, in 1997 Ed Fredkin came to Southampton to help us celebrate the 50th anniversary of our Department of Electronics and Computer Science — as far as we know the first, specifically 'electronics' department in the world. Ed gave a talk which formed the basis for his contribution to this volume, but he also told a Feynman story that does not appear in his written version. With apologies to Ed, I would like to tell it here.

The story concerns the so-called 'twin paradox' in relativity. In his book, Feynman had written "You can't make a spaceship clock, by any means whatsoever, that keeps time with the clocks at home." Now Fredkin happened to be teaching a course and this subject came up. In thinking about the paradox, Fredkin came up with a trivial way to make a spaceship clock that *did* keep time with the clock at home. Before making a fool of himself in front of his students, Fredkin thought he'd check with Feynman first. There was, of course, an ulterior motive for doing this and that was to 'sandbag' Feynman — a thing that Fredkin loved to do but rarely succeeded. The telephone conversation went something like this. Fredkin said "It says in your book that it is impossible for a clock on the spaceship to keep time with a clock at home. Is that correct?" Feynman replied "What it says in the book is absolutely correct." Having set him up, Fredkin countered "OK, but what if I made a clock this way ..." and then proceeded to describe how his proposed clock had knowledge of the whole trajectory and could be programmed to put the 'back home' time on the face of the clock. "Wouldn't that keep time with the clocks back home?" Feynman said "That is absolutely correct." Fredkin replied "Then what does that mean about what's in your book?" Feynman's instant response was "What it says in the book is absolutely wrong!"

Anyone who has had any long-term contact with Feynman will have a fund of stories such as this one. In all the things he did, Feynman was never afraid to admit he was mistaken and he constantly surprised his audience with his direct and unconventional responses. In this way, many of the Feynman stories generate themselves without any overt act of creation by Feynman himself.

Research and Teaching

What these anecdotes, and what these chapters illustrate, is how intimately research and teaching were blended in Feynman's approach to any subject. Danny Hillis remembers how Feynman worked on problems at Thinking Machines. While he was engaged in solving a problem he hated to be interrupted, but once he had found a solution 'he spent the next day or two explaining it to anyone who would listen.' Explanation and communication of his understanding were an essential part of Feynman's methodology. He also had no problem about the fact that he was sometimes recreating things that other people already knew — in fact I don't think he could learn a subject any other way than by finding out for himself.

Carver Mead, however, remembers another, more combative side to Feynman. Besides improving his skills on integrals and numerology in duels with Hans Bethe, the hot-house atmosphere of Los Alamos during the war had honed Feynman's skills in argument: 'The one who blinked first lost the argument.' As Mead says, 'Feynman learned the game well — he never blinked.' For this reason, Feynman would never say what he was working on: He preferred 'to spring it, preferably in front of an audience, after he had it all worked out.' Mead learnt to tell what problems Feynman cared about by noticing which topics made him mad when they were brought up. Furthermore, Mead goes on to say, if Feynman was stuck about something, 'he had a wonderful way of throwing up a smoke screen' which Mead calls Feynman's "proof by intimidation."

Feynman's grasp of the big picture, coupled with his love for knowing first-hand of practical details — from low-level programming to lock-picking — gave him an almost unique perspective on any subject he chose to study. It was this mastery, both of the minutiae of a subject and of its broad intellectual framework, that gave him the seemingly effortless ability to move back and forth between the two levels at will, without getting lost in the detail or losing the overall plot.

How to be an Editor

Feynman was an inspiring teacher who declined the 'easy' option of giving the same course every year. He chose to spend a large part of the last decade of his life thinking about the fundamentals of computation. Stan Williams, who works at Hewlett-Packard on nanostructures, quotes me as saying that the *Feynman Lectures on Computation* were the most important thing I have done in my career. Now I am not sure that I quite said that, but it *is* true that I am glad his last lectures have seen the light of day. Furthermore, with this volume, the links and connections with the people in the computational community that he was inspired by, or who were inspired by him, are recorded.

When I took on the job of putting together this second volume, I fondly imagined it would be easier than constructing the first from rough notes and tapes. I little knew what skills an 'editor' requires. Getting agreement in principle for a contribution is easy: Getting the contribution in reality is much more difficult. Some examples will make the point. Marvin Minsky was wonderfully encouraging about the project initially — but I felt bad at having to telephone Marvin at his home at regular intervals, badgering him for his paper. Gerry Sussman daily demonstrates an incredible breadth and depth of knowledge, on subjects ranging from programming in SCHEME to the foundations of classical mechanics. On talking with him and Tom Knight at MIT, he described their current research project by holding up his hand and saying "I want to know how to program this." It is therefore not surprising that I found it difficult to intrude on his manifold activities and persuade him to set them aside for the time required to complete his brief contribution to this volume. I'm glad he did, since his contribution to Feynman's course was certainly worthy of acknowledgement.

A special note of thanks is owing to Rolf Landauer: He not only was first to deliver his text but he was also wise enough to apply subtle pressure on me to complete the task. This he did by telling me he had no doubts about my skills to put together an exciting volume. There certainly were times when I doubted whether I would be able to persuade Charles Bennett to devote enough time to write up the talk he had given at our Southampton Electronics celebrations. Since Charles was one of those who had been responsible for educating Feynman about the field, and had participated in the original lecture course, I felt it was important to persevere. Finally, I hit on the idea of telling him that his colleague, Rolf Landauer, did not think he would make my final, final deadline ... And of course, I should thank not only those who gave me worries but delivered, but also those who delivered on time, to a tight schedule. To all of them, many thanks — I hope you like the result.

Acknowledgements

Some final acknowledgements are in order. The 1997 conference at Southampton on 'The Future of Computing', at which Charles Bennett, Ed Fredkin and Richard Hughes all spoke, was sponsored in part by the UK Engineering and Physical Sciences Research Council. My co-editor from the *Lectures on Computation*, Robin Allen, assisted with some of the figures, and David Barron gave me useful advice concerning combining Word and LaTeX documents.

In the event, in the interest of uniformity, the book has been produced entirely in LaTeX and Tim Chown, our network and LaTeX guru at Southampton, has worked tirelessly and cheerfully to achieve this goal. Thanks also to Jeff Robbins, my editor at what is now called Perseus Books (formerly Addison Wesley Longman) for his faith that I would deliver something of interest; and to Helen Tuck for the loan of $8 and for her valued friendship.

And finally, a belated acknowledgement to my wife, Jessie Hey: Some of the material in the 'Afterword' in the *Feynman Lectures on Computation* derived from an unpublished article authored by her in 1975 entitled 'The Art of Feynman's Science.' With Robin Allen, Jessie also assisted in preparing the index for this volume. Many thanks to her for this, and for her continuing patience, interest and support in the production of both these volumes.

Tony Hey

Southampton, September 1998

Part I

Feynman's Course on Computation

FEYNMAN AND COMPUTATION

John J. Hopfield

In early 1981, Carver Mead and I came up with an idea for a year-long course on computation as a physical process, trying to consider digital computers, analog computers, and the brain as having things in common. Carver thought that Richard Feynman would be interested in joining the enterprise, and indeed he was, so the three of us did a little planning. At that juncture, Feynman had one of his heroic bouts with cancer, and Mead and I went on alone.

The course was a disaster. Lectures from different visitors were not coordinated, we wandered over an immense continent of intellectual terrain without a map, the students were mystified and attendance fell off exponentially. Carver and I learned a great deal about the physics of computation. More important, we learned never, never to try such a thing again.

However, when I ran across a recovered Feynman in October 1982, he immediately asked "Hey, what happened to that course me and you and Mead were going to give?" I gave a terse and charitable interpretation of events, to which Feynman immediately asked (Mead being on leave from teaching at the time) whether he and I could try it again, saying he would really do his share and more importantly, organize the subject his way. So it was that his course on the physics of computation was born, in the winter of 1983. Mead was also dragooned into lecturing a bit. I will recount a little from this era, the class, and our lunchtime meetings to talk about the subject.

The format consisted of two lectures a week. The first of these lectures was usually from an outside lecturer. Marvin Minsky led off. The second lecture in a week was a critique by Feynman of the first lecture. He would summarize the important points from the first lecture with his own organization, integrating the lecture into the rest of science and computer science in Feynman's inimitable fashion. Occasionally, this would become a lecture on what the speaker should have said if he had really understood the essence of his subject. And of course, the students loved it.

There were three basic aspects of computers and computation which intrigued Feynman, and made the subject sufficiently interesting that he would spend time teaching it. First, what limits did the laws of physics place on computation? Second, why was it that when we wanted to do a hard problem a little better, it always seemed to need an exponentially greater amount of computing resources? Third, how did the human mind work, and could it somehow get around the second problem?

Feynman himself did the lecture on quantum computation. Charles Bennett had already prepared the way by talking about reversible computation, and Norman Margolus on the billiard ball computer, so Feynman developed quantum computation in ways which made its connection to classical reversible computers obvious. His lecture was chiefly developing the idea that an entirely reversible quantum system, without damping processes, could perform universal computation. I gave the second lecture, chiefly on the subject of restoration, since Feynman's quantum computer was built on the basis of a Hamiltonian which he had designed, and whose parameters were mathematically exact. I pressed hard on the fact that all real digital computers had noise and errors in construction, that ways of restoring the signals after such errors were essential, and that the laws of physics would not permit the building of arbitrary Hamiltonians with precision. However, in spite of his interest in the physical and the reality of nature, he found such questions of little interest in quantum computing.

While Paul Benioff had been thinking (and writing) about quantum computation before Feynman, I don't think that Feynman had ever read any of what Benioff had said. As in his approach to most other subjects, Feynman simply ignored what had been done before, easily recreating it independently if relevant, and of course not referencing what he had not read. While he is often given credit for helping originate ideas of quantum computation, my recollections of the many conversations with him on the subject contain no notion of his that quantum computers could in some sense of N scaling be better than classical computing machines. He only emphasized that the physical scale and speed of computers were not limited by the classical world, since conceptually they could be built of reversible components at the atomic level. The insight that quantum computers were really different came only later, and to others, not to Feynman.

He loved the nuts and bolts of computers and low-level coding for machines. At the time, I had a new model of associative memory, had implemented the model on a small scale, and was interested in exploring it on a much larger scale. Feynman raised the question as to whether I had considered implementing it on the Connection Machine of his friend Danny Hillis. At the time, that machine was enormous in potential, but had a very elementary processor and instruction set for each of its 64,000 nodes. I explained why the problem I was doing did not fit onto such an architecture and instruction set effectively. At the our next lunch, he sketched a different way to represent and order the operations which were needed, so that virtually the entire theoretical computing power of the Connection Machine were put into use. (My research paper reprinted in this volume is the one for which this programming was done, and is the only research paper of mine in which he ever took an interest. Strangely, he did not himself try to make mathematical models of how the brain worked.)

Throughout his life, Feynman had been an observer of his own mental processes, and enjoyed speculating on how the brain must work on the basis of his observations. For example, the socks. As a student living in the Graduate College at Princeton, most mundane worries were taken care of. However, you had to count out your shirts, socks, etc. for the laundry from time to time, which he found a bore, and which led him to be interested in what you could do at the same time as you counted. So he developed the idea of counting internally to 60, and seeing if you could judge a fixed time period (roughly a minute, but dependent on your personal internal cadence for counting). If you could do a task and also at the same time judge accurately the passing of this time interval, then the tasks did not interfere, probably because different brain hardware was being used. He found, for example, that he could read and count (by this measure) at the same time, but that he could not speak and maintain a judgement of the time interval. He elaborated on this theme at dinner one day. John Tukey took exception to it, saying he was sure that by this criterion, he could count and speak at the same time. The experiment was done, and Tukey could indeed do the both tasks at once. Feynman pursued Tukey on the issue, to ultimately learn that Tukey knew he generally counted not linguistically (as most of us do), but visually, by seeing objects group in front of him. Since he, like Feynman, should be able to do a visual task and a verbal task at the same time, he (but not Feynman!) would be able to speak and count at the same time. Feynman went on to explain to me that he might have known there would be other things going on in other minds, since for him letters and numbers had colors (as well as shapes) and when he was manipulating equations in his head, he could just use the colors. He speculated that the color-symbol link must have come from a set of blocks he had as a small child.

In Southern California, the rich sometimes give fancy parties to which they invite the accomplished, and so it was that Francis Crick and Richard Feynman met at a cocktail party in Beverly Hills. Wanting to talk, not finding the occasion, they sought a means of getting together. Finding that I seemed to be their only common link, I was ultimately assigned the task of getting them together. So there was a happy Athenaeum (Caltech Faculty Club) lunch which I hosted and at which they sparred. Once again, Feynman took up the role of observer of his own mind. He asked why it was that when you move the eyes from one spot to another in a saccade (rapid eye motion) the world does not blur during the motion. Crick replied that the visual signals to the brain are gated off during this time. Feynman responded "...but I remember when I was a kid riding my bicycle fast, looking at the front wheel, every once and a while I could see the writing on the tire very clearly, and I thought it was because my eyes were accidentally moving in such a way that the image of the tire was stationary on my retina..." Crick quickly admitted that it was only a diminution, not a complete shutting off, of the visual signals, and that Feynman was probably right about the explanation.

The Athenaeum was not always an intellectual haven. One day when the two of us were lunching, a woman who had recently read 'Surely You Are Joking...' (by Feynman, as told to Ralph Leighton) came up to our table to browbeat Feynman about his male chauvinist view of women as portrayed in that book. He put her off as best he could, then turned to me sadly, saying "It's all Ralph's fault, you know. If Ralph had been a women, I would have told him all about my relationship with my first wife (who died of tuberculosis), and everyone would have a very different notion of me. But it is not the kind of thing which men discuss with each other, so Ralph never asked, and I couldn't have said anything even if he had." He even MEANT that it was Ralph's fault.

A scientific giant who was always on stage, he went on to develop his Physics of Computation course as a regular feature of Computer Science, and was disappointed when no one stepped forward to teach it after he ceased to do so. But what human of normal accomplishment could attempt such a role?

2

NEURAL NETWORKS AND PHYSICAL SYSTEMS WITH EMERGENT COLLECTIVE COMPUTATIONAL ABILITIES

John J. Hopfield *

Abstract

Computational properties of use to biological organisms or to the construction of computers can emerge as collective properties of systems having a large number of simple equivalent components (or neurons). The physical meaning of content-addressable memory is described by an appropriate phase space flow of the state of a system. A model of such a system is given, based on aspects of neurobiology but readily adapted to integrated circuits. The collective properties of this model produce a content-addressable memory which correctly yields an entire memory from any subpart of sufficient size. The algorithm for the time evolution of the state of the system is based on asynchronous parallel processing. Additional emergent collective properties include some capacity for generalization, familiarity recognition, categorization, error correction, and time sequence retention. The collective properties are only weakly sensitive to details of the modeling or the failure of individual devices.

2.1 Introduction

Given the dynamical electrochemical properties of neurons and their interconnections (synapses), we readily understand schemes that use a few neurons to obtain elementary useful biological behavior [1–3]. Our understanding of such simple circuits in electronics allows us to plan larger and more complex circuits which are essential to large computers. Because evolution has no such plan, it becomes relevant to ask whether the ability of large collections of neurons to perform "computational" tasks may in part be a spontaneous collective consequence of having a large number of interacting simple neurons.

In physical systems made from a large number of simple elements, interactions among large numbers of elementary components yield collective phenomena such as the stable magnetic orientations and domains in a magnetic system or the vortex patterns in fluid flow. Do analogous collective phenomena in a system of simple interacting neurons have useful "computational" correlates? For example, are

*Reproduced from Proc. Natl. Acad. Sci. USA. Vol.79, pp. 2554–2558, April 1982. Biophysics.

the stability of memories, the construction of categories of generalization, or time-sequential memory also emergent properties and collective in origin? This paper examines a new modeling of this old and fundamental question [4–8] and shows that important computational properties spontaneously arise.

All modeling is based on details, and the details of neuroanatomy and neural function are both myriad and incompletely known [9]. In many physical systems, the nature of the emergent collective properties is insensitive to the details inserted in the model (e.g., collisions are essential to generate sound waves, but any reasonable interatomic force law will yield appropriate collisions). In the same spirit, I will seek collective properties that are robust against change in the model details.

The model could be readily implemented by integrated circuit hardware. The conclusions suggest the design of a delocalized content-addressable memory or categorizer using ex-tensive asynchronous parallel processing.

2.2   The general content-addressable memory of a physical system

Suppose that an item stored in memory is "H. A. Kramers & C. H. Wannier Phys. Rev. 60, 252 (1941)." A general content-addressable memory would be capable of retrieving this entire memory item on the basis of sufficient partial information. The input "& Wannier, (1941)" might suffice. An ideal memory could deal with errors and retrieve this reference even from the input "Vannier, (1941)". In computers, only relatively simple forms of content-addressable memory have been made in hardware [10, 11]. Sophisticated ideas like error correction in accessing information are usually introduced as software [10].

There are classes of physical systems whose spontaneous behavior can be used as a form of general (and error-correcting) content-addressable memory. Consider the time evolution of a physical system that can be described by a set of general coordinates. A point in state space then represents the instantaneous condition of the system. This state space may be either continuous or discrete (as in the case of N Ising spins).

The equations of motion of the system describe a flow in state space. Various classes of flow patterns are possible, but the systems of use for memory particularly include those that flow toward locally stable points from anywhere within regions around those points. A particle with frictional damping moving in a potential well with two minima exemplifies such a dynamics.

If the flow is not completely deterministic, the description is more complicated. In the two-well problems above, if the frictional force is characterized by a temperature, it must also produce a random driving force. The limit points become small limiting regions, and the stability becomes not absolute. But as long as the stochastic effects are small, the essence of local stable points remains.

Consider a physical system described by many coordinates $X_l \ldots X_N$, the components of a state vector $\mathbf{X}$. Let the system have locally stable limit points $\mathbf{X_a}, \mathbf{X_b} \ldots$ Then, if the system is started sufficiently near any $\mathbf{X_a}$, as at $\mathbf{X} = \mathbf{X_a} + \Delta$, it will proceed in time until $\mathbf{X} \approx \mathbf{X_a}$. We can regard the information stored in the system as the vectors $\mathbf{X_a}, \mathbf{X_b}, \ldots$ The starting point $\mathbf{X} = \mathbf{X_a} + \Delta$ represents a partial knowledge of the item $\mathbf{X_a}$, and the system then generates the total information $\mathbf{X_a}$.

Any physical system whose dynamics in phase space is dominated by a substantial number of locally stable states to which it is attracted can therefore be regarded as a general content-addressable memory. The physical system will be a potentially useful memory if, in addition, any prescribed set of states can readily be made the stable states of the system.

2.3 The model system

The processing devices will be called neurons. Each neuron i has two states like those of McCullough and Pitts [12]: $V_i = 0$ ("not firing") and $V_i = 1$ ("firing at maximum rate"). When neuron i has a connection made to it from neuron j, the strength of connection is defined as T_{ij}. (Nonconnected neurons have $T_{ij} \equiv 0$.) The instantaneous state of the system is specified by listing the N values of V_i , so it is represented by a binary word of N bits.

The state changes in time according to the following algorithm. For each neuron i there is a fixed threshold U_i. Each neuron i readjusts its state randomly in time but with a mean attempt rate W, setting

$$V_i \rightarrow \begin{cases} 1 & \text{if } \sum_{j \neq i} T_{ij} V_j > U_i \\ 0 & \text{if } \sum_{j \neq i} T_{ij} V_j < U_i \end{cases} \qquad (2.1)$$

Thus, each neuron randomly and asynchronously evaluates whether it is above or below threshold and readjusts accordingly. (Unless otherwise stated, we choose $U_i = 0$.)

Although this model has superficial similarities to the Perceptron [13, 14] the essential differences are responsible for the new results. First, Perceptrons were modeled chiefly with neural connections in a "forward" direction $A \rightarrow B \rightarrow C \rightarrow D$. The analysis of networks with strong backward coupling $\overrightarrow{A \rightleftarrows B \rightleftarrows C}$ proved intractable. All our interesting results arise as consequences of the strong backcoupling. Second, Perceptron studies usually made a random net of neurons deal directly with a real physical world and did not ask the questions essential to finding the more abstract emergent computational properties. Finally, Perceptron modeling required synchronous neurons like a conventional digital computer. There is no evidence for such global synchrony and, given the delays of nerve signal propagation, there would be no way to use global synchrony effectively. Chiefly computational properties which can exist in spite of asynchrony have interesting implications in biology.

2.4   The information storage algorithm

Suppose we wish to store the set of states $V^s, s = 1 \ldots n$. We use the storage prescription [15, 16]

$$T_{ij} = \sum_s (2V_i^s - 1)(2V_j^s - 1) \tag{2.2}$$

but with $T_{ii} = 0$. From this definition

$$\sum_j T_{ij} V_j^{s'} = \sum_s (2V_i^s - 1) \left[\sum_j V_j^{s'} (2V_j^s - 1) \right] \equiv H_j^{s'} \tag{2.3}$$

The mean value of the bracketed term in Eq. 2.3 is 0 unless $s = s'$, for which the mean is $N/2$. This pseudoorthogonality yields

$$\sum_j T_{ij} V_j^{s'} \equiv \langle H_i^{s'} \rangle \approx (2V_i^{s'} - 1)N/2 \tag{2.4}$$

and is positive if $V_i^{s'} = 1$ and negative if $V_i^{s'} = 0$. Except for the noise coming from the $s \neq s'$ terms, the stored state would always be stable under our processing algorithm.

Such matrices T_{ij} have been used in theories of linear associative nets [15–19] to produce an output pattern from a paired input stimulus, $S_1 \to O_1$. A second association $S_2 \to O_2$ can be simultaneously stored in the same network. But the confusing stimulus $0.6\ S_1 + 0.4\ S_2$ will produce a generally meaningless mixed output $0.6\ O_1 + 0.4\ O_2$. Our model, in contrast, will use its strong nonlinearity to make choices, produce categories, and regenerate information and, with high probability, will generate the output O_1 from such a confusing mixed stimulus.

A linear associative net must be connected in a complex way with an external nonlinear logic processor in order to yield true computation [20, 21]. Complex circuitry is easy to plan but more difficult to discuss in evolutionary terms. In contrast, our model obtains its emergent computational properties from simple properties of many cells rather than circuitry.

2.5   The biological interpretation of the model

Most neurons are capable of generating a train of action potentials — propagating pulses of electrochemical activity — when the average potential across their mem-

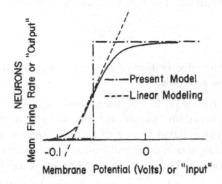

Fig. 2.1. Firing rate versus membrane voltage for a typical neuron (solid line), dropping to 0 for large negative potentials and saturating for positive potentials. The broken lines show approximations used in modeling.

brane is held well above its normal resting value. The mean rate at which action potentials are generated is a smooth function of the mean membrane potential, having the general form shown in Fig. 2.1.

The biological information sent to other neurons often lies in a short-time average of the firing rate [22]. When this is so, one can neglect the details of individual action potentials and regard Fig. 2.1 as a smooth input-output relationship. (Parallel pathways carrying the same information would enhance the ability of the system to extract a short-term average firing rate [23, 24].)

A study of emergent collective effects and spontaneous computation must necessarily focus on the nonlinearity of the input-output relationship. The essence of computation is nonlinear logical operations. The particle interactions that produce true collective effects in particle dynamics come from a nonlinear dependence of forces on positions of the particles. Whereas linear associative networks have emphasized the linear central region [14–19] of Fig. 2.1, we will replace the input-output relationship by the dot-dash step. Those neurons whose operation is dominantly linear merely provide a pathway of communication between nonlinear neurons. Thus, we consider a network of "on or off" neurons, granting that some of the interconnections may be by way of neurons operating in the linear regime.

Delays in synaptic transmission (of partially stochastic character) and in the transmission of impulses along axons and dendrites produce a delay between the input of a neuron and the generation of an effective output. All such delays have been modeled by a single parameter, the stochastic mean processing time $1/W$.

The input to a particular neuron arises from the current leaks of the synapses to that neuron, which influence the cell mean potential. The synapses are activated by arriving action potentials. The input signal to a cell i can be taken to be

$$\sum_j T_{ij} V_j \tag{2.5}$$

where T_{ij} represents the effectiveness of a synapse. Fig. 2.1 thus becomes an input-output relationship for a neuron.

Little, Shaw, and Roney [8, 25, 26] have developed ideas on the collective functioning of neural nets based on "on/off" neurons and synchronous processing. However, in their model the relative timing of action potential spikes was central and resulted in reverberating action potential trains. Our model and theirs have limited formal similarity, although there may be connections at a deeper level.

Most modeling of neural learning networks has been based on synapses of a general type described by Hebb [27] and Eccles [28]. The essential ingredient is the modification of T_{ij} by correlations like

$$\Delta T_{ij} = [V_i(t) V_j(t)]_{\text{average}} \tag{2.6}$$

where the average is some appropriate calculation over past history. Decay in time and effects of $[V_i(t)]_{\text{avg}}$ or $[V_j(t)]_{\text{avg}}$ are also allowed. Model networks with such synapses [16, 20, 21] can construct the associative T_{ij} of Eq. 2.2. We will therefore initially assume that such a T_{ij} has been produced by previous experience (or inheritance). The Hebbian property need not reside in single synapses; small groups of cells which produce such a net effect would suffice.

The network of cells we describe performs an abstract calculation and, for applications, the inputs should be appropriately coded. In visual processing, for example, feature extraction should previously have been done. The present modeling might then be related to how an entity or *Gestalt* is remembered or categorized on the basis of inputs representing a collection of its features.

2.6 Studies of the collective behaviors of the model

The model has stable limit points. Consider the special case $T_{ij} = T_{ji}$, and define

$$E = -\frac{1}{2} \sum \sum_{i \neq j} T_{ij} V_i V_j \tag{2.7}$$

ΔE due to ΔV_i is given by

$$\Delta E = -\Delta V_i \sum_{j \neq i} T_{ij} V_j \tag{2.8}$$

Thus, the algorithm for altering V_i causes E to be a monotonically decreasing function. State changes will continue until a least (local) E is reached. This case is isomorphic with an Ising model. T_{ij} provides the role of the exchange coupling, and there is also an external local field at each site. When T_{ij} is symmetric but has a random character (the spin glass) there are known to be many (locally) stable states [29].

Monte Carlo calculations were made on systems of $N=30$ and $N=100$, to examine the effect of removing the $T_{ij} = T_{ji}$ restriction. Each element of T_{ij} was chosen as a random number between -1 and 1. The neural architecture of typical cortical regions [30, 31] and also of simple ganglia of invertebrates [32] suggests the importance of 100-10,000 cells with intense mutual interconnections in elementary processing, so our scale of N is slightly small.

The dynamics algorithm was initiated from randomly chosen initial starting configurations. For $N=30$ the system never displayed an ergodic wandering through state space. Within a time of about $4/W$ it settled into limiting behaviors, the commonest being a stable state. When 50 trials were examined for a particular such random matrix, all would result in one of two or three end states. A few stable states thus collect the flow from most of the initial state space. A simple cycle also occurred occasionally — for example, $\ldots A \to B \to A \to B \ldots$

The third behavior seen was chaotic wandering in a small region of state space. The Hamming distance between two binary states A and B is defined as the number of places in which the digits are different. The chaotic wandering occurred within a short Hamming distance of one particular state. Statistics were done on the probability p_i of the occurrence of a state in a time of wandering around this minimum, and an entropic measure of the available states M was taken

$$\ln M = -\sum p_i \ln p_i \qquad (2.9)$$

A value of $M=25$ was found for $N=30$. *The flow in phase space produced by this model algorithm has the properties necessary for a physical content-addressable memory* whether or not T_{ij} is symmetric.

Simulations with $N=100$ were much slower and not quantitatively pursued. They showed qualitative similarity to $N=30$.

Why should stable limit points or regions persist when $T_{ij} \neq T_{ji}$? If the algorithm at some time changes V_i from 0 to 1 or vice versa, the change of the energy defined in Eq. 2.7 can be split into two terms, one of which is always negative. The second is identical if T_{ij} is symmetric and is "stochastic" with mean 0 if T_{ij} and T_{ji} are randomly chosen. The algorithm for $T_{ij} \neq T_{ji}$ therefore changes E in a fashion similar to the way E would change in time for a symmetric T_{ij} but with an algorithm corresponding to a finite temperature.

About 0.15 N states can be simultaneously remembered before error in recall is severe. Computer modeling of memory storage according to Eq. 2.2 was carried out for $N=30$ and $N=100$. n random memory states were chosen and the corresponding

T_{ij} was generated. If a nervous system preprocessed signals for efficient storage, the preprocessed information would appear random (e.g., the coding sequences of DNA have a random character). The random memory vectors thus simulate efficiently encoded real information, as well as representing our ignorance. The system was started at each assigned nominal memory state, and the state was allowed to evolve until stationary.

Typical results are shown in Fig. 2.2. The statistics are averages over both the states in a given matrix and different matrices. With $n=5$, the assigned memory states are almost always stable (and exactly recallable). For $n=15$, about half of the nominally remembered states evolved to stable states with less than 5 errors, but the rest evolved to states quite different from the starting points.

These results can be understood from an analysis of the effect of the noise terms. In Eq. 2.3, $H_i^{s'}$ is the "effective field" on neuron i when the state of the system is s', one of the nominal memory states. The expectation value of this sum, Eq. 2.4, is $\pm N/2$ as appropriate. The $s \neq s'$ summation in Eq. 2.2 contributes no mean, but has a rms noise of $[(n - 1)N/2]^{1/2} \equiv \sigma$. For nN large, this noise is approximately Gaussian and the probability of an error in a single particular bit of a particular memory will be

$$P = \frac{1}{\sqrt{2\pi\sigma^2}} \int_{N/2}^{\infty} e^{-x^2/2\sigma^2} dx \qquad (2.10)$$

For the case $n=10$, $N=100$, $P=0.0091$, the probability that a state had no errors in its 100 bits should be about $e^{-0.91} \approx 0.40$. In the simulation of Fig. 2.2, the experimental number was 0.6.

The theoretical scaling of n with N at fixed P was demonstrated in the simulations going between $N=30$ and $N=100$. The experimental results of half the memories being well retained at $n=0.15\ N$ and the rest badly retained is expected to be true for all large N. The information storage at a given level of accuracy can be increased by a factor of 2 by a judicious choice of individual neuron thresholds. This choice is equivalent to using variables $\mu_i = \pm 1$, $T_{ij} = \sum_s \mu_i^s \mu_j^s$, and a threshold level of 0.

Given some arbitrary starting state, what is the resulting final state (or statistically, states)? To study this, evolutions from randomly chosen initial states were tabulated for $N=30$ and $n=5$. From the (inessential) symmetry of the algorithm, if $(101110\ldots)$ is an assigned stable state, $(010001\ldots)$ is also stable. Therefore, the matrices had 10 nominal stable states. Approximately 85% of the trials ended in assigned memories, and 10% ended in stable states of no obvious meaning. An ambiguous 5% landed in stable states very near assigned memories. There was a range of a factor of 20 of the likelihood of finding these 10 states.

The algorithm leads to memories near the starting state. For $N=30$, $n=5$, partially random starting states were generated by random modification of known

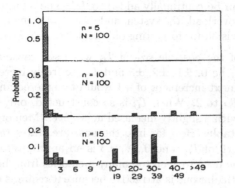

Fig. 2.2. The probability distribution of the occurrence of errors in the location of the stable states obtained from nominally assigned memories.

memories. The probability that the final state was that closest to the initial state was studied as a function of the distance between the initial state and the nearest memory state. For distance ≤ 5, the nearest state was reached more than 90% of the time. Beyond that distance, the probability fell off smoothly, dropping to a level of 0.2 (2 times random chance) for a distance of 12.

The phase space flow is apparently dominated by attractors which are the nominally assigned memories, each of which dominates a substantial region around it. The flow is not entirely deterministic, and *the system responds to an ambiguous starting state by a statistical choice* between the memory states it most resembles.

Were it desired to use such a system in an Si-based content-addressable memory, the algorithm should be used and modified to hold the known bits of information while letting the others adjust.

The model was studied by using a "clipped" T_{ij}, replacing T_{ij} in Eq. 2.3 by ± 1, the algebraic sign of T_{ij}. The purposes were to examine the necessity of a linear synapse supposition (by making a highly nonlinear one) and to examine the efficiency of storage. Only $N(N/2)$ bits of information can possibly be stored in this symmetric matrix. Experimentally, for $N=100$, $n=9$, the level of errors was similar to that for the ordinary algorithm at $n=12$. The signal-to-noise ratio can be evaluated analytically for this clipped algorithm and is reduced by a factor of $(2/\pi)^{1/2}$ compared with the unclipped case. For a fixed error probability, the number of memories must be reduced by $2/\pi$.

With the μ algorithm and the clipped T_{ij}, both analysis and modeling showed that the maximal information stored for $N=100$ occurred at about $n=13$. Some errors were present, and the Shannon information stored corresponded to about $N(N/8)$ bits.

New memories can be continually added to T_{ij}. The addition of new memories beyond the capacity overloads the system and makes all memory states irretrievable unless there is a provision for forgetting old memories [16, 27, 28].

The saturation of the possible size of T_{ij} will itself cause forgetting. Let the possible values of T_{ij} be 0, ±1, ±2, ±3 and T_{ij} be freely incremented within this range. If $T_{ij} = 3$, a next increment of +1 would be ignored and a next increment of -1 would reduce T_{ij} to 2. When T_{ij} is so constructed, only the recent memory states are retained, with a slightly increased noise level. Memories from the distant past are no longer stable. How far into the past are states remembered depends on the digitizing depth of T_{ij}, and 0,..., ±3 is an appropriate level for $N = 100$. Other schemes can be used to keep too many memories from being simultaneously written, but this particular one is attractive because it requires no delicate balances and is a consequence of natural hardware.

Real neurons need not make synapses both of $i \rightarrow j$ and $j \rightarrow i$. Particular synapses are restricted to one sign of output. We therefore asked whether $T_{ij} = T_{ji}$ is important. Simulations were carried out with only one ij connection: if $T_{ij} \neq 0$, $T_{ji} = 0$. The probability of making errors increased, but the algorithm continued to generate stable minima. A Gaussian noise description of the error rate shows that the signal-to-noise ratio for given n and N should be decreased by the factor $1/\sqrt{2}$, and the simulations were consistent with such a factor. This same analysis shows that the system generally fails in a "soft" fashion, with signal-to-noise ratio and error rate increasing slowly as more synapses fail.

Memories too close to each other are confused and tend to merge. For $N=100$, a pair of random memories should be separated by 50 ± 5 Hamming units. The case $N=100$, $n=8$, was studied with seven random memories and the eighth made up a Hamming distance of only 30, 20, or 10 from one of the other seven memories. At a distance of 30, both similar memories were usually stable. At a distance of 20, the minima were usually distinct but displaced. At a distance of 10, the minima were often fused.

The algorithm categorizes initial states according to the similarity to memory states. With a threshold of 0, the system behaves as a forced categorizer.

The state 00000... is always stable. For a threshold of 0, this stable state is much higher in energy than the stored memory states and very seldom occurs. Adding a uniform threshold in the algorithm is equivalent to raising the effective energy of the stored memories compared to the 0000 state, and 0000 also becomes a likely stable state. The 0000 state is then generated by any initial state that does not resemble adequately closely one of the assigned memories and represents positive recognition that the starting state is not familiar.

Familiarity can be recognized by other means when the memory is drastically overloaded. We examined the case $N=100$, $n=500$, in which there is a memory overload of a factor of 25. None of the memory states assigned were stable. The initial rate of processing of a starting state is defined as the number of neuron state readjustments that occur in a time $1/2W$. Familiar and unfamiliar states were distinguishable most of the time at this level of overload on the basis of the initial processing rate, which was faster for unfamiliar states. This kind of familiarity can only be read out of the system by a class of neurons or devices abstracting average properties of the processing group.

For the cases so far considered, the expectation value of T_{ij} was 0 for $i \neq j$. A set of memories can be stored with average correlations, and $\bar{T}_{ij} = C_{ij} \neq 0$ because there is a consistent internal correlation in the memories. If now a partial new state X is stored

$$\Delta T_{ij} = (2X_i - 1)(2X_j - 1) \quad i, j \leq k < N \quad (2.11)$$

using only k of the neurons rather than N, an attempt to reconstruct it will generate a stable point for all N neurons. The values of $X_{k+1} \ldots X_N$ that result will be determined primarily from the sign of

$$\sum_{j=1}^{k} C_{ij} X_j \quad (2.12)$$

and X is completed according to the mean correlations of the other memories. The most effective implementation of this capacity stores a large number of correlated matrices weakly followed by a normal storage of X.

A nonsymmetric T_{ij} can lead to the possibility that a minimum will be only metastable and will be replaced in time by another minimum. Additional nonsymmetric terms which could be easily generated by a minor modification of Hebb synapses

$$\Delta T_{ij} = A \sum_{s} (2V_i^{s+1} - 1)(2V_j^s - 1) \quad (2.13)$$

were added to T_{ij}. When A was judiciously adjusted, the system would spend a while near V_s and then leave and go to a point near V_{s+1}. But sequences longer than four states proved impossible to generate, and even these were not faithfully followed.

2.7   Discussion

In the model network each "neuron" has elementary properties, and the network has little structure. Nonetheless, collective computational properties spontaneously arose. Memories are retained as stable entities or *Gestalts* and can be correctly recalled from any reasonably sized subpart. Ambiguities are resolved on a statistical basis. Some capacity for generalization is present, and time ordering of memories can also be encoded. These properties follow from the nature of the flow in phase space produced by the processing algorithm, which does not appear to be strongly dependent on precise details of the modeling. This robustness suggests that similar effects will obtain even when more neurobiological details are added.

Much of the architecture of regions of the brains of higher animals must be made from a proliferation of simple local circuits with well-defined functions. The bridge between simple circuits and the complex computational properties of higher nervous systems may be the spontaneous emergence of new computational capabilities from the collective behavior of large numbers of simple processing elements.

Implementation of a similar model by using integrated circuits would lead to chips which are much less sensitive to element failure and soft-failure than are normal circuits. Such chips would be wasteful of gates but could be made many times larger than standard designs at a given yield. Their asynchronous parallel processing capability would provide rapid solutions to some special classes of computational problems.

The work at California Institute of Technology was supported in part by National Science Foundation Grant DMR-8107494. This is contribution no. 6580 from the Division of Chemistry and Chemical Engineering.

References

[1] A. O. D. Willows, D. A. Dorsett and G. Hoyle, *J. Neurobiol.* **4**, 207-237, 255-285 (1973).

[2] W. B. Kristan, in *Information Processing in the Nervous System*, eds. H. M. Pinsker and W.D. Willis, 241-261 (Raven, New York, 1980).

[3] B. W. Knight, *Lect. Math. Life Sci.* **5**, 111-144 (1975).

[4] D. R. Smith and C. H. Davidson, *J. Assoc. Comput. Mach.* **9**, 268-279 (1962).

[5] L. D. Harmon, in Neural Theory and Modeling, ed. R.F. Reiss, pp. 23-24 (Stanford Univ. Press, Stanford, CA, 1964).

[6] S.-I. Amari, *Biol Cybern.* **26**, 175-185 (1977).

[7] S.-I. Amari and T. Akikazu, *Biol. Cybern.* **29**, 127-136 (1978).

[8] W. A. Little, *Math. Biosci.* **19**, 101-120 (1974).

[9] J. Marr, *J. Physiol.* **202**, 437-470 (1969).

[10] T. Kohonen, *Content Addressable Memories* (Springer, New York, 1980).

[11] G. Palm, *Biol. Cybern.* **36**, 19-31 (1980).

[12] W. S. McCulloch and W. Pitts, *Bull. Math Biophys.* **5**, 115-133 (1943).

[13] M. Minsky and S. Papert, *Perceptrons: An Introduction to Computational Geometry* (MIT Press, Cambridge, MA, 1969).

[14] F. Rosenblatt, *Principles of Perceptrons* (Spartan, Washington, DC, 1962).

[15] L. N. Cooper, in *Proceedings of the Nobel Symposium on Collective Properties of Physical Systems*, eds. B. Lundqvist and S. Lundqvist, 252-264 (Academic, New York, 1973).

[16] L. N. Cooper, F. Liberman and E. Oja, *Biol. Cybern.* **33**, 9-28 (1979).

[17] J. C. Longuet-Higgins, *Proc. Roy. Soc. London Ser. B* **171**, 327-334 (1968).

[18] J. C. Longuet-Higgins, *Nature (London)* **217**, 104-105 (1968).

[19] T. Kohonen, *Associative Memory — A System-Theoretic Approach* (Springer, New York, 1977).

[20] G. Willwacher, *Biol. Cybern.* **24**, 181-198 (1976).

[21] J. A. Anderson, *Psych. Rev.* **84**, 413-451 (1977).

[22] D. H. Perkel and T. H. Bullock, *Neurosci. Res. Symp. Summ.* **3**, 405-527 (1969).

[23] E. B. John, *Science* **177**, 850-864 (1972).

[24] K. J. Roney, A. B. Scheibel and G. L. Shaw, *Brain Res. Rev.* **1**, 225-271 (1979).

[25] W. A. Little and G. L. Shaw, Math. Biosci. 39, 281-289 (1978).

[26] G. L. Shaw and K. J. Roney, *Phys. Rev. Lett.* **74**, 146-150 (1979).

[27] D. O. Hebb, *The Organization of Behavior* (Wiley, NewYork, 1949).

[28] J. G. Eccles, *The Neurophysiological Basis of Mind* (Clarendon, Oxford, 1953).

[29] S. Kirkpatrick and D. Sherrington, *Phys. Rev.* **17**, 4384-4403 (1978).

[30] V. B. Mountcastle, in *The Mindful Brain*, eds. G. M. Edelman and V. B. Mountcastle, pp 36-41 (MIT Press, Cambridge, MA, 1978).

[31] P. S. Goldman and W. J. H. Nauta, *Brain Res.* **122**, 393-413 (1977).

[32] E. R. Kandel, *Sci. Am.* **241**, 61-70 (1979).

3

FEYNMAN AS A COLLEAGUE

Carver A. Mead

Feynman and I both arrived at Caltech in 1952 — he as a new professor of physics, and I as a freshman undergraduate. My passionate interest was electronics, and I avidly consumed any material I could find on the subject: courses, seminars, books, etc. As a consequence, I was dragged through several versions of standard electromagnetic theory: **E** and **B**, **D** and **H**, curls of curls, the whole nine yards. The only bright light in the subject was the vector potential, to which I was always attracted because, somehow, it made sense to me. It seemed a shame that the courses I attended didn't make more use of it. In my junior year, I took a course in mathematical physics from Feynman — what a *treat*. This man could *think conceptually* about physics, not just regurgitate dry formalism. After one quarter of Feynman, the class was spoiled for any other professor. But when we looked at the registration form for the next quarter, we found Feynman as teaching high-energy physics, instead of our course. Bad luck! When our first class met, however, here came Feynman. "So you're not teaching high-energy physics?" I asked. "No" he replied, "low-energy mathematics." Feynman liked the vector potential too; for him it was the link between electromagnetism and quantum mechanics. As he put it "In the general theory of quantum electrodynamics, one takes the vector and scalar potentials as fundamental quantities in a set of equations that replace the Maxwell equations." I learned enough about it from him to know that, some day, I wanted to do all of electromagnetic theory that way.

By 1960 I had completed a thesis on transistor physics and had become a brand new faculty member in my own right. Fascinated by Leo Esaki's work on tunnel diodes, I started my own research on electron tunneling through thin insulating films. Tunneling is interesting because it is a purely quantum phenomenon. Electrons below the zero energy level in a vacuum, or in the forbidden gap of a semiconductor or insulator, have wave functions that die out exponentially with distance. I was working with insulators sufficiently thin that the wave function of electrons on one side had significant amplitude on the opposite side. The result was a current that decreased exponentially with the thickness of the insulator. From the results, I could work out how the exponential depended on energy. My results didn't fit with the conventional theory, which treated the insulator as though it were a vacuum. But the insulator was not a vacuum, and the calculations were giving us important information about how the wave function behaved in the forbidden gap. Feynman was enthusiastic about this tunneling work. We shared a graduate student, Karvel Thornber, who used path integral methods to work out a more detailed model of the insulator.

In 1961 Feynman undertook the monumental task of developing a completely new 2-year introductory physics course. The first year covered mechanics; although that topic wasn't of much interest to me, it would come up occasionally in our meetings on the tunneling project. When I heard that Feynman was going to do electromagnetic theory in the second year, I got very excited — finally someone would get it right! Unfortunately, it was not to be. The following quotation from the forward to the *Feynman Lectures on Gravitation* tells the story:

"It is remarkable that concurrent with this course on gravitation Feynman was also creating and teaching an innovative course in sophomore (second year undergraduate) physics, a course that would become immortalized as the second and third volumes of *The Feynman Lectures on Physics*. Each Monday Feynman would give his sophomore lecture in the morning and the lecture on gravitation after lunch. Later in the week would follow a second sophomore lecture and a lecture for scientists at Hughes Research Laboratories in Malibu. Beside this teaching load and his own research, Feynman was also serving on a panel to review textbooks for the California State Board of Education, itself a consuming task, as is vividly recounted in *Surely You're Joking, Mr. Feynman*. Steven Frautschi, who attended the lectures as a young Caltech assistant professor, remembers Feynman later saying that he was "utterly exhausted" by the end of the 1962-63 academic year."

I was another young Caltech assistant professor who attended the gravitation lectures, and I remember them vividly. Bill Wagner, with whom I still collaborate over collective electrodynamics material, took notes, and later worked out the mathematical presentation in the written version of the lectures. I also attended many of the sophomore lectures, to which I had mixed reactions. If you read volume II of *The Feynman Lectures on Physics*, you will find two distinct threads. The first is a perfectly standard treatment, like that in any introductory book on the subject. In his preface, Feynman says of this material "In the second year I was not so satisfied. In the first part of the course, dealing with electricity and magnetism, I couldn't think of any really unique or different way of doing it." There is a second thread, however, of true vintage Feynman — the occasional lectures where he waxed eloquent about the vector potential "**E** and **B** are slowly disappearing from the modern expression of physical laws; they are being replaced by **A** and ϕ." Section 15-5 contains a delightful discussion about what a field is, and what makes one field more "real" than another. He concludes "In our sense then, the **A** field is real." In Chapter 25, he develops the equations of electrodynamics in four-vector form — the approach that I have adopted in the *Collective Electrodynamics* sequence. I can remember feeling very angry with Feynman when I sat in on this particular lecture. Why hadn't he started this way in the first place, and saved us all the mess of a **B** field, which, as he told us himself, was not real anyway? When I asked him about it, he said something vague, like "There are a bunch of classical interactions that you can't get at in any simple way without Maxwell's equations.

You need the $\mathbf{v} \times \mathbf{B}$ term." I don't remember his exact words here, only the gist of the discussion. Sure enough, when volume II of the lectures was published, in table 15-1 the equation $\mathbf{F} = q(\mathbf{E} + \mathbf{v} \times \mathbf{B})$ appears in the column labelled "True Always." The equation is true for the toy electric motor he shows in Fig. 16-1. It is *not* true in general. For a real electric motor, the $\mathbf{B}$ field is concentrated in the iron, rather than in the copper where the current is flowing, and the equation gives the wrong answer by a factor of more than 100! That factor is due to the failure of $\mathbf{B}$ to be "real," precisely in Feynman's sense.

I was an active researcher in solid-state physics at that time, and I used the quantum nature of electrons in solids every day. Electrodynamics deals with how electrons interact with other electrons. The classical interactions Feynman was talking about were between electrons in metals, in which the density of electrons is so high that quantum interaction is by far the dominant effect. If we know how the vector potential comes into the phase of the electron wave function, and if the electron wave function dominates the behavior of metals, then why can't we do all of electromagnetic theory that way? Why didn't he use his knowledge of quantum electrodynamics to "take the vector and scalar potentials as fundamental quantities in a set of equations that replace the Maxwell equations," as he himself had said? I was mystified; his cryptic answer prodded me to start working on the problem. But every time I thought I had an approach, I got stuck.

Bill Fairbank from Stanford had given a seminar on quantized flux in superconducting rings that impressed me very much. The solid-state physics club was much smaller in those days, and, because I was working in electron tunneling, I was close to the people working on tunneling between superconductors. Their results were breaking in just this time frame, and Feynman gave a lecture about this topic to the sophomores; it appears as chapter 21 in volume 3. As I listened to that lecture, my thoughts finally clicked: That was how we could make the connection! A superconductor is a quantum system on a classical scale, and that fact allows us to carry out Feynman's grand scheme. But I couldn't get this approach to go all the way through at that time, so it just sat in the back of my mind all these years, vaguely tickling me.

Meanwhile my work on tunneling was being recognized, and Gordon Moore (then at Fairchild) asked me whether tunneling would be a major limitation on how small we could make transistors in an integrated circuit. That question took me on a detour that was to last nearly 30 years, but it also led me into another collaboration with Feynman, this time on the subject of computation. Here's how it happened. In 1968, I was invited to give a talk at a workshop on semiconductor devices at Lake of the Ozarks. In those days, you could get everyone who was doing cutting-edge work in one room, so the workshops were where all the action was. I had been thinking about Gordon Moore's question, and decided to make it the subject of my talk. As I prepared for this event, I began to have serious doubts about my sanity. My calculations were telling me that, contrary to all the current lore in the field, we could scale down the technology such that *everything got better*: The circuits got more complex, they ran faster, and they took less power — WOW! That's a violation of Murphy's law that won't quit! But the more I looked at the problem, the more I was convinced that the result was correct, so I went ahead and gave the talk, to hell with Murphy! That talk provoked considerable debate, and at the time most people didn't believe the result. But by the time the next workshop rolled around, a number of other groups had worked through the problem for themselves, and we were pretty much all in agreement. The consequences of this result for modern information technology have, of course, been staggering.

Back in 1959, Feynman had given a lecture entitled "There's Plenty of Room at the Bottom," in which he discussed how much smaller things can be made than we ordinarily imagine. That talk, which appears elsewhere in this volume, had made a big impression on me; I thought about it often, and it would sometimes come up in our discussions on the tunneling work. When I told him about the scaling law for electronic devices, Feynman got jazzed. He came to my seminars on the subject, and always raised a storm of good questions and comments. I was working with a graduate student, Bruce Hoeneisen, and by 1971 we had worked out the details of how transistors would look and work when they were a factor of 100 smaller in linear dimension than the limits set by the prevailing orthodoxy. Recently, I had occasion to revisit these questions, and to review the history of what has happened in the industry since those papers were published. I plotted our 1971 predictions alongside the real data; they have held up extremely well over 25 years, representing a factor of several thousand in density of integrated circuit components. That review also appears in this volume.

Because of the scaling work, I became completely absorbed with how the exponential increase in complexity of integrated circuits would change the way that we think about computing. The viewpoint of the computer industry at the time was an outgrowth of the industrial revolution; it was based on what was then called "the economy of scale." The thinking went this way. A 1000-horsepower engine cost only four times as much as a 100-horsepower engine. Therefore, the cost per horsepower became less as the engine was made larger. It was more cost effective to make a few large power plants than to make many small ones. Efficiency considerations favored the concentration of technology in a few large installations. The same was evidently true of computing. One company, IBM, was particularly successful following this strategy. The "Computing Center" was the order of the day — a central concentration of huge machines, with some bureaucrat "in charge," and plenty of people around to protect the machines from anyone who might want to use them. This model went well with the bureaucratic mindset of the time — a mindset that has not totally died out even today.

But as I looked at the physics of the emerging technology, it didn't work that way at all. The time required to move data was set by the velocity of light and related electromagnetic considerations, so it was far more effective to put whatever computing was required *where the data were located.* Efficiency considerations thus favored the distribution of technology, rather than the concentration of technology The economics of information technology were the reverse of those of mechanical technology. I gave numerous talks on this topic and, at that time, what I had to say was contrary to what the industry wanted to hear. The story is best told in George Gilder's book *Microcosm.* Feynman had started this line of thought already in his 1959 lecture, and we had a strong agreement on the general direction things were headed. He often came to my group meetings, and we had lively discussions on how to build a machine that would recognize fingerprints, how to organize many thousand little computers so they would be more efficient than one big computer, etc. Those discussions inevitably led us to wonder about the most distributed computer of all: the human brain. Years before, Feynman had dabbled in biology, and I had worked with Max Delbruck on the physics of the nerve membrane, so I knew a bit about nerve tissue. John Hopfield had delved much deeper than either Feynman or I had, and by 1982 he had a simple model — a caricature of how computation might go on in the brain.

The three of us decided to offer a course jointly, called "Physics of Computation." The first year Feynman was battling a bout with cancer, so John and I had to go it alone. We alternated lectures, looking at the topic from markedly different points of view. Once Feynman rejoined us, we had even more fun — three totally different streams of conciousness in one course. The three of us had a blast, and learned a lot from one another, but many of the students were completely mystified. After the third year, we decided, in deference to the students, that there was enough material for three courses, each with a more unified theme. Hopfield did "Neural Networks," Feynman did "Quantum Computing," and I did "Neuromorphic Systems." The material in the *Feynman Lectures on Computation* evolved during this period.

There is a vast mythology about Feynman, much of which is misleading. He had a sensitive side that he didn't show often. Over lunch one time, I told him how much he had meant to me in my student years, and how I would not have gone into science had it not been for his influence. He looked embarrassed, and abruptly changed the subject; but he heard me and that was what was important. In those days, physics was an openly combative subject — the one who blinked first lost the argument. Bohr had won his debate with Einstein that way, and the entire field adopted the style. Feynman learned the game well — he never blinked. For this reason, he would never tell you when he was working on something, but instead would spring it, preferably in front of an audience, after he had it all worked out. The only way that you could tell what he cared about was to notice what topics made him mad when you brought them up.

If Feynman was stuck about something, he had a wonderful way of throwing up a smoke screen; we used to call it "proof by intimidation." There is a good example in Vol. II of the *Lectures on Physics*, directly related to collective electrodynamics. Section 17-8 contains the following comment: "...we would expect that corresponding to the mechanical momentum p = mv, whose rate of change is the applied force, there should be an analogous quantity equal to LI, whose rate of change is V. We have no right, of course, to say that LI is the real momentum of the circuit; in fact it isn't. The whole circuit may be standing still and have no momentum." Now this passage does not mean that Feynman was ignorant of the fact that the electrical current I is made up of moving electrons, that these moving electrons have momentum, and that the momentum of the electrons does not correspond to the whole circuit moving in space. But the relations are not as simple as we might expect, and they do not correspond in the most direct way to our expectations from classical mechanics. It is exactly this point that prevented me, over all these years, from seeing how to do electrodynamics without Maxwell's equations. Feynman was perfectly aware that this was a sticking point, and he made sure that nobody asked any questions about it. There is a related comment in Vol. III, section 21-3: "It looks as though we have two suggestions for relations of velocity to momentum... The two possibilities differ by the vector potential. ...One of them... is the momentum obtained by multiplying the mass by velocity. The other is a more mathematical, more abstract momentum..."

When Feynman said that a concept was "more mathematical" or "more abstract," he was not paying it a compliment! He had no use for theory devoid of physical content. In the *Lectures on Gravitation*, he says "If there is something very slightly wrong in our definition of the theories, then the full mathematical rigor may convert these errors into ridiculous conclusions." He called that "carrying rigor to the point of *rigor mortis*." At another point he is even more explicit: "...it is the facts that matter, and not the proofs. Physics can progress without the proofs, but we can't go on without the facts... if the facts are right, then the proofs are a matter of playing around with the algebra correctly." He opened a seminar one time with the statement "Einstein was a giant." A hush fell over the audience. We all sat, expectantly, waiting for him to elaborate. Finally, he continued "His head was in the clouds, but his feet were on the ground." We all chuckled, and again we waited. After another long silence, he concluded "But those of us who are not that tall have to choose!" Amid the laughter, you could see that, not only a good joke, but also a deep point had been made.

Experiments are the ground on which physics must keep its feet — as Feynman knew well. When any of us had a new result, he was all ears. He would talk about it, ask questions, brainstorm. That was the only situation in which I have personally interacted with him without the combative behavior getting in the way. Down deep, he always wanted to do experiments himself. A hilarious account of how he was "cured" of this craving appears in *Surely You're Joking, Mr. Feynman*. In the end, he had his wish. In 1986, he was asked to join the Rodgers commission to investigate the Challenger disaster. After talking to the technical people, who knew perfectly well what the problem was, and had tried to postpone the launch, he was able to devise an experiment that he carried out on national, prime-time TV. In true

Feynman style, he sprang it full-blown, with no warning! In his personal appendix to the commission report, he concluded, "For a successful technology, reality must take precedence over public relations, for Nature cannot be fooled." The day after the report was released was Caltech graduation, and we marched together in the faculty procession. "Did you see the headline this morning?" he asked. "No," I replied "what did it say?" "It said **FEYNMAN ISSUES REPORT**." He paused, and then continued with great glee, "Not **Caltech Professor Issues Report**, not **Commission Member Issues Report**, but **FEYNMAN ISSUES REPORT**." He was a household word, known and revered by all people everywhere who loved truth. His own public relations were all about reality, and were, therefore, OK.

In 1987, one year later, his cancer came back with a vengeance, and he died in February, 1988. Al Hibbs, a former student, colleague, and friend of Feynman's, organized a wake in grand style. Bongo drums, news clips, interviews, and testimonials. It was deeply moving — we celebrated the life of this man who had, over the years, come to symbolize, not just the spirit of Caltech, but the spirit of science itself. This man had engendered the most intense emotions I have ever felt — love, hate, admiration, anger, jealousy, and, above all, a longing to share and an intense frustration that he would not. As I walked away from Feynman's wake, I felt intensely alone. He was the man who had taught me, not only what physics was, but also what science was all about: what it meant to *really understand*. He was the only person with whom I could have talked about doing electromagnetism using only the vector potential. He was the only one who would have understood why it was important. He was the one who could have related to this dream that I had carried for 25 years. This dream came direct from Feynman, from what he said, and from what he scrupulously avoided saying — from the crystal-clear insights he had, and from the topics that made him mad when I brought them up. But now he was gone. I would have to go it alone. I sobbed myself to sleep that night, but I never shared those feelings with anyone. I learned that from him too.

In 1994, I was invited to give the keynote talk at the Physics of Computation conference. That invitation gave me the kickstart I needed to get going. By a year later, I had made enough progress to ask Caltech for a year relief from teaching, so I could concentrate on the new research. In June 1997, the six graduate students working in my lab all received their doctoral degrees, and, for the first time since I joined the faculty, I was a free man. I finished the first paper on Collective Electrodynamics, which is included here.

In the end, science is all in how you look at things. Collective Electrodynamics is a way of looking at the way that electrons interact. It is a much simpler way than Maxwell's, because it is based on experiments that tell us about the electrons directly. Maxwell had no access to these experiments. The sticking point I mentioned earlier is resolved in this paper, in a way that Feynman would have liked. The paper is dedicated to him in the most sincere way I know: it opens with my favorite quotation, the quotation that defines, for me, what science is all about. In his epilogue, he tells us his true motivation for giving the *Lectures on Physics*: "I wanted most to give you some appreciation of the wonderful world, and the physicist's way of looking at it, which, I believe, is a major part of the true culture of modern times... Perhaps you will not only have some appreciation of this culture; it is even possible that you may want to join in the greatest adventure that the human mind has ever begun."

4

COLLECTIVE ELECTRODYNAMICS I

Carver A. Mead*

Abstract

Standard results of electromagnetic theory are derived from the direct interaction of macroscopic quantum systems; the only assumptions used are the Einstein-deBroglie relations, the discrete nature of charge, the Green's function for the vector potential, and the continuity of the wave function. No reference is needed to Maxwell's equations or to traditional quantum formalism. Correspondence limits based on classical mechanics are shown to be inappropriate.

> "But the real *glory* of science is that *we can find a way of thinking* such that the law is *evident.*" — R. P. Feynman

4.1 Foundations of Physics

Much has transpired since the first two decades of this century, when the conceptual foundations for modern physics were put in place. At that time, macroscopic mechanical systems were easily accessible and well understood. The nature of electrical phenomena was mysterious; experiments were difficult and their interpretation was murky. Today, quite the reverse is true. Electrical experiments of breathtaking clarity can be carried out, even in modestly equipped laboratories. Electronic apparatus pervade virtually every abode and workplace. Modern mechanical experiments rely heavily on electronic instrumentation. Yet, in spite of this reversal in the range of experience accessible to the average person, introductory treatments of physics still use classical mechanics as a starting point.

Ernst Mach wrote (p.596 in ref. [1]), "The view that makes mechanics the basis of the remaining branches of physics, and explains all physical phenomena by mechanical ideas, is in our judgement a prejudice ... The mechanical theory of Nature, is, undoubtedly, in a historical view, both intelligible and pardonable; and it may also, for a time, have been of much value. But, upon the whole, it is an artificial conception."

Classical mechanics is indeed inappropriate as a starting point for physics because it is not fundamental; rather, it is the limit of an incoherent aggregation of an enormous number of quantum elements. To make contact with the fundamental

*Reproduced from Proc. Natl. Acad. Sci. USA Vol.94, pp. 6013-6018, June 1997 Physics

nature of matter, we must work in a coherent context where the quantum reality is preserved.

R. P. Feynman wrote (p.15–8 in ref. [2]), "There are many changes in concepts that are important when we go from classical to quantum mechanics ... Instead of forces, we deal with the way interactions change the wavelengths of waves."

Even Maxwell's equations have their roots in classical mechanics. They were conceived as a theory of the ether: They express relations between the magnetic field $\mathbf{B}$ and the electric field $\mathbf{E}$, which are defined in terms of the classical force $\mathbf{F} = q(\mathbf{E} + \mathbf{v} \times \mathbf{B})$ on a particle of charge q moving with velocity $\mathbf{v}$. But it is the vector potential $\mathbf{A}$, rather than the magnetic field $\mathbf{B}$, that has a natural connection with the quantum nature of matter — as highlighted by Aharonov and Bohm [3].

Hamilton's formulation of classical mechanics was — and remains — the starting point for the concepts underlying the quantum theory. The correspondence principle would have every quantum system approach the behavior of its classical-mechanics counterpart in the limit where the mechanical action involved is large compared with Planck's constant.

Although superconductivity was discovered in 1911, the recognition that super-conductors manifest quantum phenomena on a macroscopic scale [4] came too late to play a role in the formulation of quantum mechanics. Through modern experimental methods, however, superconducting structures give us direct access to the quantum nature of matter. The superconducting state is a coherent state formed by the collective interaction of a large fraction of the free electrons in a material. Its properties are dominated by known and controllable interactions within the collective ensemble. The dominant interaction is collective because the properties of each electron depend on the state of the entire ensemble, and it is electromagnetic because it couples to the charges of the electrons. Nowhere in natural phenomena do the basic laws of physics manifest themselves with more crystalline clarity.

This paper is the first in a series in which we start at the simplest possible conceptual level, and derive as many conclusions as possible before moving to the next level of detail. In most cases, understanding the higher level will allow us to see why the assumptions of the level below were valid. In this stepwise fashion, we build up an increasingly comprehensive understanding of the subject, always keeping in view the assumptions required for any given result. We avoid introducing concepts that we must "unlearn" as we progress. We use as our starting point the magnetic interaction of macroscopic quantum systems through the vector and scalar potentials $\mathbf{A}$ and V, which are the true observable quantities. For clarity, the brief discussion given here is limited to situations where the currents and voltages vary slowly; the four-vector generalization of these relations not only removes this quasi-static limitation, but gives us electrostatics as well [5, 6].

4.2   Model System

Our model system is a loop of superconducting wire — the two ends of the loop being colocated in space and either insulated or shorted, depending on the experimental situation. Experimentally, the voltage V between the two ends of the loop is related to the current I flowing through the loop by

$$LI = \int V dt = \Phi \qquad (4.1)$$

Two quantities are defined by this relationship: Φ, called the magnetic flux [1], and L, called the inductance, which depends on the dimensions of the loop.

Current is the flow of charge: $I = dQ/dt$. Each increment of charge dQ carries an energy increment $dW = V dQ$ into the loop as it enters [2]. The total energy W stored in the loop is thus

$$W = \int V dQ = \int V I dt = L \int \frac{dI}{dt} I dt = L \int I dI = \frac{1}{2} L I^2 \qquad (4.2)$$

If we reduce the voltage to zero by, for example, connecting the two ends of the loop to form a closed superconducting path, the current I will continue to flow indefinitely: a persistent current. If we open the loop and allow it to do work on an external circuit, we can recover all the energy W.

If we examine closely the values of currents under a variety of conditions, we find the full continuum of values for the quantities I, V, and Φ, except for persistent currents, where only certain discrete values occur for any given loop [7, 8]. By experimenting with loops of different dimensions, we find the condition that describes the values that occur experimentally:

$$\Phi = \int V dt = n\Phi_0 \qquad (4.3)$$

Here, n is any integer, and $\Phi_0 = 2.06783461 \times 10^{-15}$ volt-second is called the flux quantum or fluxoid; its value is accurate to a few parts in 10^9, independent of the detailed size, shape, or composition of the superconductor forming the loop. We also find experimentally that a rather large energy — sufficient to disrupt the superconducting state entirely — is required to change the value of n.

The more we reflect on Eq. 4.3, the more remarkable the result appears. The quantities involved are the voltage and the magnetic flux. These quantities are integrals of the quantities $\mathbf{E}$ and $\mathbf{B}$ that appear in Maxwell's equations, and are therefore usually associated with the electromagnetic field. Experimentally, we

---

[1]This definition is independent of the shape of the loops, and applies to coils with multiple turns. For multiturn coils, what we call the flux is commonly referred to as the total flux linkage.
[2]We use this relation to define the voltage V.

know that they can take on a continuum of values — except under special conditions, when the arrangement of matter in the vicinity causes the flux to take on precisely quantized values. In Maxwell's theory, **E** and **B** represented the state of strain in a mechanical medium (the ether) induced by electric charge. Einstein had a markedly different view (p.383 in ref. [9]): "I feel that it is a delusion to think of the electrons and the fields as two physically different, independent entities. Since neither can exist without the other, there is only *one* reality to be described, which happens to have two different aspects; and the theory ought to recognize this from the start instead of doing things twice." At the most fundamental level, the essence of quantum mechanics lies in the wave nature of matter. Einstein's view would suggest that electromagnetic variables are related to the wave properties of the electrons. Quantization is a familiar phenomenon in systems where the boundary conditions give rise to standing waves. The quantization of flux (Eq. 4.3) is a direct manifestation of the wave nature of matter, expressed in electromagnetic variables.

4.3   Matter

To most nonspecialists, quantum mechanics is a baffling mixture of waves, statistics, and arbitrary rules, ossified in a matrix of impenetrable formalism. By using a superconductor, we can avoid the statistics, the rules, and the formalism, and work directly with the waves. The wave concept, accessible to intuition and common sense, gives us "a way of thinking such that the law is evident." Electrons in a superconductor are described by a wave function that has an amplitude and a phase. The earliest treatment of the wave nature of matter was the 1923 wave mechanics of deBroglie. He applied the 1905 Einstein postulate ($W = \hbar w$) to the energy W of an electron wave, and identified the momentum **p** of an electron with the propagation vector of the wave: $\mathbf{p} = \hbar\mathbf{k}$. Planck's constant h and its radian equivalent $\hbar = h/2\pi$ are necessary for merely historical reasons — when our standard units were defined, it was not known that energy and frequency were the same quantity.

The Einstein-deBroglie relations apply to the collective electrons in a superconductor. The dynamics of the system can be derived from the dispersion relation [10] between ω and **k**. Both ω and **k** are properties of the phase of the wave function and do not involve the amplitude, which, in collective systems, is usually determined by some normalization condition. In a superconductor, the constraint of charge neutrality is such a condition.

The wave function must be continuous in space; at any given time, we can follow the phase along a path from one end of the loop to the other: The number of radians by which the phase advances as we traverse the path is the phase accumulation φ around the loop. If the phase at one end of the loop changes relative to that at the other end, that change must be reflected in the total phase accumulation around the loop. The frequency ω of the wave function at any point in space is the rate at which the phase advances per unit time. If the frequency at one end of the loop

(ω_1) is the same as that at the other end (ω_2), the phase difference between the two ends will remain constant, and the phase accumulation will not change with time. If the frequency at one end of the loop is higher than that at the other, the phase accumulation will increase with time, and that change must be reflected in the rate at which phase accumulates with the distance l along the path. The rate at which phase around the loop accumulates with time is the difference in frequency between the two ends. The rate at which phase accumulates with distance l is the component of the propagation vector $\mathbf{k}$ in the direction $\mathbf{dl}$ along the path. Thus, the total phase accumulated around the loop is

$$\varphi = \int (w_1 - w_2)dt = \oint \mathbf{k} \cdot \mathbf{dl} \tag{4.4}$$

We can understand quantization as an expression of the single-valued nature of the phase of the wave function. When the two ends of the loop were connected to an external circuit, the two phases could evolve independently. When the ends are connected to each other, however, the two phases must match up. But the phase is a quantity that has a cyclic nature — matching up means being equal modulo 2π. Thus, for a wave that is confined to a closed loop, and has a single-valued. continuous phase, the integral of Eq. 4.4 must be $n2\pi$, where n is an integer. The large energy required to change n is evidence that the phase constraint is a strong one — as long as the superconducting state stays intact, the wave function remains intact as well.

These relations tell us that the magnetic flux and the propagation vector will be quantized for a given loop; they do not tell us how the frequency ω in Eq. 4.4 is related to the potential V in Eq. 4.1. To make this connection, we must introduce one additional assumption: The collective electron system represented by the wave function is made up of elemental charges of magnitude q_0. By the Einstein relation, the energy $q_0 V$ of an elemental charge corresponds to a frequency $\omega = q_0 V/\hbar$.

4.4 Electrodynamics

Electrodynamics is the interaction of matter via the electromagnetic field. We can formulate our first relation between the electromagnetic quantities V and Φ and the phase accumulation φ of the wave function by comparing Eq. 4.1 with Eq. 4.4:

$$\varphi = \int \omega dt = \frac{q_0}{\hbar} \int V dt = \frac{q_0}{\hbar} n\Phi_0 = n(2\pi) \tag{4.5}$$

From Eq. 4.5, we conclude that $\Phi_0 = h/q_0$. We understand that the potential V and the frequency ω refer to differences in these quantities between the two ends of the loop. Equivalently, we measure each of these quantities at one end of the loop using as a reference the value at the other end of the loop. When we substitute into Eq. 4.5 the measured value of Φ_0 and the known value of h, we obtain for q_0 a value that is exactly twice the charge q_e of the free electron. The usual explanation for this somewhat surprising result is that each state in the superconductor is occupied by a pair of electrons, rather than by an individual electron, so the elemental charge q_0 should be $2q_e$, rather than q_e. None of the conclusions that we shall reach depends on the value of q_0.

We have established the correspondence between the potential V and the frequency ω — the time integral of each of these equivalent quantities in a closed loop is quantized. The line integral of the propagation vector $\mathbf{k}$ around a closed loop also is quantized. We would therefore suspect the existence of a corresponding electromagnetic quantity, whose line integral is the magnetic flux Φ. That quantity is the well-known vector potential $\mathbf{A}$. The general relations among these quantities, whether or not the loop is closed, are

$$\left. \begin{array}{l} \text{Phase } \varphi = \int \omega \mathrm{dt} = \oint \mathbf{k} \cdot \mathbf{dl} \\ \text{Flux } \Phi = \int V \mathrm{dt} = \oint \mathbf{A} \cdot \mathbf{dl} \end{array} \right\} \quad \Phi = \frac{\hbar}{q_0} \varphi \qquad (4.6)$$

Eq. 4.6 expresses the first set of fundamental relations of collective electrodynamics.

4.5   Coupling

Up to this point, we have tentatively identified the phase accumulation and the magnetic flux as two representations of the same physical entity. We assume that "winding up" the wave function with a voltage produces a propagation vector in the superconductor related to the motion of the electrons, and that this motion corresponds to a current because the electrons are charged. This viewpoint will allow us to understand the interaction between two coupled collective electron systems. We shall develop these relations in more detail when we study the current distribution within the wire itself.

Let us consider two identical loops of superconducting wire, the diameter of the wire being much smaller than the loop radius. We place an extremely thin insulator between the loops, which are superimposed on each other as closely as allowed by the insulator. In this configuration, both loops can be described, to an excellent approximation, by the same path in space, despite their being electrically distinct. As we experiment with this configuration, we make the following observations.

(i) When the two ends of the second loop are left open, its presence has no effect on the operation of the first loop. The relationship between a current flowing in the first loop and the voltage observed between the ends of the first loop follows Eq. 4.1 with exactly the same value of L as that observed when the second loop was absent.

(ii) The voltage observed between the two ends of the second loop under open conditions is almost exactly equal to that observed across the first loop.

(*iii*) When the second loop is shorted, the voltage observed across the first loop is nearly zero, independent of the current.

(*iv*) The current observed in the second loop under shorted conditions is nearly equal to that flowing in the first loop, but is of the opposite sign.

Similar measurements performed when the loops are separated allow us to observe how the coupling between the loops depends on their separation and relative orientation.

(*v*) For a given configuration, the voltage observed across the second loop remains proportional to the voltage across the first loop. The constant of proportionality, which is nearly unity when the loops are superimposed, decreases with the distance between the loops.

(*vi*) The constant of proportionality decreases as the axes of the two loops are inclined with respect to each other, goes to zero when the two loops are orthogonal, and reverses when one loop is flipped with respect to the other.

Observation *i* tells us that the presence of electrons in the second loop does not *per se* affect the operation of the first loop. The voltage across a loop is a direct manifestation of the phase accumulation around the loop. Observation *ii* tells us that current in a neighboring loop is as effective in producing phase accumulation in the wave function as is current in the same loop. The ability of current in one location to produce phase accumulation in the wave function of electrons in another location is called magnetic interaction. Observation *vi* tells us that the magnetic interaction is vectorial in nature. After making these and other similar measurements on many configurations, involving loops of different sizes and shapes, we arrive at the proper generalization of Eqs. 4.1 and 4.6:

$$\int V_1 dt = \oint \mathbf{A} \cdot \mathbf{dl_1} = \Phi_1 = L_1 I_1 + M I_2$$
$$\int V_2 dt = \oint \mathbf{A} \cdot \mathbf{dl_2} = \Phi_2 = M I_1 + L_2 I_2 \tag{4.7}$$

Here, the line elements $\mathbf{dl_1}$ and $\mathbf{dl_2}$ are taken along the first and second loops, respectively. The quantity M, which by observation *vi* can be positive or negative depending on the configuration, is called the mutual inductance; it is a measure of how effective the current in one loop is at causing phase accumulation in the other. When $L_1 = L_2 = L$, the magnitude of M can never exceed L. Observations $i - iv$ were obtained under conditions where $M \approx L$. Experiments evaluating the mutual coupling of loops of different sizes, shapes, orientations, and spacings indicate that each element of wire of length dl carrying the current $\mathbf{I}$ makes a contribution to $\mathbf{A}$ that is proportional to $\mathbf{I}$, and to the inverse of the distance r from the current element to the point at which $\mathbf{A}$ is evaluated:

$$\mathbf{A} = \frac{\mu_0}{4\pi} \int \frac{\mathbf{I}}{r} dl \Rightarrow \mathbf{A} = \frac{\mu_0}{4\pi} \int \frac{\mathbf{J}}{r} d\text{vol} \tag{4.8}$$

The constant μ_0 is called the permeability of free space. The second form follows from the first if we visualize a distribution of current as carried by a large number of wires of infinitesimal cross section, and the current density $\mathbf{J}$ as being the number of such wires per unit area normal to the current flow. The $1/r$ form of the integrand of Eq. 4.8 is called the Green's function; it tells us how the vector potential is generated by currents everywhere in space. It is perhaps more correct to say that the vector potential is a bookkeeping device for evaluating the effect at a particular point of all currents everywhere in space. Ernst Mach wrote (p.317 in ref. [1]), "We cannot regard it as impossible that *integral laws* ... will some day take the place of the ... differential laws that now make up the science of mechanics ... In such an event, the concept of force will have become superfluous." Eqs. 4.6 and 4.8 are the fundamental integral laws for collective electromagnetic interaction. The equivalent differential equation is $\nabla_2 \mathbf{A} = -\mu_0 \mathbf{J}$ [5, 6].

We can express Eq. 4.2 in a way that gives us additional insight into the energy stored in the coil:

$$W = \int V dQ = \int V I dt = \int I d\Phi \qquad (4.9)$$

Eq. 4.9 is valid for any $\mathbf{A}$; it is not limited to the $\mathbf{A}$ from the current in the coil itself. The integrals in Eq. 4.9 involve the entire coil. From them we can take a conceptual step and, using our visualization of the current density, imagine an energy density $\mathbf{J} \cdot \mathbf{A}$ ascribed to every point in space:

$$W = \int \mathbf{I} \cdot \mathbf{A} dl = \int \mathbf{J} \cdot \mathbf{A} dvol \qquad (4.10)$$

4.6   Electrodynamic Momentum

Feynman commented on the irrelevance of the concept of force in a quantum context. At the fundamental level, we can understand the behavior of a quantum system using only the wave properties of matter. But we experience forces between currents in every encounter with electric motors, relays, and other electromagnetic actuators. How do these forces arise from the underlying quantum reality? We can make a connection between the classical concept of force and the quantum nature of matter through the concept of momentum. Using the deBroglie postulate relating the momentum $\mathbf{p}$ of an electron to the propagation vector $\mathbf{k}$ of the wave function, and identifying the two integrands in Eq. 4.6, the electrodynamic momentum of an elemental charge is

$$\mathbf{p} = \hbar \mathbf{k} = q_0 \mathbf{A} \qquad (4.11)$$

We shall now investigate the electrodynamic momentum in one of our loops of superconducting wire. There is an electric field E along the loop, the line integral of which is the voltage V between the ends. From a classical point of view, Newton's law tells us that the force $q_0 E$ on a charge should be equal to the time rate of change of momentum. From Eq. 4.11,

$$q_0 \mathbf{E} = \frac{\partial \mathbf{p}}{\partial t} = q_0 \frac{\partial \mathbf{A}}{\partial t} \Rightarrow V = \oint \mathbf{E} \cdot \mathbf{dl} = \frac{\partial \Phi}{\partial t} \qquad (4.12)$$

Integrating the second form of Eq. 4.12 with respect to time, we recover Eq. 4.6, so the classical idea of inertia is indeed consistent with the quantum behavior of our collective system. Electrodynamic inertia acts exactly as a classical mechanical inertia: It relates the integral of a force to a momentum, which is manifest as a current. We note that, for any system of charges that is overall charge neutral, as is our superconductor, the net electromagnetic momentum is zero. For the $-q\mathbf{A}$ of each electron, we have a canceling $+q\mathbf{A}$ from one of the background positive charges. The electric field that accelerates electrons in one direction exerts an equal force in the opposite direction on the background positive charges. We have, however, just encountered our first big surprise: We recognize the second form of Eq. 4.12, which came from Newton's law, as the integral form of one of Maxwell's equations!

We would expect the total momentum P of the collective electron system to be the momentum per charge times the number of charges in the loop. If there are η charges per unit length of wire that take part in the motion, integrating Eq. 4.11 along the loop gives

$$P = \eta q_0 \oint \mathbf{A} \cdot \mathbf{dl} = \eta q_0 \Phi = \eta q_0 L I \qquad (4.13)$$

The current I is carried by the η charges per unit length moving at velocity v; therefore, $I = \eta q_0 v$, and Eq. 4.13 becomes

$$P = L(\eta q_0)^2 v \qquad (4.14)$$

The momentum is proportional to the velocity, as it should be. It is also proportional to the size of the loop, as reflected by the inductance L. Here we have our second big surprise: instead of scaling linearly with the number of charges that take part in the motion, the momentum of a collective system scales as the square of the number of charges! We can understand this collective behavior as follows. In an arrangement where charges are constrained to move in concert, each charge produces phase accumulation, not only for itself, but for all the other charges as well. So the inertia of each charge increases linearly with the number of charges moving in concert. The inertia of the ensemble of coupled charges must therefore increase as the square of the number of charges.

4.7   Forces on Currents

In our experiments on coupled loops, we have already seen how the current in one loop induces phase accumulation in another loop; the relations involved were captured in Eq. 4.7. In any situation where we change the coupling of collective systems by changing the spatial arrangement, mechanical work may be involved. Our model system for studying this interaction consists of two identical shorted loops of individual inductance L_0, each carrying a persistent flux Φ. As long as the superconducting state retains its integrity, the cyclic constraint on the wave function guarantees that the flux Φ in each loop will be constant, independent of the coupling between loops. Because M enters symmetrically in Eq. 4.7, the current I will be the same in both loops. Hence, L_0 and Φ will remain constant, whereas M and I will be functions of the spatial arrangement of the loops — M will be large and positive when the loops are brought together with their currents flowing in the same direction, and will be large and negative when the loops are brought together with their currents flowing in opposite directions. From Eq. 4.7, $\Phi = (L_0 + M)I$. Substituting Φ into Eq. 4.9, and noting that the total energy of the system is twice that for a single coil,

$$W = 2 \int I d\Phi = (L_0 + M)I^2 = \frac{\Phi^2}{(L_0 + M)} \tag{4.15}$$

The force F_x along some direction x is defined as the rate of change of energy with a change in the corresponding coordinate:

$$F_x = \frac{\partial W}{\partial x} = -\left(\frac{\Phi}{L_0 + M}\right)^2 \frac{\partial M}{\partial x} \tag{4.16}$$

The negative sign indicates an attractive force because the mutual inductance M increases as the coils — whose currents are circulating in the same direction — are moved closer. It is well known that electric charges of the same sign repel each other. We might expect the current, being the spatial analog of the charge, to behave in a similar manner. However, Eq. 4.15 indicates that the total energy of the system decreases as M increases. How does this attractive interaction of currents circulating in the same direction come about?

The electron velocity is proportional to I. As M is increased, the electrons in both loops slow down because they have more inertia due to the coupling with electrons in the other loop. This effect is evident in Eq. 4.15, where $I = \Phi/(L_0+M)$. Thus, there are two competing effects: The decrease in energy due to the lower velocity, and the increase in energy due to the increase in inertia of each electron. The energy goes as the square of the velocity, but goes only linearly with the inertia, so the velocity wins. The net effect is a decrease in energy as currents in the same direction are coupled, and hence an attractive force. We can see how the classical force law discovered in 1823 by Ampère arises naturally from the collective quantum behavior, which determines not only the magnitude, but also the sign, of the effect.

4.8   Multiturn Coils

The interaction in a collective system scales as the square of the number of electrons moving in concert. Thus, we might expect the quantum scaling laws to be most clearly manifest in the properties of closely coupled multiturn coils, where the number of electrons is proportional to the number of turns. We can construct an N-turn coil by connecting in series N identical, closely coupled loops. In this arrangement, the current through all loops is equal to the current I through the coil, and the voltage V across the coil is equal to the sum of the individual voltages across the loops. If A_0 is the vector potential from the current in one loop, we expect the vector potential from N loops to be NA_0, because the current in each loop contributes. The flux integral is taken around N turns, so the path is N times the length l_0 of a single turn. The total flux integral is thus

$$\Phi = \int V\, dt = \int_0^{Nl_0} N\mathbf{A_0} \cdot \mathbf{dl} = N^2 L_0 I \qquad (4.17)$$

From Eq. 4.17 we conclude that an N-turn closely coupled coil has an inductance $L = N^2 L_0$ Once again, we see the collective interaction scaling as the square of the number of interacting charges. We remarked that collective quantum systems have a correspondence limit markedly different from that of classical mechanical systems. When two classical massive bodies, each body having a separate inertia, are bolted together, the inertia of the resulting composite body is simply the sum of the two individual inertias. The inertia of a collective system, however, is a manifestation of the interaction, and cannot be assigned to the elements separately. This difference between classical and quantum systems has nothing to do with the size scale of the system. Eq. 4.17 is valid for large as well as for small systems; it is valid where the total phase accumulation is an arbitrary number of cycles — where the granularity of the flux due to $\hbar$ is as small as might be required by any correspondence procedure. Thus, it is clear that collective quantum systems do not have a classical correspondence limit.

4.9   Total Momentum

To see why our simplistic approach has taken us so far, we must understand the current distribution within the superconductor itself. We saw that the vector potential made a contribution to the momentum of each electron, which we called the electrodynamic momentum: $\mathbf{p_{el}} = q\mathbf{A}$. The mass m of an electron moving with velocity $\mathbf{v}$ also contributes to the electron's momentum: $\mathbf{p_{mv}} = m\mathbf{v}$. The total momentum is the sum of these two contributions:

$$\hbar\mathbf{k} = \mathbf{p} = \mathbf{p_{el}} + \mathbf{p_{mv}} = q_0\mathbf{A} + m\mathbf{v} \qquad (4.18)$$

The velocity $\mathbf{v} = (\hbar\mathbf{k} - q_0\mathbf{A})/m$ is thus a direct measure of the imbalance between the total momentum $\hbar\mathbf{k}$ and the electrodynamic momentum $q_0\mathbf{A}$. When these two quatitities are matched, the velocity is zero. The current density is just the motion of $\mathcal{N}$ elementary charges per unit volume: $\mathbf{J} = q_0 \mathcal{N} \mathbf{v}$. We can thus express Eq. 4.18 in terms of the wave vector k, the vector potential $\mathbf{A}$, and the current density $\mathbf{J}$:

$$\mathbf{J} = \frac{q_0 \mathcal{N}}{m}(\hbar \mathbf{k} - q_0 \mathbf{A}) \tag{4.19}$$

4.10   Current Distribution

We are now in a position to investigate how current distributes itself inside a superconductor. If $\mathbf{A}$ were constant throughout the wire, the motion of the electrons would be determined by the common wave vector $\mathbf{k}$ of the collective electron system, and we would expect the persistent current for a given flux to be proportional to the cross-sectional area of the wire, and thus the inductance L of a loop of wire to be inversely related to the wire cross section. When we perform experiments on loops of wire that have identical paths in space, however, we find that the inductance is only a weak function of the wire diameter, indicating that the current is not uniform across the wire, and therefore that $\mathbf{A}$ is far from constant. If we make a loop of superconducting tubing, instead of wire, we find that it has exactly the same inductance as does a loop made with wire of the same diameter, indicating that current is flowing at the surface of the loop, but is not flowing throughout the bulk.

Before taking on the distribution of current in a wire, we can examine a simpler example. In a simply connected bulk superconductor, the single-valued nature of the wave function can be satisfied only if the phase is everywhere the same: $\mathbf{k} = 0$. Any phase accumulation induced through the $\mathbf{A}$ vector created by an external current will be canceled by a screening current density $\mathbf{J}$ in the opposite direction, as we saw in observations *iii* and *iv*. To make the problem tractable, we consider a situation where a vector potential $\mathbf{A}_0$ at the surface of a bulk superconducting slab is created by distant currents parallel to the surface of the slab. The current distribution perpendicular to the surface is a highly localized phenomenon, so it is most convenient to use the differential formulation of Eq. 4.8. We suppose that conditions are the same at all points on the surface, and therefore that $\mathbf{A}$ changes in only the x direction, perpendicular to the surface, implying that $\nabla^2 \mathbf{A} = \partial^2 \mathbf{A}/\partial x^2$.

$$\nabla^2 \mathbf{A} = \frac{\partial^2 \mathbf{A}}{\partial x^2} = -\mu_0 \mathbf{J} = \frac{\mu_0 q_0^2 \mathcal{N}}{m} \mathbf{A} \tag{4.20}$$

The solution to Eq. 4.20 is

$$\mathbf{A} = \mathbf{A}_0 e^{-x/\lambda} \quad \lambda^2 = \frac{m}{\mu_0 q_0^2 \mathcal{N}} \tag{4.21}$$

The particular form of Eq. 4.21 depends on the geometry, but the qualitative result is always the same, and can be understood as follows: The current is the imbalance between the wave vector and the vector potential. When an imbalance exists, a current proportional to that imbalance will flow such that it cancels out the imbalance. The resulting screening current dies out exponentially with distance from the source of imbalance. The distance scale at which the decay occurs is given by λ, the screening distance, penetration depth, or skin depth. For a typical superconductor, $\mathcal{N}$ is of the order of $10^{28}/m^3$, so λ should be a few tens of nanometers. Experimentally, simple superconductors have $\lambda \approx 50$ nanometers — many orders of magnitude smaller than the macroscopic wire thickness that we are using.

4.11   Current in a Wire

At long last, we can visualize the current distribution within the superconducting wire itself. Because the skin depth is so small, the surface of the wire appears flat on that scale, and we can use the solution for a flat surface. The current will be a maximum at the surface of the wire, and will die off exponentially with distance into the interior of the wire. We can appreciate the relations involved by examining a simple example. A 10-cm-diameter loop of 0.1-mm-diameter wire has an inductance of 4.4×10^{-7} Henry (p.193 in ref. [11]): A persistent current of 1 Ampere in this loop produces a flux of 4.4×10^{-7} volt-second, which is 2.1×10^8 flux quanta. The electron wave function thus has a total phase accumulation of 2.1×10^8 cycles along the length of the wire, corresponding to a wave vector $k = 4.25 \times 10^9$ m^{-1}. Due to the cyclic constraint on the wave function, this phase accumulation is shared by all electrons in the wire, whether or not they are carrying current.

In the region where current is flowing, the moving mass of the electrons contributes to the total phase accumulation. The 1-Ampere of current results from a current density of 6.4×10^{10} Amperes per square meter flowing in a thin "skin" $\approx \lambda$, just inside the surface. This current density is the result of the 10^{28} electrons per cubic meter moving with a velocity of $v \approx 20$ meters per second. The mass of the electron moving at this velocity contributes $mv/\hbar = 1.7 \times 10^5$m^{-1} to the total wave vector of the wave function, which is less than one part in 10^4 of that contributed by the vector potential. That small difference, existing in about 1 part in 10^6 of the cross-sectional area, is enough to bring $\mathbf{k}$ and $\mathbf{A}$ into balance in the interior of the wire.

In the interior of the wire, the propagation vector of the wave function is matched to the vector potential, and the current is therefore zero. As we approach the surface, $\mathbf{A}$ decreases slightly, and the difference between $\mathbf{k}$ and $\mathbf{A}q_0/\hbar$ is manifest as a current. At the surface, the value and radial slope of $\mathbf{A}$ inside and outside the wire match, and the value of $\mathbf{A}$ is still within one part in 10^4 of that in the center of the wire. So our simplistic view — that the vector potential and the wave vector were two representations of the same quantity — is precisely true in the center of the wire, and is nearly true even at the surface. The current $\mathbf{I}$ is not the propagation vector $\mathbf{k}$ of the wave, but, for a fixed configuration, $\mathbf{I}$ is proportional to

k by Eqs. 4.8 and 4.19. For that reason, we were able to deduce the electromagnetic laws relating current and voltage from the quantum relations between wave vector and frequency.

4.12 Conclusion

We took to heart Einstein's belief that the electrons and the fields were two aspects of the same reality, and were able to treat the macroscopic quantum system and the electromagnetic field as elements of a unified subject. We heeded Mach's advice that classical mechanics was not the place to start, followed Feynman's directive that interactions change the wavelengths of waves, and saw that there is a correspondence limit more appropriate than the classical-mechanics version used in traditional introductions to quantum theory. We found Newton's law masquerading as one of Maxwell's equations. We were able to derive a number of important results using only the simplest properties of waves, the Einstein postulate relating frequency to energy, the deBroglie postulate relating momentum to wave vector, and the discrete charge of the electron. It thus appears possible to formulate a unified, conceptually correct introduction to both the quantum nature of matter and the fundamental laws of electromagnetic interaction without using either Maxwell's equations or standard quantum formalism.

I am indebted to Richard F. Lyon, Sanjoy Mahajan, William B. Bridges, Rahul Sarpeshkar, Richard Neville, and Lyn Dupre for helpful discussion and critique of the material, and to Calvin Jackson for his help in preparing the manuscript. The work was supported by the Arnold and Mabel Beckman Foundation, and by Gordon and Betty Moore.

References

[1] F. Mach, *The Science of Mechanics* (Open Court, La Salle, IL, 1960).

[2] R. P. Feynman, R. B. Leighton and M. Sands, *The Feynman Lectures on Physics*, Vol.2, (Addison-Wesley, Reading, MA, 1964).

[3] Y. Aharonov and D. Bohm, *Phys. Rev.* **115**, 485-491 (1959).

[4] F. London, *Superfluids* (Wiley, New York, 1950).

[5] A. Sommerfeld, *Electrodynamics* (Academic, New York, 1952).

[6] P. M. Morse and H. Feshbach, *Methods of Theoretical Physics I* (McGraw-Hill New York, 1953).

[7] B. S. Deaver and W. M. Fairbank, *Phys. Rev. Lett.* **7**, 43-46 (1961).

[8] R. Doll and M. Nabauer, *Phys. Rev. Lett.* **7**, 51-52 (1961).

[9] E. T. Jaynes, in *Complexity Entropy and the Physics of Information*, ed. W. H. Zurek, pp. 381-403 (Addison-Wesley, Reading, MA, 1990).

[10] R. P. Feynman and A. R. Hibbs, *Quantum Mechanics and Path Integrals* (McGraw-Hill, New York, 1965).

[11] S. Ramo, J. R. Whinnery and T. van Duzer, *Fields and Waves in Communication Electronics* (Wiley, New York, 1994).

5

A MEMORY

Gerald Jay Sussman

I met and worked with Richard P. Feynman when I was on sabbatical leave from MIT for the year 1983-1984. At MIT I had worked in various aspects of Computer Science, including Artificial Intelligence, Computer Architecture, and High-Level Programming Languages. Hal Abelson and I had been developing an exciting and novel introductory subject for our Department entitled "Structure and Interpretation of Computer Programs." In fact, we had just produced a first draft of our book by that name as an MIT AI Laboratory Technical Report. This report came to be called the "Yellow Wizard Book", because of the picture on its yellow cover. Richard's son, Carl, had been my student at MIT.

I went to Caltech to learn about dynamical astrophysics, a perennial interest of mine. My friend and former teacher, Professor Alar Toomre of the MIT Department of Mathematics, had arranged that I would work with Peter Goldreich's group for that year. My wife, Julie, and I went to Caltech in the Spring to look into the housing situation for our stay. Mr. Feynman met us for lunch at the Athenaeum (the Caltech Faculty Club) . Apparently he had read the yellow wizard book and liked it — he said that he learned a lot from reading it. At some point Feynman, in his characteristic gruff voice, said "My son says you're a pretty good teacher. I am going to teach a course on computing — on the ultimate physical limitations of computation, and on the computational aspects of physics. Will you help me?" Now, I had been planning on learning about astrophysics for that year, so I was a bit leery of taking on a teaching assignment, but the prospect of working with Feynman was difficult to resist. I asked him when the course met and he said that it was on Monday, Wednesday and Friday, at 11:00 to 12:00, for the entire year. I replied that I would help with the subject if he would eat lunch with me afterwards. He agreed, and that was one of the best deals I ever made in my life.

So, during the 1983-1984 school year I helped Richard Feynman develop and teach a subject entitled "Potentialities and Limitations of Computing Machines" at Caltech. Our class consisted mostly of graduate students in physics and in computer science. We had a lot of fun. We taught students how to program, in Scheme, using the Yellow Wizard Book. I provided a computing laboratory component for the subject, using a Caltech computer. Richard really loved to program: he stayed up one night with me learning how to write power-series manipulations using infinite streams. We taught a bit about hardware design using MOS circuits. We investigated the limitations of size for circuits based on lithography and methods such as ion implantation and diffusion. We estimated how much variation we could expect in the parameters of an inverter as the transistors became smaller. We told students a bit of information theory (up to Shannon's theorem) and about error-correcting codes. Richard really liked the recursive application of these ideas. We examined the thermodynamic arguments and we discussed algorithmic information. We looked at the ideas of Landauer, Fredkin, Toffoli, Margolus, and Bennett, concerning the reversibility of computation and what that could mean physically. From here we were led to cellular automata. We discussed classic examples. We examined how such systems could be universal and in what sense they could be models for the physical world.

We also discussed what was then known about the quantum mechanics of computation. Richard explained his work on the computational difficulty of simulating a quantum-mechanical system with a classical computer. He also gave a few lectures in which he developed a quantum-mechanical state machine, composed of a PLA (programmed logic array) built up from operators, and a state register described in terms of spin states. We estimated the cycle time that could be expected from diffusion of the state, and how energy could be used to speed things up.

While this class was under way, I assembled a group to help me build the Digital Orrery, a special-purpose computer designed to do high-precision integrations for orbital-mechanics experiments. Using the Digital Orrery, I later worked with Jack Wisdom to discover numerical evidence for chaotic motions in the outer planets. The Digital Orrery is now retired at the Smithsonian Institution in Washington, D.C. One serious problem with long integrations is the buildup of numerical error. Before beginning the design I assumed that the software would be easy, and that the complete numerical analysis for such a venerable problem would be in the textbooks. Although he did not actually contribute to the design or to the development of the algorithms we ultimately used, Richard put out significant effort to help me to understand this problem, just how bad it was, and how little was known about its solution.

My last encounter with Richard Feynman was not long before he died. Richard and Carl ate dinner with Julie and me at our house. We spent the rest of the evening practicing lock picking with locks from my rather extensive collection. I still sorely miss my friend, Richard Feynman. When thinking about interesting ideas, from either physics or computation, I often ask myself, "What would Feynman think of this?"

6

NUMERICAL EVIDENCE THAT THE MOTION OF PLUTO IS CHAOTIC

Gerald Jay Sussman and Jack Wisdom *

Abstract

The Digital Orrery has been used to perform an integration of the motion of the outer planets for 845 million years. This integration indicates that the long-term motion of the planet Pluto is chaotic. Nearby trajectories diverge exponentially with an e-folding time of only about 20 million years.

6.1 Introduction

The determination of the stability of the solar system is one of the oldest problems in dynamical astronomy but despite considerable attention all attempts to prove the stability of the system have failed. Arnold has shown that a large proportion of possible solar systems are quasiperiodic if the masses, and orbital eccentricities and inclinations, of the planets are sufficiently small [1]. The actual solar system, however, does not meet the stringent requirements of the proof. Certainly, the great age of the solar system suggests a high level of stability, but the nature of the long-term motion remains undetermined. The apparent analytical complexity of the problem has led us to investigate the stability by means of numerical models. We have investigated the long-term stability of the solar system through an 845-million-year numerical integration of the five outermost planets with the Digital Orrery [2], a special-purpose computer for studying planetary, motion.

Pluto's orbit is unique among the planets. It is both highly eccentric ($e \approx 0.25$) and highly inclined ($i \approx 16°$). The orbits of Pluto and Neptune cross one another, a condition permitted only by the libration of a resonant argument associated with the 3:2 commensurability between the orbital periods of Pluto and Neptune. This resonance, which has a libration period near 20,000 years [3], ensures that Pluto is far from perihelion when Pluto and Neptune are in conjunction. Pluto also participates in a resonance involving its argument of perihelion, the angle between the ascending node and the perihelion, which librates about $\pi/2$ with a period of 3.8 million years [4]. This resonance guarantees that the perihelion of Pluto's orbit is far from the line of intersection of the orbital planes of Pluto and Neptune, further ensuring that close encounters are avoided.

*Reprinted from Science, Vol.241, Research Articles, pp. 433–437, 22 July 1988

We found in our 200-million-year integrations of the outer planets [5] that Pluto's orbit also undergoes significant variations on much longer time scales. The libration of the argument of perihelion is modulated with a period of 34 million years, and $h = e \sin \tilde{w}$, where e is the eccentricity and $\tilde{w}$ is the longitude of perihelion, shows significant long-period variations with a period of 137 million years. The appearance of the new 34-million-year period might have been expected, because Pluto must have two independent long-period frequencies, but the 137-million-year period was completely unexpected. It results from a near commensurability between the frequency of circulation of Pluto's ascending node and one of the principal secular frequencies of the massive planets. Pluto also participates in two other resonances involving the frequency of oscillation of the argument of perihelion and the principal secular frequencies. In our 200-million-year integration Pluto's inclination appeared to have even longer periods or possibly a secular decrease.

The similarity of Pluto's peculiar highly eccentric and inclined orbit to chaotic asteroid orbits [6], together with the very long periods, Pluto's participation in a large number of resonances, and the possible secular decline in inclination compelled us to carry out longer integrations of the outer planets to clarify the nature of the long-term evolution of Pluto. Our new numerical integration indicates that in fact the motion of the planet Pluto is chaotic.

6.2 Deterministic chaotic behavior

In most conservative dynamical systems Newton's equations have both regular solutions and chaotic solutions. For some initial conditions the motion is quasiperiodic; for others the motion is chaotic. Chaotic behavior is distinguished from quasiperiodic behavior by the way in which nearby trajectories diverge [6, 7]. Nearby quasiperiodic trajectories diverge linearly with time, on average, whereas nearby chaotic trajectories diverge exponentially with time. Quasiperiodic motion can be reduced to motion on a multidimensional torus; the frequency spectrum of quasiperiodic motion has as many independent frequencies as degrees of freedom. The frequency spectrum of chaotic motion is more complicated, usually appearing to have a broad-band component.

The Lyapunov exponents measure the average rates of exponential divergence of nearby orbits. The Lyapunov exponents are limits for large time of the quantity $\gamma = \ln(d/d_0)/(t - t_0)$ where d is the distance in phase space between the trajectory and an infinitesimally nearby test trajectory, and t is the time. For any particular trajectory of an n-dimensional system there can be n distinct Lyapunov exponents, depending on the phase-space direction from the reference trajectory to the test trajectory. In Hamiltonian systems the Lyapunov exponents are paired; for each non-negative exponent there is a non-positive exponent with equal magnitude. Thus an m-degree-of-freedom Hamiltonian system can have at most m positive exponents. For chaotic trajectories the largest Lyapunov exponent is positive; for quasiperiodic

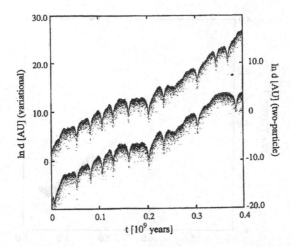

Fig. 6.1. The exponential divergence of nearby trajectories is indicated by the average linear growth of the logarithms of the distance measures as a function of time. In the upper trace we see the growth of the variational distance around a reference trajectory (left vertical axis). In the lower trace we see how two Plutos diverge with time (right vertical axis). The distance saturates near 45 AU; note that the semimajor axis of Pluto's orbit is about 40 AU. The variational method of studying neighboring trajectories does not have the problem of saturation. Note that the two methods are in excellent agreement until the two-trajectory method has nearly saturated.

trajectories all of the Lyapunov exponents are zero.

Lyapunov exponents can be estimated from the time evolution of the phase-space distance between a reference trajectory and nearby test trajectories [7, 8]. The most straightforward approach is to simply follow the trajectories of a small cloud of particles started with nearly the same initial conditions. With a sufficiently long integration we can determine if the distances between the particles in the cloud diverge exponentially or linearly. If the divergence is exponential, then for each pair of particles in the cloud we obtain an estimate of the largest Lyapunov exponent. With this method the trajectories eventually diverge so much that they no longer sample the same neighborhood of the phase space. We could fix this by periodically rescaling the cloud to be near the reference trajectory, but we can even more directly study the behavior of trajectories in the neighborhood of a reference trajectory by integrating the variational equations along with the reference trajectory. In particular, let $\mathbf{y}' = f(\mathbf{y})$ be an autonomous system of first-order ordinary differential equations and $\mathbf{y}(t)$ be the reference trajectory. We define a phase-space variational trajectory $\mathbf{y} + \delta\mathbf{y}$ and note that $\delta\mathbf{y}$ satisfies a linear system of first-order ordinary differential equations with coefficients that depend on $\mathbf{y}(t)$, $\delta\mathbf{y}' = \mathbf{J} \cdot \delta\mathbf{y}$ where the elements of the Jacobian matrix are $J_{ij} = \partial f_i / \partial y_j$.

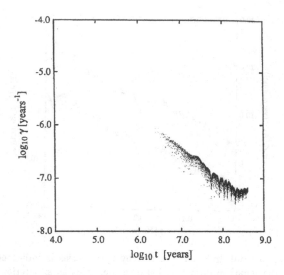

Fig. 6.2. The conventional representation of the Lyapunov exponent calculation, the logarithm of γ versus the logarithm of time. Convergence to a positive exponent is indicated by a leveling off; for regular trajectories this plot approaches a line with slope minus one.

6.3   Our numerical experiment

For many years the longest direct integration of the outer planets was the 1-million-year integration of Cohen, Hubbard, and Oesterwinter [9]. Recently several longer integrations of the outer planets have been performed [5, 10, 11]. The longest was our set of 200-million-year integrations. Our new 845-million-year integration is significantly longer and more accurate than all previously reported long-term integrations.

In our new integration of the motion of the outer planets the masses and initial conditions are the same as those used in our 200-million-year integrations of the outer planets. The reference frame is the invariable frame of Cohen, Hubbard, and Oesterwinter. The planet Pluto is taken to be a zero-mass test particle. We continue to neglect the effects of the inner four planets, the mass lost by the Sun as a result of electromagnetic radiation and solar wind, and general relativity. The most serious limitation of our integration is our ignorance of the true masses and initial conditions. Nevertheless, we believe that our model is sufficiently representative of the actual solar system that its study sheds light on the question of stability of the solar system. To draw more rigorous conclusions, we must determine the sensitivity of our conclusions to the uncertainties in masses and initial conditions, and to unmodeled effects.

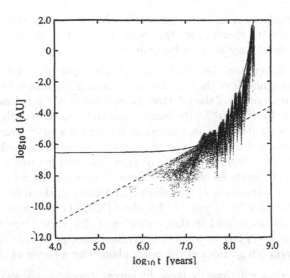

Fig. 6.3. Common logarithm of the distance between several pairs of Plutos, in AU, versus the common logarithm of the time, in years. The initial segment of the graph closely fits a 3/2 power law (dashed line). The solid line is an exponential chosen to fit the long-time divergence of Plutos. The exponential growth takes over when its slope exceeds the slope of the power law.

Our earlier integrations were limited to 100 million years forward and backward in time because of the accumulation of error, which was most seriously manifested in an accumulated longitude error of Jupiter of order 50°. In our new integrations we continue to use the 12th-order Störmer predictor [12], but a judicious choice of step size has reduced the numerical errors by several orders of magnitude. In all of our integrations the error in energy of the system varies nearly linearly with time. In the regime where neither roundoff nor truncation error is dominant the slope of energy as a function of time depends on step size in a complicated way. For some step sizes the energy level has a positive slope; for others the slope is negative. This suggests that there might be special step sizes for which there is no linear growth of energy error. By a series of numerical experiments we indeed found that there are values of the step size where the slope of the linear trend of energy vanishes. The special step sizes become better defined as the integration interval of the experiments is increased.

We chose our step size on the basis of a dozen 3-million-year integrations, and numerous shorter integrations. For our new long integration we chose the step size to be 32.7 days. This seemingly innocuous change from a step size near 40 days dramatically reduces the slope of the energy error, by roughly three orders of magnitude. If the numerical integration were truncation error-dominated, for which the accumulated error is proportional to h^n, where h is the step size and n is the order of the integrator, then this reduction of step size would improve the accumulated error by only about a factor of 10.

In our new integration the relative energy error (energy minus initial energy divided by the magnitude of the initial energy) accumulated over 845 million years is -2.6×10^{-10}; the growth of the relative energy error is still very nearly linear with a slope of -3.0×10^{-19} year^{-1}. By comparison the rate of growth of the relative energy error in our 200-million-year integrations was 1.8×10^{-16} year^{-1}. The errors in other integrations of the outer solar system were comparable to the errors in our 200-million-year integrations. The rate of growth of energy error in the 1-million-year integration of Cohen, Hubbard, and Oesterwinter was 2.4×10^{-16} year^{-1}. For the 6-million-year integration of Kinoshita and Nakai [10] the relative energy error was approximately 5×10^{-16} year^{-1}. For the LONGSTOP integration the growth of relative energy (as defined in this article) was -2.5×10^{-16} year^{-1}. Thus the rate of growth of energy error in the integration reported here is smaller than all previous long-term integrations of the outer planets by a factor of about 600.

We verified that this improvement in energy conservation was reflected in a corresponding improvement in position and velocity errors by integrating the outer planets forward 3 million years and then backward to recover the initial conditions, over a range of step sizes. For the chosen step size of 32.7 days the error in recovering the initial positions of each of the planets is of order 10^{-5} astronomical units (AU) or about 1500 km. Note that Jupiter has in this time traveled 2.5×10^{15} km.

The error in the longtitude of Jupiter can be estimated if we assume that the energy error is mainly in the orbit of Jupiter. The relative energy error is proportional to the relative error in orbital frequency so the error in longitude is proportional to the integral of the relative energy error: $\Delta\lambda \approx tn\Delta E(t)/E$, where n is the mean motion of Jupiter and t is the time of integration. Because the energy error grows linearly with time the position error grows with the square of the time. The accumulated error in the longitude of Jupiter after 100 million years is only about 4 arc minutes. This is to be compared with the 50° accumulated error estimated for our 200-million-year integrations. The error in the longitude of Jupiter after the full 845 million years is about 5°.

We have directly measured the integration error in the determination of the position of Pluto by integrating forward and backward over intervals as long as 3 million years to determine how well we can reproduce the initial conditions. Over such short intervals the round-trip error in the position of Pluto grows as a power of the time with an exponent near 2. The error in position is approximately $1.3 \times 10^{-19} t^2$ AU (where t is in years). This growth of error is almost entirely in the integration of Pluto's orbit; the round-trip error is roughly the same when we integrate the whole system and when we integrate Pluto in the field of the Sun only. It is interesting to note that in the integrations with the 32.7-day step size the

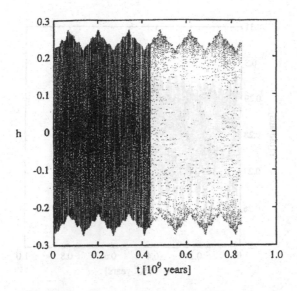

Fig. 6.4. The orbital element $h = e \sin \tilde{w}$ for Pluto over 845 million years. On this scale the dominant period (the 3.7-million-year circulation of the longitude of perihelion) is barely resolved. The most obvious component has a period of 137 million years. The sampling interval was increased in the second half of our integration.

position errors in all the planets are comparable. Extrapolation of the round-trip error for Pluto over the full 845-million-year integration gives an error in longitude of less than 10 arc minutes.

6.4   Lyapunov exponent of Pluto

We estimated the largest Lyapunov exponent of Pluto by both the variational and the phase-space distance methods during the second half of our 845-million-year run. Fig. 6.1 shows the logarithm of the divergence of the phase-space distance in a representative two-particle experiment and the growth of the logarithm of the variational phase-space distance. We measured the phase-space distance by the ordinary Euclidean norm in the six-dimensional space with position and velocity coordinates. We measured position in AU and velocity in AU/day. Because the magnitude of the velocity in these units is small compared to the magnitude of the position, the phase-space distance is effectively equivalent to the positional distance, and we refer to phase-space distances in terms of AU. For both traces in this plot the average growth is linear, indicating exponential divergence of nearby trajectories with an e-folding time of approximately 20 million years. The shapes of these graphs are remarkably similar until the two-particle divergence grows to about 1

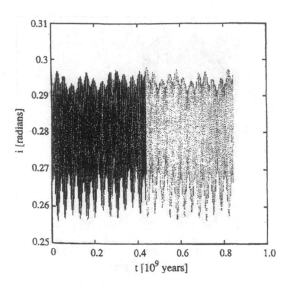

Fig. 6.5. The inclination i of Pluto over 845 million years. Besides the 34-million-year component and the 150-million-year component there appears to be a component with a period near 600 million years.

AU, verifying that the motion in the neighborhood of Pluto is properly represented. A more conservative representation of this data is to plot the logarithm of γ versus the logarithm of time (Fig. 6.2). The leveling off of this graph indicates a positive Lyapunov exponent.

To study the details of the divergence of nearby trajectories we expand the early portion of the two-particle divergence graph (Fig. 6.3). The separation between particles starts out as a power law with an exponent near 3/2. The square law we described earlier estimates the actual total error, including systematic errors in the integration process. The 3/2 power law describes the divergence of trajectories subject to the same systematic errors. Only after some time does the exponential take off. The power law is dominated by the exponential only after the rate of growth of the exponential exceeds the rate of growth of the power law. This suggests that the portion of the divergence of nearby trajectories that results only from the numerical error fits a 3/2 power law and that this error "seeds" the exponential divergence that is the hallmark of chaos. We tested this hypothesis by integrating a cloud of test particles with the orbital elements of Pluto in the field of the Sun alone. The divergence of these Kepler "Plutos" grows as $3.16 \times 10^{-17} t^{3/2}$ AU. This is identical to the initial divergence of the Plutos in the complete dynamical system, showing that two-body numerical error completely accounts for the initial divergence.

Only the second half of the integration was used in the computation of the Lyapunov exponents, because the measurement in the first half of our integration was contaminated by over-vigorous application of the rescaling method, and gave a Lyapunov exponent about a factor of 4 too large. The rescaling interval was only 275,000 years, which was far too small. The rescaling interval must be long enough that the divergence of neighboring trajectories is dominated by the exponential divergence associated with chaotic behavior rather than the power law divergence caused by the accumulation of numerical errors. In our experiment the rescaling interval should have been greater than 30 million years.

It is important to emphasize that the variational method of measuring the Lyapunov exponent has none of these problems.

6.5   Features of the orbital elements of Pluto

The largest component in the variation of h (Fig. 6.4) reflects the 3.7-million-year regression of the longitude of perihelion. The 27-million-year component we previously reported is clearly visible, as is the 137-million-year component. The change in density of points reflects a change in the sampling interval. For the first 450 million years of our integration we recorded the state of the system every 499,983 days (about 1,369 years) of simulated time. For the second 400 million years we sampled 16 times less frequently.

Besides the major 3.8-million-year component in the variation of the inclination of Pluto (Fig. 6.5) we can clearly discern the 34-million-year component we previously reported. Although there is no continuing secular decline in the inclination, there is a component with a period near 150 million years and evidence for a component with a period of approximately 600 million years.

The existence of significant orbital variations with such long periods would be quite surprising if the motion were quasiperiodic. For quasiperiodic trajectories we expect to find frequencies that are low order combinations of a few fundamental frequencies (one per degree of freedom). The natural time scale for the long-term evolution of a quasiperiodic planetary system is set by the periods of the circulation of the nodes and perihelia, which in this case are a few million years. Periods in the motion of Pluto comparable to the length of the integration have been found in all long-term integrations. This is consistent with the chaotic character of the motion of Pluto, as indicated by our measurement of a positive Lyapunov exponent.

Usually the measurement of a positive Lyapunov exponent provides a confirmation of what is already visible to the eye; that is, chaotic trajectories look irregular. In this case, except for the very long periods, the plots of Pluto's orbital elements do not look particularly irregular. However, the irregularity of the motion does manifest itself in the power spectra. For a quasiperiodic trajectory the power spectrum of any orbital element is composed of integral linear combinations of fundamental

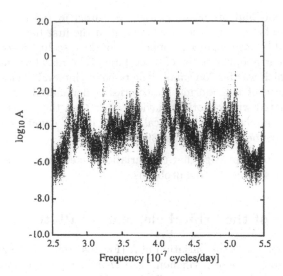

Fig. 6.6. A portion of the power spectrum of Pluto's h. In this graph A is the relative amplitude. There appears to be a broad-band component to the spectrum. This is consistent with the chaotic character of the motion of Pluto as indicated by the positive Lyapunov exponent.

frequencies, where the number of fundamental frequencies is equal to the number of degrees of freedom. The power spectrum of a chaotic trajectory usually appears to have some broad-band component.

A portion of the power spectrum of Pluto's h is shown in Fig. 6.6. For comparison the same portion of the power spectrum of Neptune's h is shown in Fig. 6.7. This portion of the spectrum was chosen to avoid confusion introduced by nearby major lines. Hanning windows have been used to reduce spectral leakage; only the densely sampled part of the run was used in the computation of the Fourier transforms. The spectrum of Neptune is quite complicated but there is no evidence that it is not a line spectrum. On the other hand the spectrum of Pluto does appear to have a broad-band component. Note that both of these spectra are computed from the same integration run, by means of the same numerical methods. They are subject to the same error processes, so the differences we see are dynamical in origin. The amplitudes in both graphs are normalized in the same way, so we can see that the broad-band components in Pluto's spectrum are mostly larger than the discrete components in Neptune's spectrum.

The lack of obvious irregularity in the orbital elements of Pluto indicates that the portion of the chaotic zone in which Pluto is currently moving is rather small. Since the global structure of the chaotic zone is not known it is not possible for us to

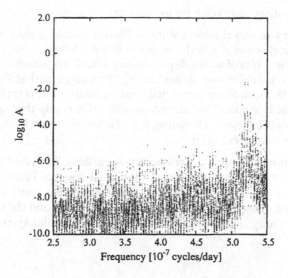

Fig. 6.7. A portion of the power spectrum of Neptune's h. In this graph A is the relative amplitude. The spectrum is apparently a quite complicated line spectrum. That we do not observe a broad-band component is consistent with the motion being quasiperiodic.

predict whether more irregular motions are likely. If the small chaotic zone in which Pluto is found connects to a larger chaotic region, relatively sudden transitions can be made to more irregular motion. This actually occurs for the motion of asteroids near the 3:1 Kirkwood gap [13].

On the other hand, the fact that the time scale for divergence is only an order of magnitude larger than the fundamental time scales of the system indicates that the chaotic behavior is robust. It is not a narrow chaotic zone associated with a high-order resonance. Even though we do not know the sensitivity of the observed chaotic behavior to the uncertainties in parameters and initial conditions, and unmodeled effects, the large Lyapunov exponent suggests that the chaotic behavior of Pluto is characteristic of a range of solar systems including the actual solar system.

6.6 Conclusions and implications of Pluto's chaotic motion

Our numerical model indicates that the motion of Pluto is chaotic. The largest Lyapunov exponent is about $10^{-7.3}$ year^{-1}. Thus the e-folding time for the divergence of trajectories is about 20 million years. It would not have been surprising to discover an instability with characteristic time of the order of the age of the solar system because such an instability would not yet have had enough time to produce apparent damage. Thus, considering the age of the solar system, 20 million years is a remarkably short time scale for exponential divergence.

The discovery of the chaotic nature of Pluto's motion makes it more difficult to draw firm conclusions about the origin of Pluto. However, the orbit of Pluto is reminiscent of the orbits of asteroids on resonant chaotic trajectories, which typically evolve to high eccentricity and inclination [6]. This suggests that Pluto might have been formed with much lower eccentricity and inclination, as is typical of the other planets, and that it acquired its current peculiar orbit purely through deterministic chaotic dynamical processes. Of course, it is also possible that Pluto simply formed in an orbit near its current orbit.

In our experiment Pluto is a zero-mass test particle. The real Pluto has a small mass. We expect that the inclusion of the actual mass of Pluto will not change the chaotic character of the motion. If so, Pluto's irregular motion will chaotically pump the motion of the other members of the solar system and the chaotic behavior of Pluto would imply chaotic behavior of the rest of the solar system.

References

[1] V. I. Arnold, paper presented at the Fourth All-Union Mathematical Congress, Leningrad (1961) [English translation in *Russ. Math. Surv.* **18**, 85 (1961)].

[2] J. Applegate et al. *IEEE Trans. on Computers* **C-34**, 822 (1985).

[3] C. J. Cohen and E. C. Hubbard, *Astron. J.* **70**, 10 (1965).

[4] J. G. Williams and G. S. Benson, ibid. **76**, 167 (1971).

[5] J. Applegate et al. ibid. **92**, 176 (1986).

[6] J. Wisdom, *Icarus* **72**, 241 (1987).

[7] B. V. Chirikov, *Phys. Rep.* **52**, 263 (1979).

[8] C. Benettin et al. *Meccanica*, March 1980, p.21 (1980).

[9] C. J. Cohen, E. C. Hubbard, C. Oesterwinter, in *Astronomical Papers of the American Ephemeris*, vol.22, pp.1-92 (Government Printing Office, Washington, DC, 1973).

[10] H. Kinoshita and H. Nakai, *Celestial Mechanics* **34**, 203 (1984).

[11] A. Nobili, in *The Few Body Problem*, M. Valtonen Ed. (Riedel, Dordrecht), pp. 147-163 (1987).

[12] P. H. Cowell and A. C. D. Crommelin, in *Greenwich Observations 1909*, appendix (Neill, Bellevue, England, 1910).

[13] J. Wisdom, *Icarus* **56**, 51 (1983).

[14] We thank W. Kahan for suggesting that a step size might be chosen that essentially eliminates the linear growth in the energy error, making this article possible; and P. Skordos for doing some of the experiments to localize the special step sizes. This research was done in part at the Artificial Intelligence Laboratory of the Massachusetts Institute of Technology. Support for the Laboratory's artificial intelligence research is provided in part by the Advanced Research Projects Agency of the Department of Defense under Office of Naval Research contract N00014-86K-0180. This work was also supported in part by the Planetary Geology and Geophysics Program of the National Aeronautics and Space Administration, under grant NAGW-706.

Part II

Reducing the Size

THERE'S PLENTY OF ROOM AT THE BOTTOM

Richard Feynman

7.1 An Invitation to Enter a New Field of Physics

I imagine experimental physicists must often look with envy at men like Kamerlingh Onnes, who discovered a field like low temperature, which seems to be bottomless and in which one can go down and down. Such a man is then a leader and has some temporary monopoly in a scientific adventure. Percy Bridgman, in designing a way to obtain higher pressures, opened up another new field and was able to move into it and to lead us all along. The development of ever higher vacuum was a continuing development of the same kind.

I would like to describe a field, in which little has been done, but in which an enormous amount can be done in principle. This field is not quite the same as the others in that it will not tell us much of fundamental physics (in the sense of, "What are the strange particles?") but it is more like solid-state physics in the sense that it might tell us much of great interest about the strange phenomena that occur in complex situations. Furthermore, a point that is most important is that it would have an enormous number of technical applications.

What I want to talk about is the problem of manipulating and controlling things on a small scale.

As soon as I mention this, people tell me about miniaturization, and how far it has progressed today. They tell me about electric motors that are the size of the nail on your small finger. And there is a device on the market, they tell me, by which you can write the Lord's Prayer on the head of a pin. But that's nothing; that's the most primitive, halting step in the direction I intend to discuss. It is a staggeringly small world that is below. In the year 2000, when they look back at this age, they will wonder why it was not until the year 1960 that anybody began seriously to move in this direction.

Why cannot we write the entire 24 volumes of the Encyclopaedia Brittanica on the head of a pin?

Let's see what would be involved. The head of a pin is a sixteenth of an inch across. If you magnify it by 25,000 diameters, the area of the head of the pin is then equal to the area of all the pages of the Encyclopaedia Brittanica. Therefore, all it is necessary to do is to reduce in size all the writing in the Encyclopaedia by

25,000 times. Is that possible? The resolving power of the eye is about 1/120 of an inch — that is roughly the diameter of one of the little dots on the fine half-tone reproductions in the Encyclopaedia. This, when you demagnify it by 25,000 times, is still 80 angstroms in diameter — 32 atoms across, in an ordinary metal. In other words, one of those dots still would contain in its area 1,000 atoms. So, each dot can easily be adjusted in size as required by the photoengraving, and there is no question that there is enough room on the head of a pin to put all of the Encyclopaedia Brittanica.

Furthermore, it can be read if it is so written. Let's imagine that it is written in raised letters of metal; that is, where the black is in the Encyclopaedia, we have raised letters of metal that are actually 1/25,000 of their ordinary size. How would we read it?

If we had something written in such a way, we could read it using techniques in common use today. (They will undoubtedly find a better way when we do actually have it written, but to make my point conservatively I shall just take techniques we know today.) We would press the metal into a plastic material and make a mold of it, then peel the plastic off very carefully, evaporate silica into the plastic to get a very thin film, then shadow it by evaporating gold at an angle against the silica so that all the little letters will appear clearly, dissolve the plastic away from the silica film, and then look through it with an electron microscope!

There is no question that if the thing were reduced by 25,000 times in the form of raised letters on the pin, it would be easy for us to read it today. Furthermore; there is no question that we would find it easy to make copies of the master; we would just need to press the same metal plate again into plastic and we would have another copy.

7.2   How do we write small?

The next question is: How do we *write* it? We have no standard technique to do this now. But let me argue that it is not as difficult as it first appears to be. We can reverse the lenses of the electron microscope in order to demagnify as well as magnify. A source of ions, sent through the microscope lenses in reverse, could be focused to a very small spot. We could write with that spot like we write in a TV cathode ray oscilloscope, by going across in lines, and having an adjustment which determines the amount of material which is going to be deposited as we scan in lines.

This method might be very slow because of space charge limitations. There will be more rapid methods. We could first make, perhaps by some photo process, a screen which has holes in it in the form of the letters. Then we would strike an arc behind the holes and draw metallic ions through the holes; then we could again use our system of lenses and make a small image in the form of ions, which would deposit the metal on the pin.

A simpler way might be this (though I am not sure it would work): We take light and, through an optical microscope running backwards, we focus it onto a very small photoelectric screen. Then electrons come away from the screen where the light is shining. These electrons are focused down in size by the electron microscope lenses to impinge directly upon the surface of the metal. Will such a beam etch away the metal if it is run long enough? I don't know. If it doesn't work for a metal surface, it must be possible to find some surface with which to coat the original pin so that, where the electrons bombard, a change is made which we could recognize later.

There is no intensity problem in these devices — not what you are used to in magnification, where you have to take a few electrons and spread them over a bigger and bigger screen; it is just the opposite. The light which we get from a page is concentrated onto a very small area so it is very intense. The few electrons which come from the photoelectric screen are demagnified down to a very tiny area so that, again, they are very intense. I don't know why this hasn't been done yet!

That's the Encyclopaedia Brittanica on the head of a pin, but let's consider all the books in the world. The Library of Congress has approximately 9 million volumes; the British Museum Library has 5 million volumes; there are also 5 million volumes in the National Library in France. Undoubtedly there are duplications, so let us say that there are some 24 million volumes of interest in the world.

What would happen if I print all this down at the scale we have been discussing? How much space would it take? It would take, of course, the area of about a million pinheads because, instead of there being just the 24 volumes of the Encyclopaedia, there are 24 million volumes. The million pinheads can be put in a square of a thousand pins on a side, or an area of about 3 square yards. That is to say, the silica replica with the paper-thin backing of plastic, with which we have made the copies, with all this information, is on an area of approximately the size of 35 pages of the Encyclopaedia. That is about half as many pages as there are in this magazine. All of the information which all of mankind has every recorded in books can be carried around in a pamphlet in your hand — and not written in code, but a simple reproduction of the original pictures, engravings, and everything else on a small scale without loss of resolution.

What would our librarian at Caltech say, as she runs all over from one building to another, if I tell her that, ten years from now, all of the information that she is struggling to keep track of — 120,000 volumes, stacked from the floor to the ceiling, drawers full of cards, storage rooms full of the older books — can be kept on just one library card! When the University of Brazil, for example, finds that their library is burned, we can send them a copy of every book in our library by striking off a copy from the master plate in a few hours and mailing it in an envelope no bigger or heavier than any other ordinary air mail letter.

Now, the name of this talk is "There is *Plenty* of Room at the Bottom" — not just "There is Room at the Bottom." What I have demonstrated is that there is room — that you can decrease the size of things in a practical way. I now want to show that there is *plenty* of room. I will not now discuss how we are going to do it, but only what is possible in principle — in other words, what is possible according to the laws of physics. I am not inventing anti-gravity, which is possible someday only if the laws are not what we think. I am telling you what could be done if the laws *are* what we think; we are not doing it simply because we haven't yet gotten around to it.

7.3   Information on a small scale

Suppose that, instead of trying to reproduce the pictures and all the information directly in its present form, we write only the information content in a code of dots and dashes, or something like that, to represent the various letters. Each letter represents six or seven "bits" of information; that is, you need only about six or seven dots or dashes for each letter. Now, instead of writing everything, as I did before, on the *surface* of the head of a pin, I am going to use the interior of the material as well.

Let us represent a dot by a small spot of one metal, the next dash, by an adjacent spot of another metal, and so on. Suppose, to be conservative, that a bit of information is going to require a little cube of atoms 5 times 5 times 5 — that is 125 atoms. Perhaps we need a hundred and some odd atoms to make sure that the information is not lost through diffusion, or through some other process.

I have estimated how many letters there are in the Encyclopaedia, and I have assumed that each of my 24 million books is as big as an Encyclopaedia volume, and have calculated, then, how many bits of information there are (10^{15}). For each bit I allow 100 atoms. And it turns out that all of the information that man has carefully accumulated in all the books in the world can be written in this form in a cube of material one two-hundredth of an inch wide — which is the barest piece of dust that can be made out by the human eye. So there is *plenty* of room at the bottom! Don't tell me about microfilm!

This fact — that enormous amounts of information can be carried in an exceedingly small space — is, of course, well known to the biologists, and resolves the mystery which existed before we understood all this clearly, of how it could be that, in the tiniest cell, all of the information for the organization of a complex creature such as ourselves can be stored. All this information — whether we have brown eyes, or whether we think at all, or that in the embryo the jawbone should first develop with a little hole in the side so that later a nerve can grow through it — all this information is contained in a very tiny fraction of the cell in the form of long-chain DNA molecules in which approximately 50 atoms are used for one bit of information about the cell.

7.4   Better electron microscopes

If I have written in a code, with 5 times 5 times 5 atoms to a bit, the question is: How could I read it today? The electron microscope is not quite good enough, with the greatest care and effort, it can only resolve about 10 angstroms. I would like to try and impress upon you while I am talking about all of these things on a small scale, the importance of improving the electron microscope by a hundred times. It is not impossible; it is not against the laws of diffraction of the electron. The wave length of the electron in such a microscope is only 1/20 of an angstrom. So it should be possible to see the individual atoms. What good would it be to see individual atoms distinctly?

We have friends in other fields — in biology, for instance. We physicists often look at them and say, "You know the reason you fellows are making so little progress?" (Actually I don't know any field where they are making more rapid progress than they are in biology today.) "You should use more mathematics, like we do." They could answer us — but they're polite, so I'll answer for them: "What you should do in order for *us* to make more rapid progress is to make the electron microscope 100 times better."

What are the most central and fundamental problems of biology today? They are questions like: What is the sequence of bases in the DNA? What happens when you have a mutation? How is the base order in the DNA connected to the order of amino acids in the protein? What is the structure of the RNA; is it single-chain or double-chain, and how is it related in its order of bases to the DNA? What is the organization of the microsomes? How are proteins synthesized? Where does the RNA go? How does it sit? Where do the proteins sit? Where do the amino acids go in? In photosynthesis, where is the chlorophyll; how is it arranged; where are the carotenoids involved in this thing? What is the system of the conversion of light into chemical energy?

It is very easy to answer many of these fundamental biological questions; you just *look at the thing*! You will see the order of bases in the chain; you will see the structure of the microsome. Unfortunately, the present microscope sees at a scale which is just a bit too crude. Make the microscope one hundred times more powerful, and many problems of biology would be made very much easier. I exaggerate, of course, but the biologists would surely be very thankful to you — and they would prefer that to the criticism that they should use more mathematics.

The theory of chemical processes today is based on theoretical physics. In this sense, physics supplies the foundation of chemistry. But chemistry also has analysis. If you have a strange substance and you want to know what it is, you go through a long and complicated process of chemical analysis. You can analyze almost anything today, so I am a little late with my idea. But if the physicists wanted to, they could also dig under the chemists in the problem of chemical analysis. It would be very easy to make an analysis of any complicated chemical substance; all one would have to do would be to look at it and see where the atoms are. The only trouble is that the electron microscope is one hundred times too poor. (Later, I would like to ask the question: Can the physicists do something about the third problem of chemistry — namely, synthesis? Is there a *physical* way to synthesize any chemical substance?)

The reason the electron microscope is so poor is that the f-value of the lenses is only 1 part to 1,000; you don't have a big enough numerical aperture. And I know that there are theorems which prove that it is impossible, with axially symmetrical stationary field lenses, to produce an f-value any bigger than so and so; and therefore the resolving power at the present time is at its theoretical maximum. But in every theorem there are assumptions. Why must the field be symmetrical? I put this out as a challenge: Is there no way to make the electron microscope more powerful?

7.5 The marvelous biological system

The biological example of writing information on a small scale has inspired me to think of something that should be possible. Biology is not simply writing information; it is *doing something* about it. A biological system can be exceedingly small. Many of the cells are very tiny, but they are very active; they manufacture various substances; they walk around; they wiggle; and they do all kinds of marvelous things — all on a very small scale. Also, they store information. Consider the possibility that we too can make a thing very small which does what we want — that we can manufacture an object that maneuvers at that level!

There may even be an economic point to this business of making things very small. Let me remind you of some of the problems of computing machines. In computers we have to store an enormous amount of information. The kind of writing that I was mentioning before, in which I had everything down as a distribution of metal, is permanent. Much more interesting to a computer is a way of writing, erasing, and writing something else. (This is usually because we don't want to waste the material on which we have just written. Yet if we could write it in a very small space, it wouldn't make any difference; it could just be thrown away after it was read. It doesn't cost very much for the material).

7.6 Miniaturizing the computer

I don't know how to do this on a small scale in a practical way, but I do know that computing machines are very large; they fill rooms. Why can't we make them very small, make them of little wires, little elements — and by little, I mean *little*. For instance, the wires should be 10 or 100 atoms in diameter, and the circuits should be a few thousand angstroms across. Everybody who has analyzed the logical theory of computers has come to the conclusion that the possibilities of computers are very interesting — if they could be made to be more complicated by several orders of magnitude. If they had millions of times as many elements, they could make judgements. They would have time to calculate what is the best way to make the calculation that they are about to make. They could select the method of analysis which, from their experience, is better than the one that we would give to them. And in many other ways, they would have new qualitative features.

If I look at your face I immediately recognize that I have seen it before. (Actually, my friends will say I have chosen an unfortunate example here for the subject of this illustration. At least I recognize that it is a *man* and not an *apple*.) Yet there is no machine which, with that speed, can take a picture of a face and say even that it is a man; and much less that it is the same man that you showed it before — unless it is exactly the same picture. If the face is changed; if I am closer to the face; if I am further from the face; if the light changes — I recognize it anyway. Now, this little computer I carry in my head is easily able to do that. The computers that we build are not able to do that. The number of elements in this bone box of mine are enormously greater than the number of elements in our "wonderful" computers. But our mechanical computers are too big; the elements in this box are microscopic. I want to make some that are *sub* microscopic.

If we wanted to make a computer that had all these marvelous extra qualitative abilities, we would have to make it, perhaps, the size of the Pentagon. This has several disadvantages. First, it requires too much material; there may not be enough germanium in the world for all the transistors which would have to be put into this enormous thing. There is also the problem of heat generation and power consumption; TVA would be needed to run the computer. But an even more practical difficulty is that the computer would be limited to a certain speed. Because of its large size, there is finite time required to get the information from one place to another. The information cannot go any faster than the speed of light — so, ultimately, when our computers get faster and faster and more and more elaborate, we will have to make them smaller and smaller.

But there is plenty of room to make them smaller. There is nothing that I can see in the physical laws that says the computer elements cannot be made enormously smaller than they are now. In fact, there may be certain advantages.

7.7   Miniaturization by evaporation

How can we make such a device? What kind of manufacturing processes would we use? One possibility we might consider, since we have talked about writing by putting atoms down in a certain arrangement, would be to evaporate the material, then evaporate the insulator next to it. Then, for the next layer, evaporate another position of a wire, another insulator, and so on. So, you simply evaporate until you have a block of stuff which has the elements — coils and condensers, transistors and so on — of exceedingly fine dimensions.

But I would like to discuss, just for amusement, that there are other possibilities. Why can't we manufacture these small computers somewhat like we manufacture the big ones? Why can't we drill holes, cut things, solder things, stamp things out, mold different shapes all at an infinitesimal level? What are the limitations as to how small a thing has to be before you can no longer mold it? How many times when you are working on something frustratingly tiny like your wife's wrist watch, have you said to yourself, "If I could only train an ant to do this!" What I would like to suggest is the possibility of training an ant to train a mite to do this. What are the possibilities of small but movable machines? They may or may not be useful, but they surely would be fun to make.

Consider any machine — for example, an automobile — and ask about the problems of making an infinitesimal machine like it. Suppose, in the particular design of the automobile, we need a certain precision of the parts; we need an accuracy, let's suppose, of 4/10,000 of an inch. If things are more inaccurate than that in the shape of the cylinder and so on, it isn't going to work very well. If I make the thing too small, I have to worry about the size of the atoms; I can't make a circle of "balls" so to speak, if the circle is too small. So, if I make the error, corresponding to 4/10,000 of an inch, correspond to an error of 10 atoms, it turns out that I can reduce the dimensions of an automobile 4,000 times, approximately — so that it is 1 mm. across. Obviously, if you redesign the car so that it would work with a much larger tolerance, which is not at all impossible, then you could make a much smaller device.

It is interesting to consider what the problems are in such small machines. Firstly, with parts stressed to the same degree, the forces go as the area you are reducing, so that things like weight and inertia are of relatively no importance. The strength of material, in other words, is very much greater in proportion. The stresses and expansion of the flywheel from centrifugal force, for example, would be the same proportion only if the rotational speed is increased in the same proportion as we decrease the size. On the other hand, the metals that we use have a grain structure, and this would be very annoying at small scale because the material is not homogeneous. Plastics and glass and things of this amorphous nature are very much more homogeneous, and so we would have to make our machines out of such materials.

There are problems associated with the electrical part of the system — with the copper wires and the magnetic parts. The magnetic properties on a very small scale are not the same as on a large scale; there is the "domain" problem involved. A big magnet made of millions of domains can only be made on a small scale with one domain. The electrical equipment won't simply be scaled down; it has to be redesigned. But I can see no reason why it can't be redesigned to work again.

7.8   Problems of lubrication

Lubrication involves some interesting points. The effective viscosity of oil would be higher and higher in proportion as we went down (and if we increase the speed as much as we can). If we don't increase the speed so much, and change from oil to kerosene or some other fluid, the problem is not so bad. But actually we may not have to lubricate at all! We have a lot of extra force. Let the bearings run dry; they won't run hot because the heat escapes away from such a small device very, very rapidly. This rapid heat loss would prevent the gasoline from exploding, so an internal combustion engine is impossible. Other chemical reactions, liberating energy when cold, can be used. Probably an external supply of electrical power would be most convenient for such small machines.

What would be the utility of such machines? Who knows? Of course, a small automobile would only be useful for the mites to drive around in, and I suppose our Christian interests don't go that far. However, we did note the possibility of the manufacture of small elements for computers in completely automatic factories, containing lathes and other machine tools at the very small level. The small lathe would not have to be exactly like our big lathe. I leave to your imagination the improvement of the design to take full advantage of the properties of things on a small scale, and in such a way that the fully automatic aspect would be easiest to manage.

A friend of mine (Albert R. Hibbs) suggests a very interesting possibility for relatively small machines. He says that, although it is a very wild idea, it would be interesting in surgery if you could swallow the surgeon. You put the mechanical surgeon inside the blood vessel and it goes into the heart and "looks" around. (Of course the information has to be fed out.) It finds out which valve is the faulty one and takes a little knife and slices it out. Other small machines might be permanently incorporated in the body to assist some inadequately-functioning organ.

Now comes the interesting question: How do we make such a tiny mechanism? I leave that to you. However, let me suggest one weird possibility. You know, in the atomic energy plants they have materials and machines that they can't handle directly because they have become radioactive. To unscrew nuts and put on bolts and so on, they have a set of master and slave hands, so that by operating a set of levers here, you control the "hands" there, and can turn them this way and that so you can handle things quite nicely.

Most of these devices are actually made rather simply, in that there is a particular cable, like a marionette string, that goes directly from the controls to the "hands." But, of course, things also have been made using servo motors, so that the connection between the one thing and the other is electrical rather than mechanical. When you turn the levers, they turn a servo motor, and it changes the electrical currents in the wires, which repositions a motor at the other end.

Now, I want to build much the same device — a master-slave system which operates electrically. But I want the slaves to be made especially carefully by modern large-scale machinists so that they are one-fourth the scale of the "hands" that you ordinarily maneuver. So you have a scheme by which you can do things at one-quarter scale anyway — the little servo motors with little hands play with little nuts and bolts; they drill little holes; they are four times smaller. Aha! So I manufacture a quarter-size lathe; I manufacture quarter-size tools; and I make, at the one-quarter scale, still another set of hands again relatively one-quarter size! This is one-sixteenth size, from my point of view. And after I finish doing this I wire directly from my large-scale system, through transformers perhaps, to the one-sixteenth-size servo motors. Thus I can now manipulate the one-sixteenth size hands.

Well, you get the principle from there on. It is rather a difficult program, but it is a possibility. You might say that one can go much farther in one step than from one to four. Of course, this has all to be designed very carefully and it is not necessary simply to make it like hands. If you thought of it very carefully, you could probably arrive at a much better system for doing such things.

If you work through a pantograph, even today, you can get much more than a factor of four in even one step. But you can't work directly through a pantograph which makes a smaller pantograph which then makes a smaller pantograph — because of the looseness of the holes and the irregularities of construction. The end of the pantograph wiggles with a relatively greater irregularity than the irregularity with which you move your hands. In going down this scale, I would find the end of the pantograph on the end of the pantograph on the end of the pantograph shaking so badly that it wasn't doing anything sensible at all.

At each stage, it is necessary to improve the precision of the apparatus. If, for instance, having made a small lathe with a pantograph, we find its lead screw irregular — more irregular than the large-scale one — we could lap the lead screw against breakable nuts that you can reverse in the usual way back and forth until this lead screw is, at its scale, as accurate as our original lead screws, at our scale.

We can make flats by rubbing unflat surfaces in triplicates together — in three pairs — and the flats then become flatter than the thing you started with. Thus, it is not impossible to improve precision on a small scale by the correct operations. So, when we build this stuff, it is necessary at each step to improve the accuracy of the equipment by working for awhile down there, making accurate lead screws, Johansen blocks, and all the other materials which we use in accurate machine work at the higher level. We have to stop at each level and manufacture all the stuff to go to the next level — a very long and very difficult program. Perhaps you can figure a better way than that to get down to small scale more rapidly.

Yet, after all this, you have just got one little baby lathe four thousand times smaller than usual. But we were thinking of making an enormous computer, which we were going to build by drilling holes on this lathe to make little washers for the computer. How many washers can you manufacture on this one lathe?

7.9 A hundred tiny hands

When I make my first set of slave "hands" at one-fourth scale, I am going to make ten sets. I make ten sets of "hands," and I wire them to my original levers so they each do exactly the same thing at the same time in parallel. Now, when I am making my new devices one-quarter again as small, I let each one manufacture ten copies, so that I would have a hundred "hands" at the 1/16th size.

Where am I going to put the million lathes that I am going to have? Why, there is nothing to it; the volume is much less than that of even one full-scale lathe. For instance, if I made a billion little lathes, each 1/4000 of the scale of a regular lathe, there are plenty of materials and space available because in the billion little ones there is less than 2 percent of the materials in one big lathe. It doesn't cost anything for materials, you see. So I want to build a billion tiny factories, models of each other, which are manufacturing simultaneously, drilling holes, stamping parts, and so on.

As we go down in size, there are a number of interesting problems that arise. All things do not simply scale down in proportion. There is the problem that materials stick together by the molecular (Van der Waals) attractions. It would be like this: After you have made a part and you unscrew the nut from a bolt, it isn't going to fall down because the gravity isn't appreciable; it would even be hard to get it off the bolt. It would be like those old movies of a man with his hands full of molasses, trying to get rid of a glass of water. There will be several problems of this nature that we will have to be ready to design for.

7.10 Rearranging the atoms

But I am not afraid to consider the final question as to whether, ultimately — in the great future — we can arrange the atoms the way we want; the very *atoms*, all the way down! What would happen if we could arrange the atoms one by one the way we want them (within reason, of course; you can't put them so that they are chemically unstable, for example).

Up to now, we have been content to dig in the ground to find minerals. We heat them and we do things on a large scale with them, and we hope to get a pure substance with just so much impurity, and so on. But we must always accept some atomic arrangement that nature gives us. We haven't got anything, say, with a "checkerboard" arrangement, with the impurity atoms exactly arranged 1,000 angstroms apart, or in some other particular pattern.

What could we do with layered structures with just the right layers? What would the properties of materials be if we could really arrange the atoms the way we want them? They would be very interesting to investigate theoretically. I can't see exactly what would happen, but I can hardly doubt that when we have some *control* of the arrangement of things on a small scale we will get an enormously greater range of possible properties that substances can have, and of different things that we can do.

Consider, for example, a piece of material in which we make little coils and condensers (or their solid state analogs) 1,000 or 10,000 angstroms in a circuit, one right next to the other, over a large area, with little antennas sticking out at the other end — a whole series of circuits. Is it possible, for example, to emit light from a whole set of antennas, like we emit radio waves from an organized set of antennas to beam the radio programs to Europe? The same thing would be to beam the light out in a definite direction with very high intensity. (Perhaps such a beam is not very useful technically or economically.)

I have thought about some of the problems of building electric circuits on a small scale, and the problem of resistance is serious. If you build a corresponding circuit on a small scale, its natural frequency goes up, since the wave length goes down as the scale; but the skin depth only decreases with the square root of the scale ratio, and so resistive problems are of increasing difficulty. Possibly we can beat resistance through the use of superconductivity if the frequency is not too high, or by other tricks.

7.11   Atoms in a small world

When we get to the very, very small world — say circuits of seven atoms — we have a lot of new things that would happen that represent completely new opportunities for design. Atoms on a small scale behave like *nothing* on a large scale, for they satisfy the laws of quantum mechanics. So, as we go down and fiddle around with the atoms down there, we are working with different laws, and we can expect to do different things. We can manufacture in different ways. We can use, not just circuits, but some system involving the quantized energy levels, or the interactions of quantized spins, etc.

Another thing we will notice is that, if we go down far enough, all of our devices can be mass produced so that they are absolutely perfect copies of one another. We cannot build two large machines so that the dimensions are exactly the same. But if your machine is only 100 atoms high, you only have to get it correct to one-half of one percent to make sure the other machine is exactly the same size — namely, 100 atoms high!

At the atomic level, we have new kinds of forces and new kinds of possibilities, new kinds of effects. The problems of manufacture and reproduction of materials will be quite different. I am, as I said, inspired by the biological phenomena in which chemical forces are used in repetitious fashion to produce all kinds of weird effects (one of which is the author). The principles of physics, as far as I can see, do not speak against the possibility of maneuvering things atom by atom. It is not an attempt to violate any laws; it is something, in principle, that can be done; but in practice, it has not been done because we are too big.

Ultimately, we can do chemical synthesis. A chemist comes to us and says, "Look, I want a molecule that has the atoms arranged thus and so; make me that molecule." The chemist does a mysterious thing when he wants to make a molecule. He sees that it has got that ring, so he mixes this and that, and he shakes it, and he fiddles around. And, at the end of a difficult process, he usually does succeed in synthesizing what he wants. By the time I get my devices working, so that we can do it by physics, he will have figured out how to synthesize absolutely anything, so that this will really be useless.

But it is interesting that it would be, in principle, possible (I think) for a physicist to synthesize any chemical substance that the chemist writes down. Give the orders and the physicist synthesizes it. How? Put the atoms down where the chemist says, and so you make the substance. The problems of chemistry and biology can be greatly helped if our ability to see what we are doing, and to do things on an atomic level, is ultimately developed — a development which I think cannot be avoided. Now, you might say, "Who should do this and why should they do it?" Well, I pointed out a few of the economic applications, but I know that the reason that you would do it might be just for fun. But have some fun! Let's have a competition between laboratories. Let one laboratory make a tiny motor which it sends to another lab which sends it back with a thing that fits inside the shaft of the first motor.

7.12 High school competition

Just for the fun of it, and in order to get kids interested in this field, I would propose that someone who has some contact with the high schools think of making some kind of high school competition. After all, we haven't even started in this field, and even the kids can write smaller than has ever been written before. They could have competition in high schools. The Los Angeles high school could send a pin to the Venice high school on which it says, "How's this?" They get the pin back, and in the dot of the "i" it says, "Not so hot."

Perhaps this doesn't excite you to do it, and only economics will do so. Then I want to do something; but I can't do it at the present moment, because I haven't prepared the ground. It is my intention to offer a prize of $1,000 to the first guy who can take the information on the page of a book and put it on an area 1/25,000 smaller in linear scale in such manner that it can be read by an electron microscope.

And I want to offer another prize — if I can figure out how to phrase it so that

I don't get into a mess of arguments about definitions — of another $1,000 to the first guy who makes an operating electric motor — a rotating electric motor which can be controlled from the outside and, not counting the lead-in wires, is only 1/64 inch cube.

I do not expect that such prizes will have to wait very long for claimants.

This transcript of the classic talk that Richard Feynman gave on December 29th 1959 at the annual meeting of the American Physical Society at the California Institute of Technology (Caltech) was first published in the February 1960 issue of Caltech's Engineering and Science, which owns the copyright. It has been made available with their kind permission.

8

INFORMATION IS INEVITABLY PHYSICAL

Rolf Landauer

Abstract

Information is inevitably tied to a physical representation, and therefore to all the possibilities and restrictions allowed by our real physical universe. The theory of computational limits is reviewed in a historical fashion. After some widespread initial errors, it was eventually understood that statistical mechanics and elementary quantum mechanics do not provide any limits. The energy requirements of the communications channel are particularly emphasized; it is an area where lower bounds, accepted for decades, are circumventable. The utility of the time-modulated potential going from monostability to bistability and back, is emphasized. Despite its use by von Neumann, Feynman, and many others, it has not received broad attention, i.e. by those not actually invoking it for their own purposes. I revisit my long-standing contention that our real universe does not permit the unlimited chain of infallible operations, envisioned in continuum mathematics, and that this has an influence on the ultimate nature of physical law. Finally, in the spirit of a volume dedicated to Richard Feynman's impact, I deplore the strong effects of fashions in science.

8.1 Information is Physical

Information is inevitably tied to a physical representation. It can be engraved on stone tablets, denoted by a spin up or down, a charge present or absent, a hole punched in a card, or many other alternative physical phenomena. It is not just an abstract entity; it does not exist except through a physical embodiment. It is, therefore, tied to the laws of physics and the parts available to us in our real physical universe. This is a viewpoint which was invoked by Szilard [1] in his analysis of Maxwell's demon. Szilard's analysis was not all that definitive as far as the demon was concerned, but his understanding of the physical nature of information was truly pioneering. Even in recent years this viewpoint is still not all that widely accepted. Penrose, [2] for example, tells us: *". . . devices can yield only approximations to a structure that has a deep and 'computer-independent' existence of its own."*

When we learned to count on our sticky little classical fingers, we were misled. We thought that an integer had to have a particular and unique value. But in the real world, which is quantum mechanical, we can have a coherent superposition of a state with two photons and one with five. This is a degree of freedom, a possibility, which has to be understood and its utility has to be assessed. The fact that its advantages are advertised too unhesitantly by the advocates of quantum parallelism (for a collection of papers on this subject, see Ref. [3]) should not blind us to the need to examine all the possibilities, as well as the restrictions, which come with the physical nature of information.

The physical nature of information leads us to an analysis of the limits imposed on information handling by the laws of physics and by the parts available in the universe, and also leads us to the attempt to exploit all the possibilities offered by physics. We will start with the discussion of limits in the next section and later, in Sec. 8.5, discuss the impact of that on the laws of physics.

8.2 Limits

The origin of the modern electronic computer can be found in diverse places; in the Jacquard loom, in Babbage's inventions, or in Hollerith's machinery for tabulating the 1890 U.S. census data. The real momentum, however, came around the end of World War II. 1996 saw the fiftieth anniversary of the ENIAC; 1995 saw the fiftieth anniversary of my own organization, IBM Research. A good many other related significant events stem from those years.

A concern with the fundamental physical limits of the computer appeared soon after the arrival of the computer. Shannon's information theory [4] had already taught us to think about the ultimate limits of information handling, and established a concern with the relation between information and entropy. Unfortunately, in its early stages, the more computer-oriented discussions were not particularly disciplined. Scientists are proud of their ability to do back of the envelope calculations; to react simply and quickly to the essence of a problem.

Unfortunately, as discussed in detail in Ref. [5], in this area most of the early attempts turned out to be wrong. The zig-zag pattern emphasized in the title of [5] continues to the present. For example, the possibility of computation by totally coherent quantum mechanical machinery [6] was appreciated at a late stage. It took several years after that for the invention of quantum parallelism [7]. The widespread understanding of the need to handle errors in such machinery has come only in the last few years.

Brillouin's book [8] typifies the early thinking. Without deprecating the many earlier accomplishments of this major scientist, the book left me with the reaction: "There must be a better way to think about that." This reaction was shared by my local colleague and collaborator, John Swanson. One result of that concern was Ref. [9] pointing out that the computer operations which inevitably demand a minimal unavoidable energy dissipation are those that discard information. That history has been discussed in Refs. [5] and [10]. It may be worthwhile, however, to point out quite how much Ref. [9] defied the prevailing wisdom. It was "known" in 1961 that it takes $kTln2$ to ship a bit, and in a computer we do lots of bit transmission, even when we do not discard information. (Example: A shift register where we simply move bits along.) It was also "known" that it takes energy to make a measurement, and a circuit can be presumed to measure its inputs. It took a long time, however, after Ref. [9] to rectify the prevalent, but incorrect, notions about measurement and communication.

The fact that computing can be accomplished through a sequence of 1:1 mappings, which do not discard information, was pointed out in Ref. [9]. The consistent and clear understanding and utilization of that for reversible computation had to wait for Bennett [11]. Ref. [11] showed that computation in classical systems, with noise, and with frictional forces proportional to velocity, can be done with arbitrarily little dissipation, per step. Bennett's work, showing that computation can be done by a sequence of 1:1 mappings, in turn, allowed the way for the appreciation that a totally coherent quantum mechanical time evolution (which inevitably has to be 1:1) can cause interacting bits to change with time as needed in a computer [6]. Benioff's insight, in Ref. [6], is one that, today, may also be hard to appreciate. The prevailing belief, at the time, was that the uncertainty principle was a problem for computation. Indeed, there were prevalent casual assertions (see Ref. [12] for a listing of some of these) that the uncertainty principle specified a minimal energy expenditure required by a fast switching event.

The notion of reversible computation arose out of discussions concerned with purely conceptual questions. Eventually, however, it was realized that the energy stored in capacitors in CMOS logic circuits need not be discarded, but could - to a large extent - be returned to a suitably designed power supply [13]. An early version of this approach, *hot-clocking*, generally ascribed to Feynman's Caltech colleague Charles Seitz, is described in Chapter 7 of Feynman's Lectures on Computation [14].

In a book closely related to Feynman, it is appropriate to comment on his role in understanding quantum mechanical computers. Ref. [15] was generated in connection with, and after a 1981 conference, where Benioff was also present, and where Benioff presented his emerging notions about a totally quantum mechanical computational process. Quite independently, Feynman's paper pointed to the difficulty that classical computers had in following events in the large Hilbert space of quantum mechanical time evolution. Feynman appreciated the greater power of a quantum computer, but provided no details suggesting how that might be accomplished. Later on, Feynman [16] enlarged on the work of Benioff and supplied his own very appealing and useful description of a quantum computer, without citing Benioff. Feynman's computer is not clocked; there is no external intervention. The computation proceeds under its initial kinetic energy, and is launched onto its computational track much like an electron wave packet can be sent down along a one-dimensional periodic lattice. Remarkably, Feynman never put his two contributions, [15] and [16], together. He, thus, failed to anticipate Deutsch's invention of quantum parallelism [7]. Quite possibly that was obvious to Feynman, and the connection was not pointed out because Feynman gave only very informal lectures, leaving it to others to record these in written form. Incidentally, my own discussion in Sec. 4 of Ref. [12], triggered by Benioff's work, anticipated in a crude way, the Feynman computer [16]. Unfortunately, it was a somewhat incorrect anticipation, assuming that the variable data structure during the computational progress, could cause quantum mechanical reflection, i.e. reversal of the computation. In a properly designed perfect computer it need not do so. The subject of reflections was eventually revisited by Benioff [17]. While Feynman has been influential in the development of our understanding of quantum computation, his teacher, John Wheeler, has had an equally significant role, even if a somewhat more indirect one. Wheeler's papers will be mentioned in Sec. 8.5. But Wheeler's influence via his students, associates, and active participation in conferences, has helped to shape the field.

In Ref. [18] John Swanson asked how much memory can be obtained from a given quantity of storage material. He had the typical cooperative phenomena such as ferromagnetism, ferroelectricity and superconductivity in mind. The theory was not totally independent of model details; his conclusions do not have universal applicability. If we take a limited amount of material and use it to make a few large bistable memory elements, they will be very immune to decay by thermal activation or tunneling, but will not provide a great deal of storage. If we make very small elements, we have a great deal of storage initially. But we lose it very quickly, even if we allow redundancy, as invoked by Swanson, to protect against errors. Redundancy is effective against small error probabilities, but not very useful once most elements have already lost their initial status. Thus, there is an intermediate size for memory elements which optimizes the number of stored bits. The optimum size is one where the individual storage element is relatively reliable.

Swanson's calculation did not describe fundamental and unavoidable limits; his result depends on device details and has two further shortcomings. He paid no attention to the machinery addressing the bits. Furthermore, the calculation had to assume that there was a given period, T, over which information had to be preserved. This paper was written long before modern random access dynamic memory was invented. (Memories with regular refresh operation, e.g. delay lines, did exist well before Ref. [18].) The calculation did not allow for frequent intermediate refresh operations, which read out the bits before too many have changed, and reset them all, accordingly. Indeed, the resetting can be done at two different levels. First of all, at a systems level, invoking redundancy, to reset those memory elements which have become erroneous. But it can also be done at the level of the individual element, before it has drifted too far from its intended state. Indeed, in the case of tunneling, we now understand that measurements not only allow resetting; they actually slow down the tunneling process [19]. If the information loss occurs by thermal noise, or other noise sources, and information is held in a local state of stability, a similar reduction can be achieved. But in a classical system, separate measurement and resetting operations are needed. In such a system the two local states of stability, denoting *0* and *1*, respectively, will typically lie in potential valleys separated by a saddle point, over which escape from one state to the other occurs. A bit on its way to the saddle point, as a result of earlier noise pulses, can be restored to its original valley, through a measurement which determines on which side of the saddle point the particle is located, followed by a return to the appropriate valley. Whether we actually deal with potential valleys, or with more active dynamic systems which have local states of stability, does not matter.

Despite its limits, Swanson's paper was perceptive and pioneering in other ways. It was, very likely, the first broad perception of the use of redundancy in memory, even though its practical use was already understood at that time. The paper was certainly the first place that tunneling, as a source of information loss for a bit utilizing more than one particle or crystal cell, was considered. Tunneling of a small ferromagnet from one state of stability to another has become a subject of active concern in the last decade, associated with the label *Macroscopic Quantum Tunneling*. This literature rarely acknowledges Swanson's work. Swanson's work stems from the same period as Feynman's [20] famous: *There's plenty of room at the bottom*. Feynman's paper, with its proposal of small machines making still smaller machines, was that of a supremely gifted visionary and amateur; Swanson's that of a professional in the field. Feynman, for example, foresaw the entire Caltech library of 120,000 volumes recorded on just one library card, ten years after his paper. Instead, since the time of Feynman's paper, most university libraries have expanded their space. A large library on a card poses a number of secondary problems. How many users can have access to it simultaneously? Can it be updated every day as new acquisitions appear?

Swanson's paper [18] appeared after his untimely death in a diving accident, and I prepared it for publication. Swanson, in collaboration with me, also addressed the escape rate from a metastable state, activated by thermal equilibrium noise [21], generalizing an earlier one-dimensional case treated by Kramers [22]. John Swanson saw the possibility of this theory; the details were left to me. This paper went only modestly beyond an earlier one by Brinkman [23], whose work, unfortunately, was not known to us at the time. This process of unintended rediscovery was to repeat itself several more times. The most significant of these later contributions is recorded in [24].

Bistable information holding systems need not, like ferromagnets, be dissipationless systems in their steady information holding state. They can be active dissipative circuits built out of relays, vacuum tubes, or transistors. How small can these be and still hold information effectively, protected against fluctuations? Ref. [25] answered this for bistable Esaki diode circuits. Even more than Swanson's memory element theory [18], this was a specialized theory, far from universal applicability. Nevertheless, it was a prototype of a calculation which started to become fashionable in areas unrelated to computation, about a dozen years later. For a history of that field see Ref. [26].

Science is very much a matter of fashionability. Some topics come into public focus, others are ignored. The track started by Refs. [18] and [25] never really caught on. Concern with the interface of physics and computation, at a fundamental level, has achieved visibility [27]. At the same time, concern with the noise activated escape from a state of local stability is also a subject of widespread concern [28]. We hope that the intersection of these fields will be rediscovered. At a minimum the extension of Swanson's theory, sketched qualitatively above, needs to be stated more precisely.

8.3 Energy Requirements of the Communications Channel

This subject, like almost all branches of our field, has had a convoluted history, which I will not recapitulate in detail. The fact that in practice electromagnetic waves are used has misled many to assume that is essential, and to go even further and assume that the energy in the signal *has* to be dissipated. Actually, of course, we can use mail or floppy disks, and have many other ways to ship information.

Essentially no energy has to be consumed to send a bit. That answer, for classical bits sent by classical machinery, was given a decade ago [29, 30]. The answer for classical bits sent by quantum mechanical machinery was given more recently [31], and after that for quantum bits (qubits) [32].

The quantum communications channel has received a great deal of attention, and we cite here only the most recent investigations [33]. But that work, despite its very real contribution and importance, fails to ask about ultimate unavoidable energy requirements. Actually, if we are interested in minimizing energy requirements, we may well want to avoid the quantum channel altogether. For a given signal energy (ignoring for the moment whether it needs to be consumed) we can send more bits if we divide the energy between several channels [34]. That requires a lower bit rate in each channel, and therefore the quanta associated with the message have a lower energy. If we go far enough in that direction, then the channels become classical where kT, rather than $\hbar\omega$, matters.

This author's proposals for low energy communication have all been of a conceptual nature, and very far from practicality. Are there more practical embodiments? Even if there are, is energy so important that it is worth going to the extra complexity that is likely to be involved? The history of reversible computation provides an interesting lesson in this context. It, too, was originally a purely conceptual innovation. Nevertheless, as already stated in Sec. 8.1, closely related proposals for saving energy in CMOS logic circuits have appeared [13]. These proposals were not oriented at saving an energy of order kT per step, but only at cutting down on some of the much larger power requirements of real circuits. The real utility of these proposals is still far from clear. Perhaps, the communications channel can

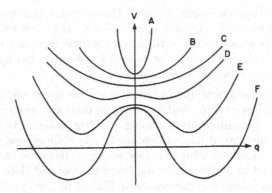

Fig. 8.1. Time-dependent potential well going from single minimum at A to a deeply bistable state at F, and later returning to A. The curves are displaced vertically relative to one another for clarity. The variable q gives the position of the particle in the well.

follow a similar history.

It is also worth noting that all of the conceptual proposals for low energy communication channels put forth, so far, involve mechanical motion. Can it, possibly be done by invoking nonlinear photon interactions? There, too, we have a historical precedent. For a number of years all of the detailed embodiments of apparatus for reversible computation were mechanical or chemical in form. I asked myself whether there were deep reasons for the absence of electrical versions. But Likharev [35] invented a Josephson junction version; there were no deep reasons, only a lack of invention.

8.4   Time-Modulated Potential

Particles in a classical time-dependent potential going from a monostable state to a deeply bistable state and back to the monostable state can be used to carry out logic. The time-dependent potential is illustrated in Fig. 8.1, and will be taken as heavily damped. A particle in the well, approaching the bifurcation point (where the well is flat, as in curve C of Fig. 8.1) from the monostable state, is easily influenced, pushing it toward one of the two developing pockets or the other. The biasing influence comes from other particles which are already locked into the deeply bistable state and coupled to the one under consideration. The time dependence of the potential, *per se*, is not a source of dissipation. That can only come from motion of the particle against the viscous forces associated with the potential. We take these to be proportional to the particle velocity. If the particle's motion is kept slow, then very little energy is consumed per event. The time-dependent potential can be generated, for example in the case of a charged particle, by moving charges to or away from the information bearing particle. I have invoked this approach repeatedly in my papers. For example, a linear sequence of such wells, each influencing the next, can be a communication channel with arbitrarily little dissipation per transmitted bit [30]. The method is not immune to noise-induced errors, but by proper "design" choices; these can be made as small as desired.

The time-modulated potential approach is not my invention; its history will be described. I have, however, invoked it more often than anyone else. Despite the fact that a long chain of investigators, including von Neumann and Feynman, have paid attention to this method, it has received amazingly little general recognition. You will not find it in textbooks or broad review articles. The notion stems from simultaneous and independent inventions by von Neumann [36] and Goto [37]. These come from a time when junction transistors were limited to low speeds and alternative high speed logic approaches were sought. The inventions use the development of a subharmonic through parametric excitation of a non-linear resonant circuit, driven at twice its approximate resonant frequency. This is a bistable system; the subharmonic can develop with two possible phases, 180° apart. Attempts to develop this approach took place at several different U. S. laboratories. In Japan actual working computers using the non-linear susceptibility of magnetic cores were built. The basic tool for this approach is majority logic. Three input circuits influence a subsequent one, which is controlled by the majority phase of its three inputs. The approach is beset by two problems when considered as a serious technological candidate. First of all, majority logic is demanding on tolerances, i.e. on the deviation allowed for signals from their supposed ideal value. Furthermore, the approach requires a precise clock signal delivered to every stage. Logic proposals which need to clock every stage have never been successful. Goto, subsequently, adapted the approach to Esaki diode circuits [38].

The invention was adapted to time-dependent potentials by Keyes and Landauer [39], but at a time when reversible computation was not yet understood. The fact that the approach could be used in reversible computation, and could be carried out with Josephson junction circuits, was pointed out by Likharev [35]. There has been some tendency to think of reversible computation as "Brownian" computers, i.e. computers which move back and forth diffusively along their trajectory, but with a net drift velocity due to an applied driving force. Likharev's invention showed that reversible computers could be clocked, moving forward at a steady and predictable rate.

Bennett used the approach to point out that copying can be done reversibly, with arbitrarily little energy dissipation [40]. In Bennett's case a ferromagnet was taken up through its Curie point, and then taken back down under the influence of another magnetized bit. Feynman, in connection with Figs. 5.15 and 5.16 of Chapter 5 in Ref. [14] once again explains the copying process, invoking time-modulated potentials, with reference to Bennett. Merkle [41] used the approach for mechanically buckled cards, under an oscillating compressive force, bending out one way or the other. Recently, Lent and Tougaw [43] used the method in connection with their Quantum Cellular Automata, which have two electrons at opposite corners of a square. The field due to one of these bistable cells polarizes an adjacent cell. In the most recent version of this approach the tunneling barriers, which allow electrons to move around the square, are externally controlled. Despite the utilization of tunneling in the proposal of Ref. [43], the time-modulated potential scheme is intrinsically a dissipative scheme, invoking relaxation to the ground state as the bifurcation develops. It is not directly suitable for totally coherent quantum mechanical computation. That can be recognized from the majority logic operation, which is not 1:1, whereas a Hamiltonian method must be 1:1. A totally coherent quantum mechanical proposal for controlled tunneling has been put forth [32], but is very different in character. As is true of all the existing quantum logic proposals, it does not provide *exactly* the required unitary transformation, and does not provide any natural error immunity, allowing for small departures from the design specifications. Likharev revisited the time-modulated potential with Korotkov in Ref. [42]. The shift register proposed there is an example of a reversible communications link discussed in Sec. 8.3. As in Ref. [43], the motion from one pocket to another involves quantum mechanical tunneling, but the overall process is not quantum mechanically coherent. Relaxation to the ground state is essential.

8.5   Broader Implication of Limits

How far can technology improve? Limits are of interest as declarations of boundaries for that, or as a declaration of their absence. Computational limits, however, also have a more fundamental scientific significance to be discussed in this section. The execution of all information handling operations, including all mathematical operations, has to take place in our physical world. We have all been indoctrinated by the mathematicians, given $\in \exists N$, stating that with enough successive operations, any accuracy requirement can be met. It is questionable whether this can be carried out in our real physical world, where it needs to be done. This is a theme which I have visited in a number of papers, and here cite only an early [44] and a recent item [45].

Arbitrary precision requires an unlimited memory, and this is unlikely to exist in a finite universe. Even if the universe is unlimited, it seems unlikely that we can collect arbitrarily large parts of it into an organized memory structure, and even if we grant the availability of unlimited memory, it still seems unlikely that we can have an unlimited sequence of operations, each guaranteed to be totally free of error. Thus, we are questioning that the mathematicians' continuum and real number system reflect executable operations. That is, of course, not a totally unprecedented reaction, and related to many other existing views, though probably not exactly equivalent to any of them. Ref. [45] stressed the relationship to John Wheeler's work. Feynman, in Ref. [15] comes close to suggesting that a bounded volume of space and time is associated with a limited amount of information, but does not quite say that. It is my recollection that in his actual lecture in 1981 that was stated much more clearly and unconditionally.

I have already alluded to quantum parallelism, utilizing qubits which can be in a quantum superposition of a *0* and *1* state. My published papers give a very conservative appraisal of the realizability of quantum parallelism. Nevertheless, I welcome it, as a celebration of the physical nature of information. Even if quantum parallelism never comes to pass, those who prove theorems about the minimum number of steps required by an algorithm, must take its possibility into account.

What can replace the real number system and allow for the limited precision of real physical operations? This author does not know. It does not necessarily mean algorithms with a limited number of bits. The limits are likely to appear in a more statistical fashion. The laws of physics are algorithms for calculation, and as stressed by this author, in their proper form must respect the limited information handling capabilities of the real universe. I am often asked: "Does our limited ability to describe what the physical world is doing prevent that world from doing its own thing more precisely? Isn't it just our knowledge that is limited?" My answer to that: "Behavior which cannot be followed, described, or observed is not a matter of science. If I am told that seven angels are on the head of a pin and that angels are not detectable, I cannot call that an erroneous assertion. But it is not a matter for science."

What is the likely impact of limited precision on science? In earlier papers I have emphasized that this may be related to the ultimate source of irreversibility and fluctuation in the real world, a world where we can readily observe departure from Hamiltonian behavior. This is admittedly a very speculative conjecture. But limited precision may also underlie the apparent classical behavior manifested by the events around us. There are a great many theories which claim to explain that [46], and we can cite only a few, including some skeptical reactions. Some of these explanations may well be correct, and there may not be a need to say more. Nevertheless, a totally quantum mechanical behavior in systems with some complexity and followed for some time, requires a precise evaluation of phases for the competing histories, for the competing Feynman paths. In a world with limited precision relative phases will, eventually, get lost and this will lead to classical behavior. As in the discussion of noise and irreversibility, the limited precision acts as if the universe had an unpredictable environment with which it interacts.

At a minimum, however, we caution those who invoke the wave function of the whole universe. How can that wave function be recorded, unless you have a second and separate universe available for that?

I am suggesting that, contrary to our prevailing views, the laws of physics did not precede the universe and control it, but are part of it. Once again, this is not a view totally orthogonal to that of others. Wheeler [47] has stated that the laws of physics result from quantum measurement on the universe. More recently, Smolin has suggested that the laws evolved with the universe [48]. Paul Benioff [49] stresses that the quantum mechanical physics of our computational apparatus is part of a totally consistent picture of physics. Indeed, this interest in self-consistency was, apparently, the motivation for Benioff's original concern with quantum computation [6].

8.6   Fashions in Science

Fashions are not totally unreasonable. We cannot each go our separate way; we need communities for interaction. After a perceptive pioneer sees a new concept, it takes time for the building of a consensus that it is important. Without that consensus we cannot expect grants, conference invitations, promotions, or acceptance by Physical Review Letters. Fashions, therefore, to some extent, simply represent the fact that it takes time to develop the conviction that an area is ripe for exploitation. Nevertheless, it seems totally apparent that the positive feedback involved in the formation of fashions has gotten out of hand. This paper is not aimed at the sociology of science and will not try to analyze the causes, nor the possible and unlikely cures. Feynman, of course, was very far from a follower of fashions, in science, or in other ways. Indeed, the many Feynman anecdotes picture someone who enjoyed an unconventional role. His *There's Plenty of Room at the Bottom* [20] may not have been a totally correct vision in all its details, but it was hardly typical of someone with Feynman's range of research activities. His later work on quantum computation anticipated what was to become a fashionable field. It was far from that when he generated Refs. [15] and [16].

The physics of computation, viewed at a fundamental level, was an almost invisible field until a 1981 conference at MIT, which included Refs. [12, 15, 35, 40] as well as papers by Benioff and Wheeler. Dyson also was a participant, as were a number of noted computer scientists. One of these, Konrad Zuse, presented his view that the universe is a digital computer [50]; a view also espoused by Ed Fredkin, one of the session's organizers.

The field as a whole has achieved visibility, without having become a really major fashion. But the subfields within the area have not developed equally. Almost all of the visibility, in recent years, has come from the study of quantum information, and particularly from quantum parallelism. It is a history we have seen elsewhere. Fractals, chaos, self-organized criticality are just a few examples of fields and concepts which received belated recognition through the path-breaking insights of pioneers. But then they went on to become industries. Industries which are fueled at the expense of other ideas which deserve to be nourished. Not all fashionable fields are the outgrowth of key new insights; some represent public relations efforts with little substance.

When we organized the 1981 MIT session, we intentionally encouraged a diverse participation. That, inevitably, brought in some contributions which were recognizably faulty. If we had not done that, it would have been hard to collect a reasonably sized group. But additionally, we knew that what was nonsense to some would be visionary to others. I cannot help but contrast the diversity of that session to some of the many recent sessions which have concentrated on the role of quantum mechanical entanglement in information processing. The narrowing represents real progress; we now understand more about what counts. But, at the same time, the narrowing strengthens fashionability in science. A carefully selected group reinforces its existing values, and declares to science journalists that their stuff is what really counts. I hope that quantum information can receive the attention it deserves, without eclipsing the other questions we have touched in the preceding sections.

References

[1] L. Szilard, Über die Entropieverminderung in einem thermodynamischen System bei Eingriff intelligenter Wesen, *Z. Phys.* **53**, 840-856 (1929); English translation: On the Decrease of Entropy in a Thermodynamic System by the Intervention of Intelligent Beings, in *Quantum Theory and Measurement*, edited by J. A. Wheeler and W. H. Zurek, pp. 539-548 (Princeton University Press, Princeton, NJ, 1983).

[2] R. Penrose, *The Emperor's New Mind* (Oxford University Press, Oxford, 1989).

[3] D. Divincenzo et al., eds. *Quantum coherence and decoherence, Proc. R. Soc. Lond. A,* **454**, 257-486 (1998).

[4] C. E. Shannon, A Mathematical Theory of Communication, *Bell Syst. Tech. J.* **27**, 379-423, Part I, II; 623-656, Part III (1948).

[5] R. Landauer, Zig-Zag Path to Understanding, in *Proceedings Workshop on Physics and Computation, PhysComp'94*, pp. 54-59 (IEEE Comp. Soc. Press, Los Alamitos, 1994).

[6] P. Benioff, The Computer as a Physical System: A Microscopic Quantum Mechanical Hamiltonian Model of Computers as Represented by Turing Machines, *J. Stat. Phys.* **22**, 563-591 (1980); P. Benioff, Quantum Mechanical Hamiltonian Models of Turing Machines, *J. Stat. Phys.* **29**, 515-546 (1982); P. Benioff, Quantum Mechanical Models of Turing Machines that Dissipate No Energy, *Phys. Rev. Lett.* **48**, 1581-1585 (1982).

[7] D. Deutsch, Quantum Theory, the Church-Turing Principle and the Universal Quantum Computer, *Proc. R. Soc. Lond. A* **400**, 97-117 (1985).

[8] L. Brillouin, *Science and Information Theory* (Academic Press, New York, 1956).

[9] R. Landauer, Irreversibility and Heat Generation in the Computing Process, *IBM J. Res. Dev.* **5**, 183-191 (1961); reprinted in *Maxwell's Demon*, edited by H. S. Leff and A. F. Rex (Princeton U. P., Princeton, NJ, 1990) pp. 188-196.

[10] C. H. Bennett, Notes on the History of Reversible Computation, *IBM J. Res. Dev.* **32**, 16-23 (1988).

[11] C. H. Bennett, Logical Reversibility of Computation, *IBM J. Res. Dev.* **17**, 525-532 (1973).

[12] R. Landauer, Uncertainty Principle and Minimal Energy Dissipation in the Computer, *Int. J. Theor. Phys.* **21**, 283-297 (1982).

[13] D. Frank and P. Solomon, in *Proceedings of the International Symposium on Low Power Design*, pp. 197-202 (Association for Computing Machinery, New York, 1995); A. Kramer et al., Adiabatic Computing with the 2N-2N2D Logic Family, in *Proceedings of 1994 IEEE Symposium on VLSI Circuits Digest of Technical Papers*, pp. 25-26 (IEEE, New York, 1994); A. De Vos, Introduction to r-MOS Systems, in *Proceedings of Fourth Workshop on Physics and Computation, PhysComp'96*, edited by T. Toffoli, M. Biafore and J. Leão, pp. 92-96 (New England Complex Systems Institute, Cambridge, MA, 1997).

[14] R. P. Feynman, *Feynman Lectures on Computation*, edited by A. J. G. Hey and R. W. Allen (Addison-Wesley Publishing Company, Inc., Reading, MA, 1996).

[15] R. P. Feynman, Simulating Physics with Computers, *Int. J. Theor. Phys.* **21**, 467-488 (1982).

[16] R. P. Feynman, Quantum Mechanical Computers, *Optics News* **11**, 11-20 (1985); reprinted in *Found. Phys.* **16**, 507-531 (1986), and in Ref. [14].

[17] P. Benioff, Transmission and Spectral Aspects of Tight-binding Hamiltonians for the Counting Quantum Turing Machine, *Phys. Rev. B* **55**, 9482-9494 (1997).

[18] J. A. Swanson, Physical Versus Logical Coupling in Memory Systems, *IBM J. Res. & Dev.* **4**, 305-310 (1960).

[19] W. H. Zurek, Pointer Basis, and Inhibition of Quantum Tunneling by Environment-Induced Superselection, *Proc. Int. Symp. Foundations of Quantum Mechanics*, edited by S. Kamefuchi, pp. 181-189 (Physical Society of Japan, Tokyo, 1983).

[20] R. P. Feynman, There's Plenty of Room at the Bottom, in *Miniaturization*, edited by H. D. Gilbert, pp. 282-296 (Reinhold Publishing Corporation, New York, 1961).

[21] R. Landauer and J. A. Swanson, Frequency Factors in the Thermally Activated Process, *Phys. Rev.* **121**, 1668-1674 (1961).

[22] H. A. Kramers, Brownian Motion in a Field of Force and the Diffusion Model of Chemical Reactions, *Physica* **7**, 284-304 (1940).

[23] H. C. Brinkman, Brownian Motion in a Field of Force and the Diffusion Theory of Chemical Reactions II, *Physica* **22**, 149-155 (1956).

[24] J. S. Langer, Theory of Nucleation Rates, *Phys. Rev. Lett.* **21**, 973-976 (1968).

[25] R. Landauer, Fluctuations in Bistable Tunnel Diode Circuit, *J. Appl. Phys.* **33**, 2209-2216 (1962); reprinted in *Dynamic Patterns in Complex Systems*, edited by J.A. S. Kelso, A. J. Mandell, M. F. Schlesinger, pp. 103-111 (World Scientific, Singapore, 1988).

[26] R. Landauer, Noise Activated Escape from Metastable States: an Historical View, in *Noise in Nonlinear Dynamical Systems*, edited by F. Moss and P. V. E. McClintock, pp. 1-15 (Cambridge U. Press, Cambridge, 1989).

[27] *Proceedings of Workshop on Physics and Computation PhysComp'92* (IEEE Computer Society Press, Los Alamitos, CA, 1993); *Proceedings of Workshop on Physics and Computation PhysComp'94* (IEEE Computer Society Press, Los Alamitos, CA, 1994); *Proceedings of Fourth Workshop on Physics and Computation, PhysComp'96*, edited by T. Toffoli, M. Biafore and J. Leão (New England Complex Systems Institute, Cambridge, MA, 1997).

[28] *Fluctuations and Order: The New Synthesis*, edited by M. Millonas (Springer, New York, 1996).

[29] R. Landauer, Computation, Measurement, Communication and Energy Dissipation, in *Selected Topics in Signal Processing*, edited by S. Haykin, pp. 18-47 (Prentice-Hall, Englewood Cliffs, NJ, 1989).

[30] R. Landauer, Energy Requirements in Communication, *Appl, Phys. Lett.* **51**, 2056-2058 (1987).

[31] R. Landauer, Minimal Energy Requirements in Communication, *Science* **272**, 5270, 1914-1918 (1996).

[32] R. Landauer, Energy Needed to Send a Bit, *Proc. R. Soc. Lond. A*, **454**, 305-311 (1998).

[33] B. Schumacher and M. D. Westmoreland, Sending Classical Information via Noisy Quantum Channels, *Phys. Rev. A* **56**, 131-138 (1997); A. S. Holevo, The Capacity of the Quantum Channel with General Signal States, *IEEE Transactions on Information Theory*, 44, 269-273 (1998); C. A. Fuchs, Nonorthogonal Quantum States Maximize Classical Information Capacity, *Phys. Rev. Lett.* **79**, 1162-1165 (1997).

[34] R. Landauer and J. F. Woo, Cooperative Phenomena in Data Processing, in *Synergetics*, edited by H. Haken, pp. 97-123 (B. G. Teubner, Stuttgart, 1973).

[35] K. K. Likharev, Classical and Quantum Limitations on Energy Consumption in Computation, *Int. J. Theor. Phy.* **21**, 311-326 (1982).

[36] J. von Neumann, Non-linear Capacitance or Inductance Switching, Amplifying and Memory Organs, U. S. Patent 2,815,488, filed 4/28/54, issued 12/3/57, assigned to IBM.

[37] E. Goto, On the Application of Parametrically Excited Nonlinear Resonators (in Japanese), *J. Electr. Commun. Eng.*, Japan, **38**, 770-775 (1955).

[38] E. Goto, K. Murata, K. Nakazawa, K. Nakagawa, T. Motooka, Y. Matsuoka, Y. Ishibashi, II. Ishida, T. Soma and E. Wada, Esaki Diode High-Speed Logical Circuit, *IRE Trans. Electronic Computer* **EC-9**, 25-29 (1960).

[39] R. W. Keyes and R. Landauer, Minimal Energy Dissipation in Logic, *IBM J. Res. & Dev.* **14**, 152-157 (1970).

[40] C. H. Bennett, The Thermodynamics of Computation –A Review, *Int. J. Theor. Phys.* **21**, 905-940 (1982).

[41] R. C. Merkle, Two Types of Mechanical Reversible Logic, *Nanotech.* **4**, 114-131 (1993).

[42] K. K. Likharev and A. N. Korotkov, "Single-Electron Parametron": Reversible Computation in a Discrete-State System, *Science* **273**, 763-765 (1996).

[43] C. S. Lent, and P. D. Tougaw, A Device Architecture for Computing with Quantum Dots, *IEEE* **85**, 541-557 (1997).

[44] R. Landauer, Wanted: a Physically Possible Theory of Physics, *IEEE Spectrum* **4**, 105-109 (1967); reprinted in *Speculations in Science and Technology* **10**, 292-302 (1987).

[45] R. Landauer, The Physical Nature of Information, *Phys. Lett. A* **217**, 188-193 (1996).

[46] R. B. Griffiths, Consistent Histories and the Interpretation of Quantum Mechanics, *J. Stat. Phys.* **36**, 219-272 (1984); M. Gell-Mann and J. B. Hartle, Quantum Mechanics in the Light of Quantum Cosmology, in *Complexity, Entropy, and the Physics of Information*, edited by W. Zurek, pp.425-458 (Addison-Wesley, Reading, MA, 1990); R. Omnès, *The Interpretation of Quantum Mechanics* (Princeton University Press, Princeton, NJ, 1994); R. Omnès, *Rev. Mod. Phys.* **64**, 339 (1992); M. Gell-Mann and J. B. Hartle, in *Proceedings of the 3rd International Symposium on the Foundations of Quantum Mechanics in the Light of New Technology*, edited by S. Kobayashi, H. Ezawa, Y. Murayama, and S. Nomura (Physical Society of Japan, Tokyo, 1990); N. Yamada, Probabilities for Histories in Nonrelativistic Quantum Mechanics, *Phys. Rev. A* **54**, 182-203 (1996); W. H. Zurek and J. P. Paz, Decoherence, Chaos, the Quantum and the Classical, *Il Nuovo Cimento* **110B**, 611-624 (1995); W. H. Zurek, Preferred States, Predictability, Classicality, and Environment-Induced Decoherence, in *Physical Origins of Time Asymmetry*, edited by J. J. Halliwell, J. Pérez-Mercader and W. H. Zurek, pp. 175-213 (Cambridge University Press, Cambridge, UK, 1994); W. H. Zurek, *Phys. Today* **46**, 81 (1993); C. H. Woo, Merging Histories and the Second Law, *unpublished;* F. Dowker and A. Kent, On the Consistent Histories Approach to Quantum Mechanics, *J. Stat. Phys.* **82**, 1575-1646 (1996).

[47] J. A. Wheeler, in *Problems in Theoretical Physics*, edited by A. Giovanni, F. Mancini and M. Marinaro, p. 121 (University of Salerno Press, Salerno, Italy, 1984); in *Frontiers of Nonequilibrium Statistical Physics*, edited by G. T. Moore and M. O. Scully (Plenum, New York, 1986).

[48] L. Smolin, The Life of the Cosmos (Oxford University Press, New York, 1997).

[49] P. Benioff, Quantum Robots and Quantum Computers, *Phys. Rev. A*, **58**, 893-904 (1998).

[50] K. Zuse, The Computing Universe, *Int. J. Theor. Phys.* **21**, 589-600 (1982).

9

SCALING OF MOS TECHNOLOGY TO SUBMICROMETER FEATURE SIZES

Carver A. Mead *

Abstract

Industries based on MOS technology now play a prominent role in the developed and the developing world. More importantly, MOS technology drives a large proportion of innovation in many technologies. It is likely that the course of technological development depends more on the capability of MOS technology than on any other technical factor. Therefore, it is worthwhile investigating the nature and limits of future improvements to MOS fabrication. The key to improved MOS technology is reduction in feature size. Reduction in feature size, and the attendant changes in device behaviour, will shape the nature of effective uses of the technology at the system level. This paper reviews recent, and historical, data on feature scaling and device behavior, and attempts to predict the limits to this scaling. We conclude with some remarks on the system-level implications of feature size as the minimum size approaches physical limits.

9.1 Introduction

It is always difficult to predict the future; few attempts to do so have met with resounding success. One remarkable example of successful prediction is the exponential increase in complexity of integrated circuits, first noted by Gordon E. Moore. As we contemplate the ongoing evolution of this great technology, many questions arise: Can the trend continue? Will single-chip systems attain levels of complexity that render present system architectures unworkable [1]? Will digital techniques completely replace analog methods [2]? The answers to these questions depend critically on the properties of the individual transistors that provide the essential active functions, without which no interesting system behavior is possible. Integrated-circuit density is increased by a reduction in the size of elementary features of the underlying structures; therefore, any discussion of the capabilities of future technologies must rely on an understanding of how the properties of transistors evolve as the transistors' dimensions are made smaller.

Elsewhere [3], we described the factors that limit how small an MOS transistor

---

*Reproduced from Journal of VLSI Signal Processing, 8, 9-25 (1994) Kluwer Academic Publishers, Boston. Manufactured in The Netherlands.

can be and still operate properly. That discussion will not be repeated here, but I will outline the major issues:

1. For the device current to be primarily controlled by the gate, the device should not be punched through; that is, the sum of the source and drain depletion layers should be less than the geometric channel length. As a direct consequence of this requirement, the bulk doping must increase as dimensions are decreased.

2. Increasing the bulk doping has two important consequences:

   a. Junction breakdown voltage is lowered.

   b. A larger electric field is required in the gate oxide to obtain a given change in surface potential.

Because of 2a, the operating voltage must be reduced. So that sufficient electric field can be obtained with a lower operating voltage, the gate oxide must be made thinner. Thus, it is inevitable that, as the minification process is continued, both drain depletion layer and gate oxide will become thin enough that electron tunneling through them will become comparable with other device currents. In 1971, when our original study [3] was written, we described a device of 0.15 micrometer (μ) channel length, having a 50 Angstrom ($\overset{\circ}{A}$) gate oxide. Although we were confident that a device of this size could be made to work, we were not at all sure that smaller devices could be made viable.

Over the ensuing 22 years, feature sizes have evolved from 6 to 0.6 μ and the trend shows no sign of abating [4–10]. In this paper, I shall examine what we have learned from the past 22 years of technology evolution, and shall discuss to what extent these same trends may continue into the future. I shall conclude that we can safely count on at least one more order of magnitude of scaling, with a concomitant increase in both density and performance. Several of the conclusions of this study were reached independently by Hu [11].

9.2   Scaling Approach

In Figure 9.1, I have plotted the historic trend of gate-oxide thickness t_{ox} as a function of l, the minimum feature size of the process. The trend can be expressed accurately as

$$t_{ox} = 210l^{0.77}$$

where the feature size is in μ, and the gate-oxide thickness is in $\overset{\circ}{A}$. This observation suggests that it may be fruitful to express all important process parameters as

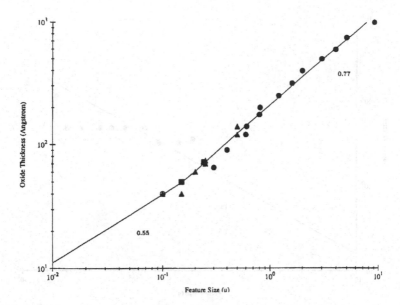

Fig. 9.1. Gate-oxide thickness as a function of feature size. The solid circles are production processes in silicon-gate technology, starting in 1970. Triangles are processes reported in the literature. Solid squares are the two most advanced devices described in our previous study [3]. The solid line is the analytic expression used in this study (Equation 9.1).

powers of the feature size, and to determine whether there is a scaling of this form that allows sensible process evolution to dimensions well below 0.1 μ. To prevent the gate oxide thickness from becoming thinner than a single atomic layer, I have chosen a scaling of the form

$$t_{\text{ox}} = \max(210l^{0.77}, 140l^{0.55}) \tag{9.1}$$

This expression is plotted as the solid line in Figure 9.1. In reviewing the historic trend, it is clear that we expressed previously [3] more concern with gate-oxide tunneling than has been justified by the experience accumulated through the intervening years. It is conceivable that I am repeating the same bit of paranoia here. In any case, if oxide thickness continues to decrease at the present rate, the resulting devices will be somewhat more capable than those I present.

The oxide thickness and feature size together determine the gate-oxide capacitance C_g of a minimum-sized device:

$$C_g = \epsilon_{\text{ox}} \frac{l^2}{t_{\text{ox}}}$$

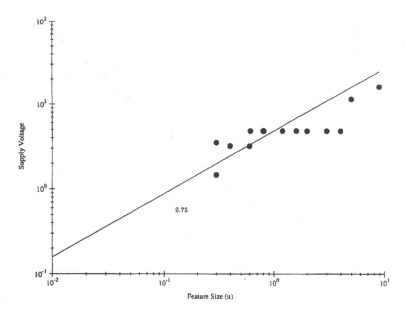

Fig. 9.2. Power-supply voltage as a function of feature size. The solid line is the analytic expression used in this study (Equation 9.2).

The historic trend in supply voltage V is shown in Figure 9.2. This trend is not as smooth as the trend in oxide thickness, due to the long period of standardization at 5 volts (V). It is clear, however, that modern submicrometer devices operate better on lower voltages [7, 12], and that this trend to lower voltages must continue. The scaling I use in this study is

$$V = 5l^{0.75} \tag{9.2}$$

This expression is plotted as the solid line in Figure 9.2.

Once we have the gate-oxide capacitance and supply voltage, we can estimate the energy W_g stored on the gate of a minimum-sized transistor at any given feature size. I have slightly overestimated the stored energy as

$$W_g = \frac{1}{2}C_g V^2 \tag{9.3}$$

For the scaling laws given here, the stored energy (in Joules) works out to be

$$W_g = 2.2 \times 10^{-14} l^{2.75} \tag{9.4}$$

This expression is plotted as the long solid line in Figure 9.3. Even with the slight "kink" introduced by Equation 9.1, this expression is a good abstraction of

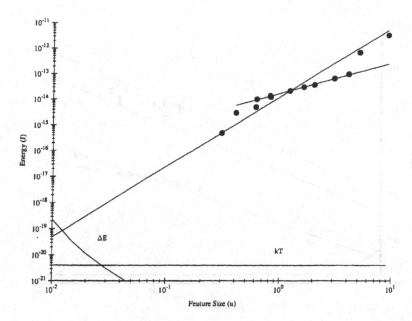

Fig. 9.3. Energy stored on the gate of a minimum-sized transistor as a function of feature size. We compute the points from Equation 9.3 using oxide-thickness values from figure 1 and the supply-voltage values from figure 2. The solid line is the analytic expression used in this study (Equation 9.4). Also shown for reference are the thermal energy kT at room temperature, and the quantum-level spacing for electrons in the channel with momenta in the direction of current flow.

the actual energy over the entire range of the plot. In the central section of historic data, however, the constant 5-V power-supply voltage has established a trend with much less dependence on feature size.

This shorter trend is well represented by the expression

$$W_5 \times 2 \times 10^{-14} l^{1.22} \qquad (9.5)$$

Also shown for reference on Figure 9.3 is the thermal energy kT, and the spacing of levels in the channel with momenta in the direction of current flow. It is clear that the stored energy is more than 10 kT even at feature sizes of 0.01μ.

The minimum stored energy is an interesting quantity because it sets the scale for the switching energy dissipated in a digital system. The energy per operation of computation-intensive digital chips is compared with the minimum stored energy in Figure 9.4. The system energy per operation is four to six orders of magnitude higher than the minimum stored energy, and can be bounded by the two solid trend

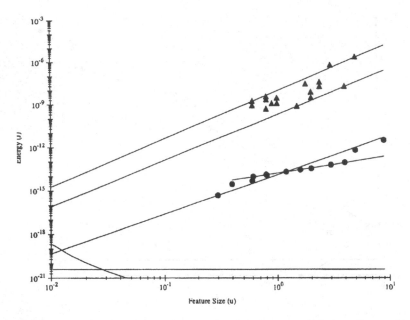

Fig. 9.4. Energy dissipated per operation at the chip level. Filled circles are data taken from the literature and from manufacturers' data sheets. Examples are all computation-intensive single chips, such as multipliers, digital signal processors, and similar devices. So that the data could be plotted on a single scale, all values were normalized to 8 x 8 multiply-add operations, assuming that the energy is proportional to the product of the word lengths of the multiplicand and multiplier. Minimum and maximum trend lines shown are Equations 9.5 and 9.6. Also shown for reference are the data of Figure 9.3.

lines:

$$W_{max} = 1.15 \times 10^{-8} l^{3.4} \tag{9.6}$$

$$W_{min} = 2.5 \times 10^{-10} l^{3.25} \tag{9.7}$$

The overall system trend is steeper than that for minimum stored energy, presumably because designers have become more skilled over the years, and processes have an ever increasing set of features on which designers can draw (multiple levels of metal, for example). A 5-V subtrend is clearly discernible in the system data as well.

With the information on hand, we can determine the tunneling current density J_{ox} through the gate oxide [13–15], making the worst-case assumption that the

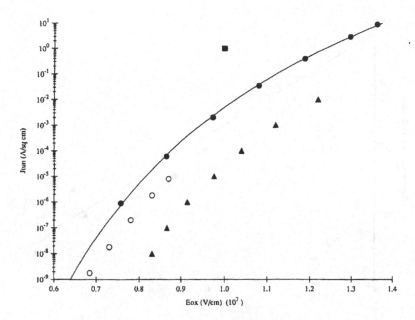

Fig. 9.5. Oxide tunneling current as a function of electric field. The open circles are from the original work of Lenzlinger and Snow [13]. Filled circles are from the recent work of Suñé et al. [15]. Filled triangles are from Hori et al. [14]. The solid line is the analytical expression used in this study (Equation 9.7). The filled square is inferred from Iwase et al. [10], but is not directly comparable with the other data because it was taken from a transistor drain characteristic, and may be corrupted with other effects such as gate-enhanced drain tunneling. The gate current was not reported separately, so this value shown represents a worst-case estimate.

entire supply voltage appears across the entire gate area:

$$J_{ox} = J_0 E_{ox}^2 e^{-kt_{ox}} \tag{9.8}$$

where $J_0 = 6.5 \times 10^{10}$ A/V/cm^2 was adjusted to match experimental data, as shown in Figure 9.5. The imaginary part of the wave vector k is given by

$$k = \frac{2k_0}{3} \frac{\phi}{V} \left[1 - \left(1 - \min\left(1, \frac{V}{\phi} \right) \right)^{3/2} \right] \tag{9.9}$$

These expressions are valid for voltages both above and below the barrier potential ϕ which was taken to be 3.2 V. The preexponential constant $k_0 = 1.2 \overset{\circ}{A}^{-1}$ was used. It is comforting to note that oxide tunneling data are available over the entire range of electric fields that will be encountered down to the smallest dimensions

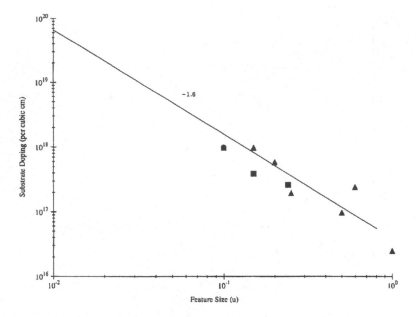

Fig. 9.6. Substrate doping as a function of feature size. The solid line is the analytical expression used in this study (Equation 9.8). Filled triangles represent processes reported in the literature. The two solid squares are the two smallest transistor designs shown in our earlier work [3].

studied here. It will be helpful, however to have actual experimental data in the 10 $\overset{\circ}{A}$ range. For these extremely thin oxides, it will be essential to take into account the quantum corrections discussed in Suñé et al. [15].

The other major source of parasitic current is tunneling through the drain junction. The junction-tunneling current density J_j is critically dependent on the substrate acceptor concentration n, which must be increased to avoid punch-through as device dimensions are decreased [16–22]. The scaling law used in this study is plotted in Figure 9.6:

$$n = 4 \times 10^{16} l^{-1.6} \tag{9.10}$$

Given the doping density n, we can compute the depletion-layer thickness x for any potential ψ relative to substrate using the usual step-junction approximation:

$$x = \sqrt{\frac{2\epsilon_{si}\psi}{qn}} \tag{9.11}$$

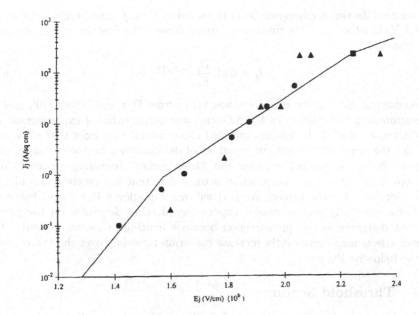

Fig. 9.7. Junction-tunneling current density as a function of peak electric field in the junction. The filled triangles are from alloy tunnel diodes, which were reported as step junctions by Chynoweth et al. [16]. The filled circles are from diffused emitter-base junctions reported as graded junctions by Fair and Wivell [19]. These were the only references that I was able to locate for electric fields in the range encountered in the finest feature sizes considered in this study. Some data are shown by Reisch [22], but not enough information is given to allow direct comparison with the other data. For reference, the solid square represents the parameters encountered in the 0.03-μ device described in this study. The solid line is the analytical expression used in this study (Equation 9.10).

The corresponding depletion-layer capacitance C is given by

$$C = \frac{\epsilon_{si}}{x}$$

We can determine the maximum electric field in the drain junction, from the junction voltage, which in the worst case will be the supply voltage plus the built-in voltage:

$$E_j = \sqrt{\frac{2qn(V + V_b)}{\epsilon_{si}}}$$

We could alternatively use a graded-junction approximation, such as that used by Fair and Wivell [19]. For our purposes, the two approaches are nearly equivalent, so

I have used the simpler step-junction expression with the junction built-in voltage V_b = 1.1 V. In either case, the tunneling current density is a function of the maximum electric field:

$$J_j = G_0 V \frac{E_j}{E_0} e^{-E_0/E}, \tag{9.12}$$

The constant $E_0 = 2.9$ x 10^7 V/cm was taken from Fair and Wivell [19], and the preexponential factor G0 = 3 x 10^9 A/V cm^2 was chosen to fit the experimental data plotted in Figure 9.7. It is significant that experimental data exist that allow us to predict the tunneling currents in junctions of devices down to 0.03-μ feature sizes. Previously [3], we pointed out that the "drain corner" tunneling occurs at lower voltage than that across the junction area, a fact that has received considerable attention [23]. For the present study, I will use Equation 9.10 for area tunneling, both for simplicity and because I expect considerable cleverness on the part of process designers as this phenomenon becomes limiting. Caution, however, that corner effects may significantly increase the drain tunneling over the values shown in the following Figures.

9.3   Threshold Scaling

To determine the detailed properties of small devices, we must take into account the short-channel properties, most notable of which are carrier-velocity saturation and drain-induced barrier lowering (the precursor to punch-through). Previously [2], we developed a model that gives closed-form expressions for the current in short-channel devices, including the effects of velocity saturation. To apply the model, we need some abstraction of the vertical doping profile under the gate. The most widely used such abstraction is the threshold voltage V_t. We therefore proceed by choosing a nominal threshold voltage of the form

$$V_t = 0.55l^{0.23} \tag{9.13}$$

The actual threshold voltage will be lower than the nominal one by the amount of drain-induced barrier lowering (DIBL) [24–27]. In this study, I use the expression given by Fjeldly and Shur [28]:

$$\text{DIBL} = V \frac{x_c}{\lambda} \frac{\sinh(x_s/\lambda}{\cosh((l - x_d)/\lambda) - \cosh(x_s/\lambda)} \tag{9.14}$$

where x_s and x_d are the classical depletion-layer thicknesses of the source and drain junctions. I have used a surface potential of 0.5 V in Equation 9.9 to compute x_c, the thickness of the depletion layer under the channel. The distance scale λ is given by

$$\lambda = x_c \left(1 + \frac{C_{\text{ox}}}{C - C}\right)^{-1/2}$$

where the depletion-layer capacitance per unit area C_c from channel to substrate is

$$C_c = \frac{\epsilon_{si}}{x_c}$$

and the oxide capacitance per unit area C_{ox} from gate to channel is

$$C_{ox} = \frac{\epsilon_{ox}}{t_{ox}}$$

The nominal threshold voltage; the actual threshold voltage, including DIBL; and the supply voltage are plotted as a function of feature size in Figure 9.8. For the scaling parameters used in this study, DIBL does not become a serious problem until feature sizes are less than 0.03 μ.

9.4 Device Characteristics

Threshold is defined as the gate voltage at which mobile charge Q_s at the source end of the channel changes the surface potential by kT/q [2]. The channel charge at threshold is

$$Q_t = \frac{kt}{q}(C_{ox} + C_c) \qquad (9.15)$$

For higher gate voltages, essentially all charge on the gate attracts equal and opposite countercharge of mobile carriers in the channel. Thus, we can form an excellent estimate of the channel charge Q_s at the source end of the channel:

$$Q_s = C_{ox}(V - V_t) \qquad (9.16)$$

For gate voltages below V_t, channel current decreases exponentially with decreasing gate voltage. At zero gate voltage, the channel charge is:

$$Q_s = Q_t e^{-q\kappa V_t/kT} \qquad (9.17)$$

where

$$\kappa = \frac{C_{ox}}{C_c + C_{ox}}$$

Given Q_t and Q_s, we can compute the saturated channel current for a minimum-sized transistor of any given channel length using Equation (B.28) from [2]:

$$I_{sat} = Q_s \nu_0 + Q_t \nu_0 \left(\frac{l}{l_0} + 1\right) \left(1 - \sqrt{1 + \frac{2Q_s l}{Q_t l_0}\left(\frac{l}{l_0} + 1\right)^{-2}}\right) \qquad (9.18)$$

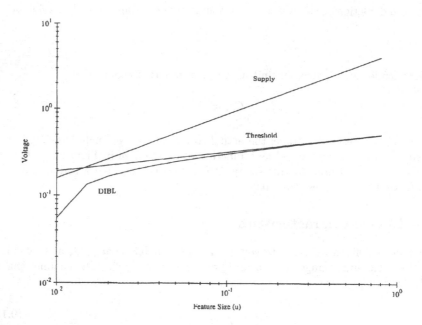

Fig. 9.8. Threshold voltage used in this study. The middle curve is the nominal threshold voltage, given by Equation 9.11. The bottom curve is the actual threshold voltage, which is lowered from the nominal value by drain-induced barrier lowering (DIBL), given by Equation 9.12. The top curve is the nominal supply voltage from Equation 9.2.

where ν_0, the saturated velocity of electrons in silicon, is taken to be 10^7 cm/s [29], and $l_0 = D/\nu_0$ can be thought of as the mean free path of the carrier, which is taken to be $0.007\ \mu$ [2].

We obtain the threshold current I_t by substituting $Q_s = Q_t$ from Equation 9.13 into Equation 9.16. We obtain the on current I_{on} by substituting Q_s from Equation 9.14 into Equation 9.16, using the threshold voltage lowered only by the built-in junction voltage, rather than by the total junction voltage. We obtain the off current (15) I_{off} by substituting Q_s from Equation 9.15 into Equation 9.16, using the threshold voltage as lowered by DIBL. These expressions thus represent a conservative characterization of the transistor performance, since the on current will be somewhat underestimated.

The several currents associated with a minimum-sized transistor are shown as a function of feature size in Figure 9.9. The trade-offs mentioned in the introduction are immediately apparent in this plot. As features become smaller, substrate doping must increase to prevent punch-through. The increase in substrate doping increases the junction electric field, thereby increasing drain-junction tunneling current into the substrate. To limit the tunneling current to a reasonable value, we reduce the supply voltage, thereby reducing the ratio of channel on current to channel off current. The most remarkable conclusion from Figure 9.9 is that transistors of 0.03-μ channel length still function essentially as do present-day devices. With proper scaling of all parameters of the process, device miniaturization is alive and well. Many issues will arise in the development of ever-finer-scale fabrication, but, in the end, the endeavor will prevail.

Given that devices at least one order of magnitude smaller than today's are feasible, we may enquire what their characteristics may be. Figure 9.10 shows several quantities of interest. It is clear that discreteness of all quantities will become increasingly important at smaller feature sizes — particularly that of doping ions in the substrate. We have given elsewhere a simple discussion of the effects of discrete substrate charge [3]; a recent analysis is presented by Nishinohara et al. [30].

Perhaps the single most important aspect of device performance is the speed of logic fabricated from any particular technology. We can estimate the time τ required for an elementary logic element to drive another like it:

$$\tau = \frac{V C_{tot}}{I_{on}} \tag{9.19}$$

where the total capacitance C_{tot} is taken to be three times the sum of the oxide and drain junction capacitances. This delay should correspond rather directly to the delay per stage measured for ring oscillators in any given process, and is plotted along with several experimental points in Figure 9.11. It is remarkable that, despite the reduction in supply voltage at small feature sizes, logic performance continues to improve. Several authors have emphasized the improvement in speed that we can make available by reducing threshold and power-supply voltages [12, 31–33].

The primary effect behind this somewhat counterintuitive trend is velocity saturation, an excellent recent account of which can be found in Noor Mohammad [29]. We gave an early treatment of the effect of velocity saturation on device characteristics [34]; an extended analysis appears in Appendix B of a previous work [2].

The supply voltage V affects the performance of standard CMOS digital logic in three ways:

1. The channel charge is proportional to $V - V_t$.

2. The electric field in the channel is proportional to V.

3. The logic swing is proportional to V.

For long-channel devices, the carrier velocity is proportional to the electric field in the channel. The channel current is the product of the channel charge and the carrier velocity. Therefore, the device current has a quadratic dependence on the

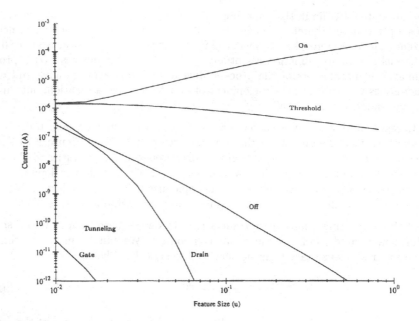

Fig. 9.9. Currents characteristic of minimum-sized devices as a function of feature size. We obtain the threshold current I_t by substituting $Q_s = Q_t$ from Equation 9.13 into Equation 9.16. We obtain the on current I_{on} by substituting Q_s from Equation 9.14 into Equation 9.16, using the threshold voltage lowered only by the built-in junction voltage, rather than by the total junction voltage. We obtain the off current Ioff by substituting Q_s from Equation 9.15 into Equation 9.16, using the threshold voltage as lowered by the full supply voltage. The junction tunneling current was computed from Equation 9.10, assuming the drain area is the square of the feature size. The gate-oxide tunneling current was computed from Equation 9.7, assuming that the full supply voltage is present across the full gate area (the square of the feature size).

supply voltage. This current must charge the load capacitance to approximately one-half of the supply voltage to achieve a logic transition. This factor cancels one of the V terms in the current, leaving the circuit speed linear in the supply voltage.

Once the carrier velocity is saturated, however, increasing the electric field in the channel no longer increases the channel current. Both the charge in transit and the voltage to be traversed by the output are increased by the same factor. In this regime, the only effect of increased supply voltage is an increase in the switching energy, with virtually no increase in performance. Just how close devices of the present day come to this limit can be seen in the delay-versus-voltage plots in the recent literature; see, for example, [6, 10, 14].

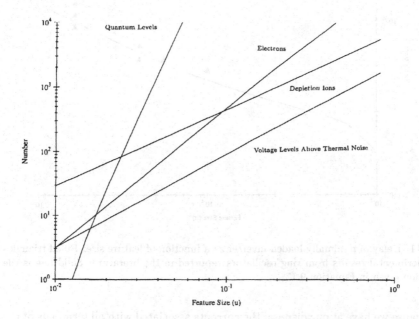

Fig. 9.10. Number of signal levels resolvable by a minimum-sized device according to the scaling laws used in this study. Thermal noise limits the analog depth representable by a single voltage. The number of voltage levels above thermal noise was taken to be the square root of the minimum stored energy shown in figure 3, expressed in units of kT. The quantum-level separation was taken to be the energy spacing of states in a one-dimensional box of length $l - x_s - x_d$. The number of electrons under the gate was taken to be the on-value of Qs multiplied by the gate area (a slight overestimate). The number of depletion ions was taken to be the doping density n given by Equation 9.8, multiplied by the gate area and the depletion depth x from Equation 9.9, using 1 V for ψ. As the number of depletion ions becomes smaller, the range of threshold voltages encountered across a single chip increases. In analog systems, adaptation techniques can mitigate or eliminate the variation among transistors.

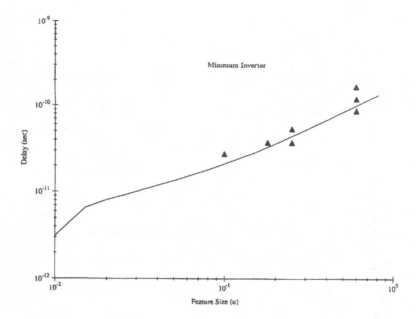

Fig. 9.11. Delay of minimally loaded inverter as a function of feature size. Filled triangles are experimental results from ring oscillators reported in the literature. Solid line is the expression given in Equation 9.17.

Because we have at our disposal the currents associated with all terminals of the transistor, we can evaluate the conductances associated with these currents. For logic devices to function properly, it is necessary that an elementary logic circuit have a gain greater than unity, which in turn requires that the transconductance G_m of the transistor be larger than the sum of all contributions to the drain conductance. As feature size decreases below 0.1 μ, both DIBL and drain-junction tunneling make rapidly increasing contributions to the drain conductance, as can be seen in Figure 9.12. In spite of these parasitic effects, the device is still capable of providing greater than unity gain down to the smallest feature sizes investigated.

9.5   System Properties

The enormous effect of device scaling on computational capability becomes apparent only when viewed from the system level. We can estimate the system-level capabilities of digital chips fabricated with advanced processes by extrapolation from present-day systems. The first such extrapolation is the number of devices per unit area. If every transistor in a modern digital chip were to be shrunk to minimum size, the entire active area would cover approximately 2% of the chip area. If we assume that this coverage factor can be maintained in future designs, the density of active elements scales with feature size, as shown in Figure 9.13. The system clock period in today's processors is approximately 100τ. Even today, it is becoming more economical to break each chip into several processors that can operate in parallel, than it is to merely build larger "dinosaur" processors. For purposes of extrapolation, we can assume that each processor contains 106 transistors. The computation available under these clearly oversimplified assumptions is plotted versus feature size in Figure 9.14. If we further assume that all devices are in fact of minimum size, and that they are clocked at the system-clock frequency, we can estimate the power that will be dissipated by chips built in these advanced technologies. The power attributable to useful switching, and the dissipations of various parasitic currents that do not depend on clock speed, are shown in Figure 9.15. Down to about 0.03 μ feature size, most of the energy supplied to the chip is dissipated in real, useful computation. Only below this scale do the parasitic currents overwhelm the energy consumed in performing real computation.

9.6 Conclusions

The MOS transistor has become the workhorse of modern microelectronics; it has survived many generations of process scaling to finer feature sizes. In this study, I have explored the extent to which the MOS device, as we know it today, can be scaled to still smaller dimensions. We have data available to provide experimental support for the tunneling currents that will be encountered in the heavily doped source and drain junctions of devices down to 0.03 μ. Neither do we have comparable data to support the theory for oxides in the 10 $\overset{\circ}{A}$ range, nor do we have direct experimental verification of the effect of statistical fluctuations on very small structures built in heavily doped material. As such data become available, we will be better able to chart the course of future minification, of which the present study is only an outline. It is already clear that MOS circuits will be integrated to upward of 10^9 devices per square centimeter merely by scaling, without any major change in the conceptual framework that we use today. There are many challenges involved in this technology evolution [4], but I do not expect any show-stoppers. The prospect of very high levels of integration was daunting in 1971 when our earlier study was written, and is far more daunting today. Whereas massive parallelism is possible in present-day technology, it will clearly become mandatory if we are to realize even a fraction of the potential of more highly evolved technology. Even as this study is written, there is far more potential in a square centimeter of silicon than we have developed the paradigms to use, as has often been the case in periods of rapid technological evolution.

I should clarify the "limits" considered in this study. It is clear that devices much smaller than those treated here can be made to show useful characteristics. Conventional MOS devices can be fabricated on insulating substrates (SOI-SOS),

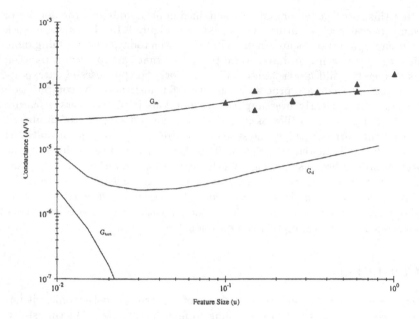

Fig. 9.12. Several conductances associated with minimum-sized transistors, as a function of feature size. The top curve is the transconductance. The filled triangles are experimental values given in the literature, normalized to a minimum-sized device at the reported dimension. The second curve is the drain conductance due to DIBL, computed by evaluating Equation 9.16 at a drain voltage equal to V and at 0.9 V, and dividing the difference by 0.1 V. The current through this conductance flows from drain to source. The bottom curve is the drain conductance due to drain junction tunneling. Current through this conductance flows from drain to substrate.

thereby removing the constraint imposed by substrate tunneling. Much smaller devices are possible at molecular scale. The most obvious example of an extremely small device is an electron-transfer reaction occurring along an amino acid path, the potential of which is determined by the charge on a nearby atomic site. Such arrangements are thought to occur in many biological systems. The physics of such a transfer corresponds directly to that of an MOS transistor operating in weak inversion (below threshold). Imagining a device that functions is easy; building a device that works is much harder; and having a process by which billions of devices can be constructed in a single physical structure is many orders of magnitude harder still. I have limited this study to the consideration of direct extensions to existing technology.

Finally, I emphasize that I have considered only the properties of transistors themselves, and have not even touched many other important aspects of the tech-

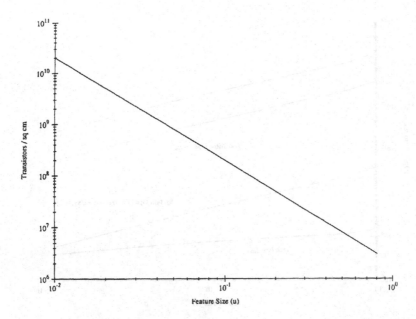

Fig. 9.13. Assumed number of active devices per square centimeter of chip area. If all devices are of minimum size, active (transistor channel) area is 2% of total area.

nology. Of the latter, interconnect — both within a single chip and across chip boundaries — is obviously a key concern. We have given elsewhere preliminary discussion of the global scaling properties of a single-chip interconnect network for ultradense technology [1]. The topic of interconnect, along with many other issues, such as the fabrication technology itself, deserve a great deal of consideration as the technology evolves. Whatever complications arise, however, it is clear that the technology will evolve. It will evolve because that evolution is possible, because there is so much to be gained at the system level by that evolution, and because the same energy and will on the part of bright, energetic, devoted people that has overcome enormous obstacles in the past will overcome those that lie ahead.

References

[1] C. Mead and L. Conway, *Introduction to VLSI Systems* (Addison-Wesley: Reading, MA, 1980).

[2] C. Mead, *Analog VLSI and Neural Systems* (Addison-Wesley: Reading, MA, 1989).

[3] B. Hoeneisen and C. A. Mead, "Fundamental limitations in microelectronics. I: MOS technology," *Solid-State Electron.*, Vol. 15, pp.819-829 (1972).

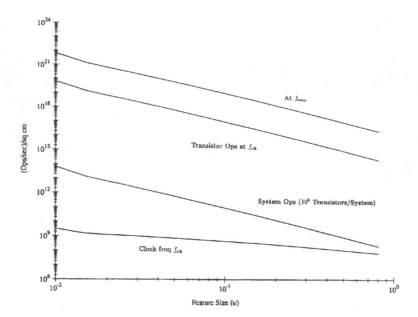

Fig. 9.14. Several measures of computation capability per unit area as a function of feature size. The bottom curve is a typical processor clock frequency, the clock period assumed to be 100 times the inverter delay shown in Figure 9.11. The second curve is the number of systems (of 10^6 transistors each) per square centimeter multiplied by the clock frequency. The third curve is the number of transistors per square centimeter shown in Figure 9.13 multiplied by the clock frequency. The top curve is the number of transistors per square centimeter multiplied by the reciprocal of the inverter delay shown in Figure 9.11.

[4] M. Nagata, "Limitations, innovations, and challenges of circuits and devices into a half micrometer and beyond," *IEEE J Solid-State Circuits*, Vol.27, pp.465-472 (1992).

[5] B. Davari, W.-H. Chang, K. F. Petrillo, CY. Wong, D. May, Y. Taur, M. R. Wordeman, J. Y-C. Sun, C. C-H. Hsu, and M. R. Polcari, "A high-performance 0.25-μm CMOS technology. II: Technology," *IEEE Trans. Electron Dev.*, Vol.39, pp.967-975 (1992).

[6] W.-H. Chang, B. Davari, M. Wordeman, Y. Taur, CC-H. Hsu, and M. D. Rodriguez, "A high-performance 0.25-μm CMOS technology. I: Design and characterization," *IEEE Trans. Electron Dev.*, Vol.39, pp.959-966 (1992).

[7] A. Bryant, B. El-Kareh, T. Furukawa, W. P Noble, E. J. Nowak, W. Schwittek, and W. Tonti, "A fundamental performance limit of optimized 3.3-V sub-quarter-micrometer fully overlapped LDD MOSFETs," *IEEE Trans. Electron Dev.*, Vol.39, pp.1208-1215 (1992).

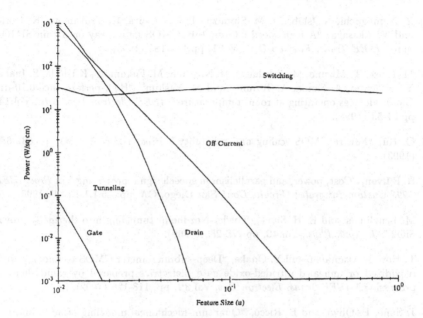

Fig. 9.15. Several contributions to the power dissipated by typical digital systems as a function of feature size. I obtained the curve labeled **Switching** by multiplying the number of transistors per unit area shown in Figure 9.13 by the switching energy shown in Figure 9.3 and by the clock frequency shown in Figure 9.14. This power contributes to the performance of computation: It scales directly with clock frequency. In addition to the switching power, there are several parasitic mechanisms by which power is wasted, each being the result of one of the parasitic currents shown in Figure 9.9. These parasitic mechanisms are present even at zero clock frequency, and perform no useful work. The values shown assume that all devices are of minimum size, and have the full voltage V on their drains. All values depend critically on the assumptions embodied in the scaling laws of Equations 9.1, 9.2, 9.8, and 9.11. Even slightly different scaling can lead to substantially different results for the smallest feature sizes. The particular laws that I have put forth in this study were fine tuned to produce reasonable results down to 0.02 μ. For example, a slight increase in doping density markedly decreases the off current by reducing DIBL, while dramatically increasing the drain-junction tunneling current. Similar trade-offs can be made with other parameters.

[8] R. H. Yan, K. E Lee, DY Jeon, YO. Kim, D. M. Tennant, E. H. Westerwick, G. M. Chin, M. D. Morris, K. Early, and P Mulgren, "High-performance deep-submicrometer Si MOSFETs using vertical doping engineering," *IEEE Trans. Electron Dev.*, Vol.**39**, p.2639 (1992).

[9] Y. Yamaguchi, A. Ishibashi, M. Shimizu, T. Nishimura, K. Tsukamoto, K. Horie, and Y. Akasaka, "A high-speed 0.6-μm 16K CMOS gate array on a thin SIMOX film," *IEEE Trans. Electron Dev.*, Vol.**1**, pp.179-185 (1993).

[10] M. Iwase, T. Mizuno, M. Takahashi, H. Niiyama, M. Fukumoto, K Ishida, S. Inaba, Y. Takigami, A. Sanda, A. Toriumi, and M. Yashimi, "High-performance 0.10-μm CMOS devices operating at room temperature," *IEEE Electron Dev. Lett.*, Vol.**14**, pp.51-53 (1993).

[11] C. Hu, "Future CMOS scaling and reliability," *Proc. IEEE*, Vol. **81**, pp.682-689 (1993).

[12] R. F Lyon, "Cost, power, and parallelism in speech signal processing," in *Proc. IEEE 1993 Custom Integrated Circuits Conf*, San Diego, CA, pp. 15.1.1-15.1.9, 1993.

[13] M. Lenzlinger and E. H. Snow, "Fowler-Nordheim tunneling into thermally grown SiO2," *J. Appl. Phys.*, Vol.**40**, pp.278-283 (1969).

[14] T. Hori S. Akamatsu, and Y. Odake, "Deep-submicrometer CMOS technology with reoxidized or annealed nitrided-oxide gate dielectrics prepared by rapid thermal processing," *IEEE Trans. Electron Dev*, Vol.**39**, pp. 118-126 (1992).

[15] J. Suñé, P. Olivo, and B. Riccò, "Quantum-mechanical modeling of accumulation layers in MOS structure," *IEEE Trans. Electron Dev.*, Vol.**39**, pp. 1732-1739 (1992).

[16] A. G. Chynoweth, W. L. Feldmann, C. A. Lee, R. A. Logan, and G. L. Pearson, "Internal field emission at narrow silicon and germanium $p - n$ junctions," *Phys. Rev*, Vol.**118**, pp.425-434 (1960).

[17] R. A. Logan and A. G. Chynoweth, "Effect of degenerate semiconductor band structure on current-voltage characteristics of silicon tunnel diodes," *Phys. Rev.*, Vol. **131**, pp.89-95 (1963).

[18] J. B. Krieger, "Theory of electron tunneling in semiconductors with degenerate band structure," *Ann. Phys.*, Vol.**36**, pp.1-60 (1966).

[19] R. B. Fair and R. W. Wivell, "Zener and avalanche breakdown in as-implanted low-voltage Si n-p junctions," *IEEE Trans. Electron Dev*, Vol. **ED-23**, pp. 512-518 (1976).

[20] J. M. C. Stork and R. D. Isaac, "Tunneling in base-emitter junctions," *IEEE Trans. Electron Devi.*, Vol. **ED-30**, pp. 1527-1534 (1983).

[21] F. Hackbarth and D.-L. Tang, "Inherent and stress-induced leakage in heavily doped silicon junctions," *IEEE Trans. Electron Dev.*, Vol. **35**, pp. 2108-2118 (1988).

[22] M. Reisch, "Tunneling-induced leakage currents in pn junctions," AEÜ, Band **44**, pp.368-376 (1990).

[23] G. P Li, F. Hackbarth, and T.-C. Chen, "Identification and implication of a perimeter tunneling current component in advanced self-aligned bipolar transistors," *IEEE Trans. Electron Dev*, Vol. **ED-35**, pp.89-95 (1988).

[24] R. R. Troutman, "VLSI limitations from drain-induced barrier lowering," *IEEE Trans. Electron Dev.*, Vol. **ED-26**, pp.461-469 (1979).

[25] M. J. Deen and Z. X. Yan, "DIBL in short-channel NMOS devices at 77 K," *IEEE Trans. Electron Dev*, Vol.**39**, pp.908-915 (1992).

[26] M. J. Van der Tol and S. G. Chamberlain, "Drain-induced barrier lowering in buried-channel MOSFETs," *IEEE Trans. Electron Dev*, Vol. **40**, pp.741-749 (1993).

[27] J. G. C. Bakker, "Simple analytical expressions for the fringing field and fringing-field-induced transfer time in charge-coupled devices," *IEEE Trans. Electron Dev.*, Vol. **38**, pp.1152-1161 (1991).

[28] T. A. Fjeldly and M. Shur, "Threshold voltage modeling and the subthreshold regime of operation of short-channel MOSFETs," *IEEE Trans. Electron Dev.*, Vol.**40**, pp.137–145 (1993).

[29] S. Noor Mohammad, "Unified model for drift velocities of electrons and holes in semiconductors as a function of temperature and electric field," *Solid-State Electron.*, Vol.**35**, pp. 1391-1396 (1992).

[30] K. Nishinohara, N. Shigyo, and T. Wada, "Effects of microscopic fluctuations in dopant distributions on MOSFET threshold voltage," *IEEE Trans. Electron Dev.*, Vol.**39**, pp.634-639 (1992).

[31] J. B. Burr and A. M. Peterson, "Energy considerations in multichip-module based multiprocessors," in *IEEE Int. Conf, Computer Design*, pp.593-600 (1991).

[32] D. Liu and C. Svensson, "Trading speed for low power by choice of supply and threshold voltages," *IEEE J. Solid-State Circuits*, Vol.**28**, pp.10-17 (1993).

[33] B. T. Murphy, "Minimization of transistor delay at a given power density," *IEEE Trans. Electron Dev.*, Vol.**40**, 414-420 (1993).

[34] B. Hoeneisen and C. A. Mead, "Current-voltage characteristics of small size MOS transistors," *IEEE Trans. Electron Dev.*, Vol. **19**, pp.382-383 (1972).

10

RICHARD FEYNMAN AND CELLULAR VACUUM

Marvin Minsky

10.1 Richard Feynman

He was not only a physicist — but also a great psychologist. Whatever Feynman thought about, he reflected upon his thinking, too — about how his mind computed itself. His explanations more than explained the subjects he was talking about. They also made you think of new ways for you to make yourself think about things.

Feynman loved computation. He loved every aspect of it: Algorithms, computing machines, abstract theories about computations, and all. I had an old Marchant Calculator in my basement. Richard and I tried to fix it once, but never got it unjammed. I was somewhat annoyed, but Richard thought this was very funny. Of course, he thought everything was funny. Especially, he liked to discover novel ways to simplify calculations.

Why did he like computation so much? Partly of course, he liked everything that had deep intellectual content. (And no one ever more despised ideas that only pretended to depth.) In particular, though, it usually was computation that showed the relations between a theory and an experiment. One thing that drew us together was a shared interest in extending the concept of a calculating machine. Earlier in my career, I had explored and promoted the idea that we could program computers, not merely to calculate numerically, but to perform "symbolic computations" — that is, to manipulate not only the numbers but the symbols of mathematical expressions. This led to the first programs that could compute formal derivatives, integrals, and finally, formal solutions to differential equations. This was the sort of thing that Feynman needed for expanding the symbolic power series that arise in QED. When large such expressions were expanded and then 'simplified' by humans, you could never tell when there might have been a symbolic (rather than numerical) error.

Would computers eventually also be able to "compute" good new ideas? Certainly, Feynman seemed sure of that; if we knew enough about how minds/brains work, then surely we could simulate them. The trouble is that we still don't know much about such things. None of us are conscious yet — in the sense of understanding ourselves enough to have good ideas about how we think. (In that particular sense Richard Feynman was perhaps the most conscious person I've met.) Once Feynman showed me an impressive trick. He asked how many people were in the

room — and before I could barely start to count he said that there were thirty-one.
Here's how he did it: Without ceasing to speak, he had scanned the room, while
thinking a rhythm of visual groups — three, three, three one. This enabled him
to count ten people per second, so at the end of three seconds he'd counted three
tens and there was only one person left. He said he did it by using his bongo-drum
expertise.

10.2   Prospects of a Final Theory

In one of his lectures, Feynman asked whether physics would always continue to
evolve, with new "fundamental" laws — or, as he preferred to say — corrections
to deficiencies in earlier discovered laws? Will we always find new phenomena that
are not explained by our old ideas? Or will there someday come a time when the
physicists finally find themselves with a theory that would never again need to be
changed or augmented?

I've asked many physicists whether they expected we would ever find such a
"theory of everything". Most of them seemed inclined to assume that there always
would be new theories to come, perhaps for increasingly high energies. Only a few
preferred the view that some final theory would be found. As for Feynman's opinion,
I didn't make notes, but what I recall is that he saw no reason why physics should
perpetually change. Of course there was no way to prove that wrong and (as he
said in "Character" [1]) the only thing we can be sure of is that we can't be sure of
anything. In particular, even if we did reach a final theory, we'd have no way to be
sure that we'd found it — so one must always be open to new evidence. However,
he cautioned against worrying about this too much, lest it promote a discouraged
attitude.

Some of those physicists did not even like the prospect of a final theory. "That
would be a terrible thing, because then there would be no mysteries left!" But
Feynman had no patience for the idea that mysteries ought to be preserved, like
endangered species. After all, even if you had all the basic laws, you'd still need
new ways to compute their implications. The last time I saw him he was trying to
find better ways to compute the predictions of QCD.

10.3   Locally finite information theories

What could be a possible form of such a final theory? Here's one idea that spread
around in the 1950s, along with the emergence of computer science. Consider that
most of the deficiencies of classical physics were related to extremes of size and
speed. I recall several friends suggesting that this could be because the universe is
"supposed to be" Newtonian — but, because it is actually only being simulated on
some very large computer, there are some problems of precision. Perhaps, for very
small quantities, it is round-off errors in the lowest bits that give rise to quantum

phenomena. As for larger, more cosmological quantities, perhaps it is the limited word-length that prohibits transluminal velocities. This fantasy does not work well when one tries to fill in more details. However, it does suggest some other ideas that seem to be more promising: What if space-time is "discrete" — that is, composed of separate points, with no continuum between them? Feynman did like that idea because he felt that there might be something wrong with the old concept of continuous functions. How could there possibly be an infinite amount of information in any finite volume? Why do all those new particles appear when we try to put too much energy in one place? Could that be because there just isn't room for that much information?

Ed Fredkin pursued the idea that information must be finite in density. One day, he announced that things must be even more simple than that. He said that he was going to assume that information itself is conserved. "You're out of your mind, Ed," I pronounced. "That's completely ridiculous. Nothing could happen in such a world. There couldn't even be logical gates. No decisions could ever be made." But when Fredkin gets one of his ideas, he's quite immune to objections like that; indeed, they fuel him with energy. Soon he went on to assume that information processing must also be reversible — and invented what's now called the Fredkin gate.

When Bennett and Landauer wrote their paper showing that reversible, non-dissipative computation was indeed possible in principle, I was asked to be a referee for the journal they had submitted it to. I read the paper over and over — every day for a solid month. Finally I sent a note to the journal. "This result just doesn't seem possible. However, I have read it very carefully, and cannot find where might be the mistake. I suppose you'll just have to publish it!"

Feynman too was skeptical. However, instead of merely checking their proof, he set out to prove it for himself. He came up with a different and simpler proof, which convinced many others that the discovery was believable. Soon he started to design the first quantum computers.

10.4 The Idolatry of Uncertainty

For generations the public has been told that modern physics has changed our view of the universe. Our teachers tell our children that "In the world of classical mechanics, everything worked like clockwork, with deterministic certainty. But Quantum Theory has shown us that things are indeterminate. The mechanical, deterministic world of Newton has been replaced by one in which everything is uncertain and unpredictable." This view has become quite popular — but it actually puts things upside down. Uncertainty lay in the classical view — and it was quantum theory that actually showed why things could be depended on. It is true that Newton's laws were replaced by a scheme in which such quantities as place and time are separately indeterminate. But the implications of this are not what they seem — but almost exactly the opposite. For it was the planetary orbits of classical mechanics that were truly undependable, because of chaotic interactions. In contrast, the "orbits" of electrons in atoms, according to quantum mechanics, are extremely stable — and it is these that enable us to have certainty!

To explain this seeming paradox, let's contrast two systems. The first is a classical Newtonian solar system that has a heavy object in the middle, and several planets surrounding it. In the particular case of the Solar System in which we live, Gerald Sussman and Jack Wisdom [2] have shown that the orbit of Pluto is chaotic. Eventually, Pluto may eventually be hurled out of the solar system. We earth-people might not consider this a serious loss. However, our large outer planets have more than enough angular momentum that, given suitable coupling, they could also throw Earth itself into outer space. (This should not be our chief concern, however, because simulations appear to show that it won't happen before the Sun itself becomes a red giant.) Solar systems are unstable. So also would be molecules if their atoms behaved in accord with classical laws. Even if each atom were stable by itself, when they approach one another, the electron orbits inside them would soon be perturbed, and one or both atoms would soon break up. As Feynman said in his 1965 Lectures on Physics, "It is true classically that if we knew the position and velocity of every particle in the world, or in a box of gas, we could predict exactly what would happen. And therefore the classical world is deterministic. Suppose, however, that we have a finite accuracy and do not know exactly where just one atom is, say to one part in a billion. Then as it goes along it hits another atom, and because we did not know the position better than to one part in a billion, we find an even larger error in the position after the collision. And that is amplified, of course, in the next collision, so that if we start with only a tiny error it rapidly magnifies to a very great uncertainty" [3].

The Newtonian world was inadequate. If the atoms of our universe moved only according to Newton's laws, there could exist no molecules, but only drifting, featureless clouds. In contrast, chemical atoms are actually extremely stable because their electrons are constrained by quantum laws to occupy only certain separate levels of energy and momentum. Furthermore, combinations of atoms can combine to form configurations, called molecules, that are also confined to have definite states. Although the internal state of a molecule can change suddenly and unpredictably, such events may not happen for billions of years — during which there is absolutely no change at all. Our stability comes from those quantum fields, by which everything is locked into place, except during moments of clean, sudden change.

In contrast, consider what happens in quantum theory, where each electron level remains unchanged until there occurs a transition jump. The result is that we can have molecules with covalent bonds, which can remain stable for billions of years. Thus, contrary to what our teachers say, it was in that classical world that everything was unstable and indeterminate, whereas it is those stable quantum levels that make possible chemistry, life, and nanotechnology. It is because of quantum states that you can remember what you had for breakfast; this is because new neural connections made in your brain can persist throughout your day. When something more important occurs, you'll remember it as long as you live. Everything we can know depends on that "Quantum Certainty."

10.5   Quantum Psychology

Early in this century, some physicists began to speculate that the uncertainty principle of quantum mechanics left room for the freedom of will. What attracted those physicists to such views? As I see it, they still believed in freedom of will as well as in quantum uncertainty — and these subjects had one thing in common: They both confounded those scientists' conceptions of causality. To be sure, the "uncertainty principle" must affect (to whatever small extent) the tiny synapses in the brain; therefore those structures must act, to that extent, unpredictably. But mere probabilistic uncertainty offers no genuine freedom. It but merely adds some capriciouness to a system based on lawful rules. Two generations of philosophers — and retired, burned-out physicists — have celebrated uncertainty in romantic, nonsensical terms, not only regarding physics itself, but also the mind and its "freedom of will."

What connects the mind to the world? This problem has always caused conflicts between physics, psychology, and religion. In the world of Newton's mechanical laws, every event was entirely caused by what had happened earlier. There was simply no room for anything else. Yet commonsense psychology said that events in the world were affected by minds: People could decide what occurred by using their freedom of will. Most religions concurred in this, although some preferred to believe in schemes involving divine predestination. Most theories in psychology were designed to support deterministic schemes, but those theories were usually too weak to explain enough of what happens in brains. In any case, neither physical nor psychological determinism left a place for the freedom of will.

It is only because of quantum laws that what we call "things" can exist at all. It is why there can be bodies with separate cells, so that there can be animals with synapses, nerves, and memories... It is why we can have genes to specify brains in which memories can be maintained — so that we can have our illusions of will. Richard Feynman said, "It is therefore not fair to say that from the apparent freedom and indeterminacy of the human mind, we should have realized that classical 'deterministic' physics could not ever hope to understand it, and to welcome quantum mechanics as a release from a 'completely mechanistic' universe. For already in classical mechanics there was indeterminability from a practical point of view" [3].

We have heard two generations of philosophers and retired, burned-out physicists speaking about uncertainty in romantic, nonsensical terms — not only about basic physics itself, but about the mind and its "freedom of will." Next time you hear this, ask them if they realize that only Quantum Certainty makes anything we know persist.

10.6   Cellular Vacuum Revisited

This section was originally published in "Cellular Vacuum" in 1982 [4]. Here I've revised most of the original, mainly by deleting some unsound speculations.

This fantasy about conservation in cellular arrays was inspired by the first conference on computation and physics. The "cellular array" idea had already emerged in such forms as Ising models, renormalization theories, the "game of life," and Von Neumann's work on self-reproducing machines. In the 1980's the subject became more popular with the work of Tommaso Toffoli, Norman Margolus, Edward Fredkin and several others. Richard Feynman was interested too, and encouraged my writing this essay (without approving its contents)...

Imagine a crystalline space-time world of separate "cells" in which each volume of space contains only a finite (and bounded) amount of information. Assume also that the state of each cell is determined by the previous states of itself and its neighbors. (Time also comes in discrete moments, which synchronize the entire machine.) Over some range of size and speed, could the mechanics of such a world be approximately classical? To answer this, we'll construct analogs of particles and fields, and ask what it would mean for these to satisfy constraints like conservation of momentum. In each case classical mechanics will break down on scales both small and large — and strange phenomena will emerge: A maximal velocity, a slowing of internal clocks, limits of simultaneous measurements, and quantumlike effects in very weak or intense fields.

10.6.1   Cellular Arrays

Envision space as a cubic array of finite-state "cells." At any moment, each cell is in one of a few possible "states" — and the rules for how states change from one moment to the next are the "vacuum field equations" of this universe. These rules are starkly local, each cell's state is determined only by its own and neighbors' states of the preceding moment. A one-dimensional example illustrates a simple moving "packet": there are just four states: 1, P, "0" and Q. Initially all cells are "0" except that somewhere appears this pattern:

$$- \ - \ - \ 0 \ 0 \ 0 \ 1 \ 1 \ 1 \ 1 \ P \ 0 \ 0 \ 0 \ 0 \ 0 \ 0 \ 0 \ 0 \ 0 \ - \ - \ -$$

A typical state-change rule has the form of $1 : Q : P \to 1$, which means when a cell in state Q sees a 1 to its left and a P to its right, it switches to state 1. Now consider this set of state-change rules:

$$1, 1, P \to P \qquad 0, 1, P \to 0 \qquad 0, P, P \to Q$$
$$Q, 0, 0 \to P \qquad Q, P, X \to Q \qquad X, Q, X \to 1$$

where X means the transition does not depend on that neighbor's state. Unless otherwise specified, each cell remains in its previous state. This initial configuration reproduces itself one unit over to the right, repeating this forever:

$$
\begin{array}{ccccccccccccccc}
t = & 0 & 0 & 0 & 0 & 1 & 1 & 1 & 1 & P & 0 & 0 & 0 & 0 & 0 \\
& 1 & 0 & 0 & 0 & 1 & 1 & 1 & P & P & 0 & 0 & 0 & 0 & 0 \\
& 2 & 0 & 0 & 0 & 1 & 1 & P & P & P & 0 & 0 & 0 & 0 & 0 \\
& 3 & 0 & 0 & 0 & 1 & P & P & P & P & 0 & 0 & 0 & 0 & 0 \\
& 4 & 0 & 0 & 0 & 0 & P & P & P & P & 0 & 0 & 0 & 0 & 0 \\
& 5 & 0 & 0 & 0 & 0 & Q & P & P & P & 0 & 0 & 0 & 0 & 0 \\
& 6 & 0 & 0 & 0 & 0 & 1 & Q & P & P & 0 & 0 & 0 & 0 & 0 \\
& 7 & 0 & 0 & 0 & 0 & 1 & 1 & Q & P & 0 & 0 & 0 & 0 & 0 \\
& 8 & 0 & 0 & 0 & 0 & 1 & 1 & 1 & Q & 0 & 0 & 0 & 0 & 0 \\
& 9 & 0 & 0 & 0 & 0 & 1 & 1 & 1 & 1 & P & 0 & 0 & 0 & 0
\end{array}
$$

We can write state interaction rules to do almost anything one can imagine. There even exist "universal" sets of state-change rules with which a single cellular array can "simulate" any computation. The trick is to encode, into the universal array's initial conditions, another set of state-change rules. The universal rules then "interpret" those other rules. Roger Banks has described a remarkably simple universal scheme in which each cell has only two states, depending only on four neighbors [5].

Size and Precision. In the example above the size of a packet is inverse to its speed. More generally, there must be an absolute constraint between the amount of information in any packet and the volume of that packet! Just as in Heisenberg's principle, it is not so much a parameter's value that determines packet size, as its precision — the number of "bits of information" needed to specify its properties.

If the information carried in a packet were "optimally encoded" then the packet's size would depend on the base-2 logarithm of its precision. Then why is there no logarithm in Heisenberg's principle? This suggests the conjecture that most physical information (particularly in photons) is encoded not in base 2, but in the less dense base-1 form. Later we'll argue that particles with rest mass may employ denser codes. This argument relates position not only to velocity but also to any other property, so this does not lead directly to the particular commutators of quantum theory.

Uniform Motion. Any bounded packet that moves within a regular lattice must have an eventually repeating trajectory. Such trajectories will appear perfectly straight, on any large enough scale — so we can deduce Newton's law of inertia (for compact particles) directly from the regularity of the lattice.

Maximal Speed. Since no effect can propagate faster than the basic lattice speed of one cell per "moment," there is a largest possible speed. Let's identify this with the speed of light. It is easy to design small light-speed packets, but more machinery is needed for sub-light-speed propagation. There are fundamental differences between light-speed things (call them "photons") and slower ones. Information behind a photon's wave front can never catch up — hence the information mechanics of photons must be relatively simple; they cannot do "three-dimensional" computations. However, if data can propagate diagonally, then some computation can proceed along the wavefront of a light-speed packet, and thus have a cyclic phase, or other more complex properties.

Time Contraction. The faster a non-photon moves, the slower must proceed its internal computations. Imagine that some computation inside a packet has to send information back and forth through some number L of cells; then when the packet is 'at rest' the round-trip will take time $2L$. Now, make the packet move in L's direction at $(N-1)/N$ the speed of light. The retrograde time remains of order L, but now it takes at least $(N-1)L$ moments for data to advance L spaces (relative to the packet) so the round-trip time has then the order of NL. Therefore the speed of internal computations, relative to the fixed frame, must slow down by $N/2$. (This is not what Lorentz invariance requires. I complained to Feynman about not seeing how to make those ideas relativistic. He said not to worry about that because any good physicist should be able to fix that. I wish I'd asked for a few more hints.)

Divergence and Aperture. In real optics it takes twice the aperture to halve a beam's divergence. However, an "optimal" encoding of the angle should only need a single extra "bit." This would suggest Nature uses "base-1" codes for photons — perhaps because only base-1 codes could let a discrete mechanism "add" quickly enough to make things linear at light speed.

Frequency and Time. To maintain a beam's divergence while shrinking both aperture and wavelength, the wavelength information must be spread out transversally. This suggests a constraint resembling the energy-time version of Heisenberg's principle.

10.6.2   Spherical Symmetry

No regular lattice is invariant under rotation, Euclidean or Lorentz, since it needs different information to move along different axes. So, just as waves in crystals show Bragg diffraction, "discrete vacuums" must show angular anisotropies at some extremes of size or energy. Such problems already lurk beneath the surface of all other modern theories — but here our main concern is seeing how a discrete model could have other ordinary properties on ordinary scales.

Liquid Lattice Model. One could imagine cell connections so randomly irregular that, in the large, the space is isotropic — like water, which is almost crystalline from each atom to the next, but isotropic on the larger scale. But to build our packets into such a world, we would have to find transition rules insensitive to local cell-connection fluctuations.

Continuous Creation. Instead of starting with a liquid vacuum, we could randomly insert new cells from time to time. This would red shift cosmically old photons (by lengthening their unary frequency counters) and uniformly expand the universe. As Richard Stallman has pointed out, this would require an amorphous (rather than a regular lattice) cellular system — making it harder to invent adequate sets of transition rules.

Spherical Propagation. How can we approximate isotropic propagation in a regular lattice? In the original essay, I posed this as the problem of how to produce asymptotically spherical expanding wave fronts in regular lattices. Since then, Margolus, Wolfram, and others discovered some surprisingly simple solutions. Another approach would be to make each particle emit showers of randomly oriented "force pellets." This would produce an isotropic inverse-square force field — but raises the problem of how to approximate a uniform spherical distribution of such pellets. Another approach might be to fill the universe with a gas of light-speed momentum pellets whose "shadows" cause inverse-square forces. This transfers the isotropy burden to the universe as a whole but, as Richard Feynman pointed out in 1963, this eventually drags everything to rest within a distinguished inertial frame.

Curvature. Suppose a spherical force field were known to have emerged from a "unit charge." Now represent that field by marking space itself as a family of equipotential surfaces. These markings need no further local information at all, because the field intensity at any point can be determined just from local curvature. However, for such a field to act on any particle, when that curvature is very small, the particle must probe correspondingly large distances. How could such a particle respond as though the interaction works at light speed? We discuss this in the next section.

10.6.3   Fields

The idea of a field abandons long range forces, and only asks the vacuum to constrain some local quantity — for example, by a differential equation. Classical 'continuous' theories assume that the vacuum can use computational schemes methods that are infinitely rapid and precise. This conceals many questions about the nature of distance and the character of partial derivatives — how can the vacuum measure and compute these? We have all become so comfortable with "real" numbers that we have come to think they are really real. Then we grumble when our theories give us series that make us pick and choose which terms to keep or throw away! The cellular model avoids all derivatives and real numbers. Here I'll try to show how we could make the state-change laws control a family of surfaces to act like a potential field on a charged particle. The resulting field has many peculiarities. Some of those may be just plain wrong — but perhaps some others could be related to other physical phenomena.

To represent the potential field, we will simply "mark" those vacuum cells that are close to equipotential surfaces. In a classical field, the force on a particle will depend on the gradient, which requires computing the surface normal — but in the cubic cellular model, there are no curved surfaces. Instead, imagine that the surfaces are entirely composed of axis-parallel polygons. Then we'll approximate the effect of the gradient by interacting simply with those polygons. Whenever a particle crosses a surface it will add a signed unit vector normal to that surface — so that as the particle moves along its trajectory, those polygons will separately supply vector components.

Now, how could the make the field maintain the right shapes for those surfaces? Some sort of physical activity must represent the information in the field's Laplacian derivatives. What we'll do is to populate the vacuum with a gas of light-speed "exchange" photons — call them "ghotons." We assign one ghoton to each oriented surface element. Each ghoton bounces repeatedly between one surface and the next one, along their common normal axis. The result is that the frequency of a ghoton's reflections is in inverse proportion to the (axis-parallel) distances between those surfaces. Each collision moves the surface element one unit away — and this has the effect of a pressure that tend to push successive potential surfaces apart. Now the corresponding field-gradient component along the x-axis is (approximately) inverse to x-distance between consecutive surfaces, so each ghoton collides once in every $2/Ex$ moments; hence the impact "pressure" is proportional to the vector E. This way, we can approximate a field in spite of local finiteness, by using something like "exchange forces." Thus, these pressures will be in equilibrium when the intersurface spacings are approximately those of the corresponding field equipotentials. (So far as I know, no one has simulated such a model to see how well it could approximate solutions to a wave equation.)

Are such ideas worth further pursuit? I think they could lead to good new ideas. Classically, we tend to view exchange-force models as mere approximations. But it is at least conceivable that some such finite-based model might turn out to actually yield exact results. Then we might be led to conclude that all those infinite series and analytic integrals were merely artifacts that came from using continuous approximations for things that by Nature should have been discrete!

10.6.4   Interactions with non-local "Particles"

Consider a collision between two bodies A and B whose momenta are very precisely specified — hence their packets must be large in size. What happens when A and B exchange momentum — if some of that information lies far away? If they are to approximate a classical interaction, this will have to happen in a process that locally first "estimates" the dispersed particles' momenta and then later repairs those estimates. In order that they interact at all, the particles must work with less than all the information classically required. It would seem that any such scheme that approximates prompt, conservative interaction must lead to some quantumlike phenomena. I have a vague idea of how this might work, but it's so complicated that I doubt it could lead to a sound theory.

Step 1. An "event" occurs at some space-time locus in which the incoming momenta are "estimated," and the outgoing momenta are determined by applying classical rules to these estimates. Because of estimation errors, the scattered momentum sums will not exactly equal the initial sums, so we need an "error correction" mechanism:

Step 2. Each scattered particle deposits a "receipt" with the other particle, recording how much momentum was actually removed. These receipts are combined with those from previous interactions, and are later "discharged" in the course of subsequent interactions. (An alternative would be to discharge the receipt momentum along the trajectory, but that would lead to curved trajectories.)

Such systems would show some quantumlike, mixed-state properties. When we measure a particle's momentum in event 1, we cannot yet observe its "receipt" momentum, because that cannot have any effect until some subsequent event 2 at which time it will be already mixed with another estimate. And so on. One can never simultaneously measure both estimates and receipts, though it all adds up eventually. This involves no probabilities, but results from the temporary inaccessibility of information.

Estimates and receipts would also permit tunneling-interactions that temporarily require more momentum than available. All is repaid eventually when receipts return their information in new estimates, but at every moment every particle carries invisible receipts not yet observable. Such models would show some qualitative features of quantum interference; if ever two particles were involved in the same interaction, they may still share receipt information that gives them some coherent "same random" properties in later interactions.

10.6.5 Rest Mass as a product of "vacuum saturation."

Why do we have particles with rest mass? I will argue that they are needed whenever a field becomes so intense that neighboring planes have not enough space. Then, some information would have to be destroyed if we continue to use the base-1 representations. However, there is a possible loophole: we can provide that at some certain threshold of intensity, the vacuum state rules impose a more compact coding — for example, by compression to base-2. That must sound silly, but it has some interesting consequences. First, this "abbreviation" process must be almost instantaneous, or information will be lost. But the light-speed limitation means that abbreviations must be done very locally, by compression into standard units, and that could lead to quantizations of momenta, orientations, spins, and energy levels. However, each such "abbreviation" must still carry enough information that, when the packet interacts again (or decays back into the field) the conserved quantities can be reproduced.

If an abbreviation carries a compressed momentum vector which can later interact in a computation, then that abbreviation must move at less than light speed. This suggests the conjecture that particles with rest mass are compressed, densely encoded representations of fragments of unary-coded fields. Presumably, then, a particle's rest mass represents the potential energy of the field consumed to create it. So, in this fantasy, we must suffer the creation of particles purely to conserve the energy and momentum of strong fields.

What else can we say about massive particles? There are many possible encodings between unary codes — that can propagate at the speed of light — and the binary codes which are most possibly compact? The closer the compression approaches that ultimate density, the fewer ways will remain for different particles to share the same space — so we must see either stronger exclusion rules, or interactions in which particles change. So we can conclude, at least qualitatively, the following:

1. Particles with rest mass have strong, short-range forces. Ultimately some information must be lost, at some threshold of intensity. If conservation of energy has the top priority, then geometric information has to be sacrificed first. That is, unless the basic state rules themselves are reversible. Fredkin has shown local time reversibility to be compatible with many cellular array computations, so it would seem of physical interest to consider time-reversible vacuum-state rules.

2. When part of a field is compressed into a particle, this must relieve the compression of the field's other surfaces, in an effect that propagates at luminal velocity. Hence the creation of particles must be accompanied by radiation.

3. A massive particle moves slower than its field and takes time to "decay." This suggests that particle creation cannot conserve all of a field's topology. When later that particle returns its information to its field, this will happen at some other place — and the global configuration of the field there will change. Conjecture: Properties like charge represent relics of the original field's topology.

Can we pursue this down to the very lattice elements? Previously we argued that three unit base vectors would suffice to abbreviate a surface normal of a field. But when things become too much compressed, there might not be room enough for all those ghotons and surfaces. If a surface were disrupted, or two surfaces were merged, this would lead to long range effects. There are eight different types of trihedral vertices and twelve kinds of edges, and various of these might combine and/or annihilate one another. The state-transition rules would need to specify what should happen in such cases. For example, we might program the vacuum to attach "axis-objects" to disrupted surface-edges. Presumably these could not be "observed" until enough of them combine to constitute or modify some other more "genuine" particle. In the meantime there would also be "radiation" as the rest of the field finds a new equilibrium configuration. It is easy to imagine various ways in which this could give rise to various six- and eight-fold "ways" that the field could use to restore itself.

10.6.6   Conclusions

In a cellular array, no field can work at light speed except by using "base-1" information codes. To approximate a Coulomb field may require something like an exchange force. Objects with "rest mass" must emerge from suitably intense fields — to conserve information before it is squeezed to death — and these must move with subluminal speed. These illustrate how, starting with simple finitistic ideas, we can end up in a world cluttered with sluggish, complicated objects with queer interactions and exclusion rules, and peculiar short-range forces.

Conservation also caused "uncertainty" to invade our simple world, because local finiteness requires that information to be dispersed. We also found some need to "balance the books" by using a complex system of "events," "receipts," and "estimates." Surely we can find simpler schemes to approximate classical physics. The ones proposed here are much too weird — and probably wouldn't work anyway. In spite of all these problems, the informational and computational clarity of such models could stimulate new insights. It remains to be seen whether this sort of approach can lead to good physical theories.

In any case, I'll argue next, beings like us could not exist in a classical sort of universe. Only worlds with firmer constraints can evolve beings like us that make theories like this. For example, unless some conservation laws hold, too many things would tend to explode. Quantum-like states seem essential for life. In the popular view, quantum states lead to uncertainty. However, on the contrary, it is they that support our stability.

This essay exploits many ideas I got from Edward Fredkin [6]. The ideas about field and particle are original; Richard Feynman persuaded me to consider fields instead of forces, but he's not responsible for my compromise on potential surfaces. I also thank Danny Hillis and Richard Stallman for other ideas.

References

[1] R. P. Feynman, *The Character of Physical Law* (MIT Press, Cambridge, MA, 1965).

[2] G. Sussman and J. Wisdom, *Science*, **22**, 433 (1988); reprinted in this volume.

[3] R. P. Feynman, *Vol.III of the Feynman Lectures on Physics*, p.2-9 and p2-10 (Addison-Wesley, Reading, MA 1965).

[4] M. L. Minsky, *International Journal of Theoretical Physics*, **21**, 537 (1982).

[5] E. R. Banks, "Information processing and transmission in cellular automata", Ph.D. Thesis, MIT, Cambridge, Massachusetts (1971).

[6] E. Fredkin, *International Journal of Theoretical Physics*, **21**, 219 (1982).

Part III

Quantum Limits

SIMULATING PHYSICS WITH COMPUTERS

Richard P. Feynman *

11.1 Introduction

On the program it says this is a keynote speech — and I don't know what a keynote speech is. I do not intend in any way to suggest what should be in this meeting as a keynote of the subjects or anything like that. I have my own things to say and to talk about and there's no implication that anybody needs to talk about the same thing or anything like it. So what I want to talk about is what Mike Dertouzos suggested that nobody would talk about. I want to talk about the problem of simulating physics with computers and I mean that in a specific way which I am going to explain. The reason for doing this is something that I learned about from Ed Fredkin, and my entire interest in the subject has been inspired by him. It has to do with learning something about the possibilities of computers, and also something about possibilities in physics. If we suppose that we know all the physical laws perfectly, of course we don't have to pay any attention to computers. It's interesting anyway to entertain oneself with the idea that we've got something to learn about physical laws; and if I take a relaxed view here (after all I'm here and not at home) I'll admit that we don't understand everything.

The first question is, "What kind of computer are we going to use to simulate physics?" Computer theory has been developed to a point where it realizes that it doesn't make any difference; when you get to a *universal computer*, it doesn't matter how it's manufactured, how it's actually made. Therefore my question is, "Can physics be simulated by a universal computer?" I would like to have the elements of this computer *locally interconnected*, and therefore sort of think about cellular automata as an example (but I don't want to force it). But I do want something involved with the locality of interaction. I would not like to think of a very enormous computer with arbitrary interconnections throughout the entire thing.

Now, what kind of physics are we going to imitate? First, I am going to describe the possibility of simulating physics in the classical approximation, a thing which is usually described by local differential equations. But the physical world is quantum mechanical, and therefore the proper problem is the simulation of quantum physics — which is what I really want to talk about, but I'll come to that later. So what kind of simulation do I mean? There is, of course, a kind of approximate simulation

---

*Reprinted with permission from International Journal of Theoretical Physics, Vol. 21, Nos. 6/7, 1982.

in which you design numerical algorithms for differential equations, and then use the computer to compute these algorithms and get an approximate view of what physics ought to do. That's an interesting subject, but is not what I want to talk about. I want to talk about the possibility that there is to be an *exact* simulation, that the computer will do *exactly* the same as nature. If this is to be proved and the type of computer is as I've already explained, then it's going to be necessary that *everything* that happens in a finite volume of space and time would have to be exactly analyzable with a finite number of logical operations. The present theory of physics is not that way, apparently. It allows space to go down into infinitesimal distances, wavelengths to get infinitely great, terms to be summed in infinite order, and so forth; and therefore, if this proposition is right, physical law is wrong.

So good, we already have a suggestion of how we might modify physical law, and that is the kind of reason why I like to study this sort of problem. To take an example, we might change the idea that space is continuous to the idea that space perhaps is a simple lattice and everything is discrete (so that we can put it into a finite number of digits) and that time jumps discontinuously. Now let's see what kind of a physical world it would be or what kind of problem of computation we would have. For example, the first difficulty that would come out is that the speed of light would depend slightly on the direction, and there might be other anisotropies in the physics that we could detect experimentally. They might be very small anisotropies. Physical knowledge is of course always incomplete, and you can always say well try to design something which beats experiment at the present time, but which predicts anistropies on some scale to be found later. That's fine. That would be good physics if you could predict something consistent with all the known facts and suggest some new fact that we didn't explain, but I have no specific examples. So I'm not objecting to the fact that it's anistropic in principle, it's a question of how anistropic. If you tell me it's so-and-so anistropic, I'll tell you about the experiment with the lithium atom which shows that the anistropy is less than that much, and that this here theory of yours is impossible.

Another thing that had been suggested early was that natural laws are reversible, but that computer rules are not. But this turned out to be false; the computer rules can be reversible, and it has been a very, very useful thing to notice and to discover that. This is a place where the relationship of physics and computation has turned itself the other way and told us something about the possibilities of computation. So this is an interesting subject because it tells us something about computer rules, and *might* tell us something about physics.

The rule of simulation that I would like to have is that the number of computer elements required to simulate a large physical system is only to be proportional to the space-time volume of the physical system. I don't want to have an explosion. That is, if you say I want to explain this much physics, I can do it exactly and I need a certain-sized computer. If doubling the volume of space and time means I'll need an *exponentially* larger computer, I consider that against the rules (I make up

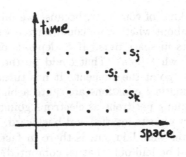

Fig. 11.1.

the rules, I'm allowed to do that). Let's start with a few interesting questions.

11.2   Simulating Time

First I'd like to talk about simulating time. We're going to assume it's discrete. You know that we don't have infinite accuracy in physical measurements so time might be discrete on a scale of less than 10^{-27} sec. (You'd have to have it at least like this to avoid clashes with experiment — but make it 10^{-41} sec. if you like, and then you've got us!) One way in which we simulate time — in cellular automata, for example — is to say that "the computer goes from state to state." But really, that's using intuition that involves the idea of time — you're going from state to state. And therefore the time (by the way, like the space in the case of cellular automata) is not simulated at all, it's imitated in the computer.

An interesting question comes up: "Is there a way of simulating it, rather than imitating it?" Well, there's a way of looking at the world that is called the space-time view, imagining that the points of space and time are all laid out, so to speak, ahead of time. And then we could say that a "computer" rule (now computer would be in quotes, because it's not the standard kind of computer which operates in time) is: We have a state s_i at each point i in space-time. (See Figure 11.1) The state s_i at the space time point i is a given function $F_i(s_j, s_k, ...)$ of the state at the points j, k in some neighborhood of i:

$$s_i = F_i(s_j, s_k, ...)$$

You'll notice immediately that if this particular function is such that the value of the function at i only involves the few points behind in time, earlier than this time i, all I've done is to redescribe the cellular automaton, because it means that you calculate a given point from points at earlier times, and I can compute the next one and so on, and I can go through this in that particular order. But just let's think of a more general kind of computer, because we might have a more general function. So let's think about whether we could have a wider case of generality of interconnections of points in space-time. If F depends on *all* the points both in the future and the past, what then? That could be the way physics works. I'll mention how our theories go at the moment. It has turned out in many physical theories that the mathematical equations are quite a bit simplified by imagining such a thing — by imagining positrons as electrons going backwards in time, and other things that connect objects forward and backward. The important question would be, if this computer were laid out, is there in fact an organized algorithm by which a solution could be laid out, that is, computed? Suppose you know this function F_i and it is a function of the variables in the future as well. How would you lay out numbers so that they automatically satisfy the above equation? It may not be possible. In the case of the cellular automaton it is, because from a given row you get the next row and then the next row, and there's an organized way of doing it. It's an interesting question whether there are circumstances where you get functions for which you can't think, at least right away, of an organized way of laying it out. Maybe sort of shake it down from some approximation, or something, but it's an interesting different type of computation.

Question: "Doesn't this reduce to the ordinary boundary value, as opposed to initial-value type of calculation?"

Answer: "Yes, but remember this is the computer itself that I'm describing."

It appears actually that classical physics is causal. You can, in terms of the information in the past, if you include both momentum and position, or the position at two different times in the past (either way, you need two pieces of information at each point) calculate the future in principle. So classical physics is local, causal, and reversible, and therefore apparently quite adaptable (except for the discreteness and so on, which I already mentioned) to computer simulation. We have no difficulty, in principle, apparently, with that.

11.3 Simulating Probability

Turning to quantum mechanics, we know immediately that here we get only the ability, apparently, to predict probabilities. Might I say immediately, so that you know where I really intend to go, that we always have had (secret, secret, close the doors!) we always have had a great deal of difficulty in understanding the world view that quantum mechanics represents. At least I do, because I'm an old enough man that I haven't got to the point that this stuff is obvious to me. Okay, I still get nervous with it. And therefore, some of the younger students ... you know how it always is, every new idea, it takes a generation or two until it becomes obvious that there's no real problem. It has not yet become obvious to me that there's no real problem. I cannot define the real problem, therefore I suspect there's no real problem, but I'm not sure there's no real problem. So that's why I like to

investigate things. Can I learn anything from asking this question about computers
— about this may or may not be mystery as to what the world view of quantum
mechanics is? So I know that quantum mechanics seems to involve probability —
and I therefore want to talk about simulating probability.

Well, one way that we could have a computer that simulates a probabilistic
theory, something that has a probability in it, would be to calculate the probability
and then interpret this number to represent Nature. For example, let's suppose
that a particle has a probability $P(x,t)$ to be at x at a time t. A typical example
of such a probability might satisfy a differential equation, as, for example, if the
particle is diffusing:

$$\frac{\partial P(x,t)}{\partial t} = -\nabla^2 P(x,t)$$

Now we could discretize t and x and perhaps even the probability itself and
solve this differential equation like we solve any old field equation, and make an
algorithm for it, making it exact by discretization. First there'd be a problem
about discretizing probability. If you are only going to take k digits it would mean
that when the probability is less that 2^{-k} of something happening, you say it doesn't
happen at all. In practice we do that. If the probability of something is 10^{-700},
we say it isn't going to happen, and we're not caught out very often. So we could
allow ourselves to do that. But the real difficulty is this: If we had many particles,
we have R particles, for example, in a system, then we would have to describe the
probability of a circumstance by giving the probability to find these particles at
points $x_1, x_2, \ldots x_R$ at the time t. That would be a description of the probability
of the system. And therefore, you'd need a k-digit number for every configuration
of the system, for every arrangement of the R values of x. And therefore if there
are N points in space, we'd need N^R configurations. Actually, from our point of
view that at each point in space there is information like electric fields and so on,
R will be of the same order as N if the number of information bits is the same as
the number of points in space, and therefore you'd have to have something like N^N
configurations to be described to get the probability out, and that's too big for our
computer to hold if the size of the computer is of order N.

We emphasize, if a description of an isolated part of Nature with N variables
requires a general function of N variables and if a computer simulates this by
actually computing or storing this function then doubling the size of nature ($N \to$
$2N$) would require an exponentially explosive growth in the size of the simulating
computer. It is therefore impossible, according to the rules stated, to simulate by
calculating the probability.

Is there any other way? What kind of simulation can we have? We can't expect
to compute the probability of configurations for a probabilistic theory. But the
other way to simulate a probabilistic Nature, which I'll call $\mathcal{N}$ for the moment,
might still be to simulate the probabilistic Nature by a computer $\mathcal{C}$ which itself is

probabilistic, in which you always randomize the last two digits of every number, or you do something terrible to it. So it becomes what I'll call a probabilistic computer, in which the output is not a unique function of the input. And then you try to work it out so that it simulates Nature in this sense: that C goes from some state — initial state if you like — to some final state with the same probability that N goes from the corresponding initial state to the corresponding final state. Of course when you set up the machine and let Nature do it, the imitator will not do the same thing, it only does it with the same probability. Is that no good? No it's okay. How do you know what the probability is? You see, Nature's unpredictable; how do you expect to predict it with a computer? You can't — it's unpredictable if it's probabilistic. But what you really do in a probabilistic system is repeat the experiment in Nature a large number of times. If you repeat the same experiment in the computer a large number of times (and that doesn't take any more time than it does to do the same thing in Nature of course), it will give the frequency of a given final state proportional to the number of times, with approximately the same rate (plus or minus the square root of n and all that) as it happens in Nature. In other words, we could imagine and be perfectly happy, I think, with a probabilistic simulator of a probabilistic Nature, in which the machine doesn't exactly do what Nature does, but if you repeated a particular type of experiment a sufficient number of times to determine Nature's probability, then you did the corresponding experiment on the computer, you'd get the corresponding probability with the corresponding accuracy (with the same kind of accuracy of statistics).

So let us now think about the characteristics of a local probabilistic computer, because I'll see if I can imitate Nature with that (by "Nature" I'm now going to mean quantum mechanics). One of the characteristics is that you can determine how it behaves in a local region by simply disregarding what it's doing in all other regions. For example, suppose there are variables in the system that describe the whole world (x_A, x_B) — the variables x_A you're interested in, they're "around here"; x_B are the whole result of the world. If you want to know the probability that something around here is happening, you would have to get that by integrating the total probability of all kinds of possibilities over x_B. If we had *computed* this probability, we would still have to do the integration

$$P_A(x_A) = \int P(x_A, x_B)dx_B$$

which is a hard job! But if we have *imitated* the probability, it's very simple to do it: you don't have to do anything to do the integration, you simply disregard what the values of x_B are, you just look at the region x_A. And therefore it does have the characteristic of Nature: if it's local, you can find out what's happening in a region not by integrating or doing an extra operation, but merely by disregarding what happens elsewhere, which is no operation, nothing at all.

The other aspect that I want to emphasize is that the equations will have a form, no doubt, something like the following. Let each point $i = 1, 2, \ldots, N$ in space be in a state s_i chosen from a small state set (the size of this set should be reasonable, say, up to 2^5). And let the probability to find some configuration $\{s_i\}$ (a set of values of the state s_i at each point i) be some number $P(\{s_i\})$. It satisfies an equation such that at each jump in time

$$P_{t+1}(\{s\}) = \sum_{\{s'\}} \left[\prod_i m(s_i | s'_j, s'_k \ldots) \right] P_t(\{s'\})$$

where $m(s_i | s'_j, s'_k \ldots)$ is the probability that we move to state s_i at point i when the neighbors have values $s_j, s_k \ldots$, where j, k etc. are points in the neighborhood of i. As j moves far from i, m becomes ever less sensitive to s'_j. At each change the state at a particular point i will move from what it was to a state s with a probability m that depends only upon the states of the neighborhood (which may be so defined as to include the point i itself). This gives the probability of making a transition. It's the same as in a cellular automaton; only, instead of its being definite, it's a probability. Tell me the environment, and I'll tell you the probability after a next moment of time that this point is at state s. And that's the way it's going to work, okay? So you get a mathematical equation of this kind of form.

Now I explicitly go to the question of how we can simulate with a computer — a universal automaton or something — the quantum mechanical effects. (The usual formulation is that quantum mechanics has some sort of a differential equation for a function ψ.) If you have a single particle, ψ is a function of x and t, and this differential equation could be simulated just like my probabilistic equation was before. That would be all right and one has seen people make little computers which simulate the Schröedinger equation for a single particle. But the full description of quantum mechanics for a large system with R particles is given by a function $\psi(x_1, x_2, \ldots, x_R, t)$ which we call the amplitude to find the particles at $x_1, \ldots, x_R$, and therefore, because it has too many variables, it *cannot be simulated* with a normal computer with a number of elements proportional to R or proportional to N. We had the same troubles with the probability in classical physics. And therefore, the problem is, how can we simulate quantum mechanics? There are two ways that we can go about it. We can give up on our rule about what the computer was, we can say: Let the computer itself be built of quantum mechanical elements which obey quantum mechanical laws. Or we can turn the other way and say: Let the computer still be the same kind that we thought of before — a logical, universal automaton; can we imitate this situation? And I'm going to separate my talk here, for it branches into two parts.

11.4 Quantum Computers — Universal Quantum Simulators

The first branch, one you might call a side-remark, is, "Can you do it with a new kind of computer — a quantum computer?" (I'll come back to the other branch in a moment.) Now it turns out, as far as I can tell, that you can simulate this with a quantum system, with quantum computer elements. It's not a Turing machine, but a machine of a different kind. If we disregard the continuity of space and make it discrete, and so on, as an approximation (the same way as we allowed ourselves in the classical case), it does seem to be true that all the various field theories have the same *kind* of behavior, and can be simulated in every way, apparently, with little latticeworks of spins and other things. It's been noted time and time again that the phenomena of field theory (if the world is made in a discrete lattice) are well imitated by many phenomena in solid state theory (which is simply the analysis of a latticework of crystal atoms, and in the case of the kind of solid state I mean, each atom is just a point which has numbers associated with it, with quantum mechanical rules). For example, the spin waves in a spin lattice imitating Bose particles in the field theory. I therefore believe it's true that with a suitable class of quantum machines you could imitate any quantum system, including the physical world. But I don't know whether the general theory of this intersimulation of quantum systems has ever been worked out, and so I present that as another interesting problem: To work out the classes of different kinds of quantum mechanical systems which are really intersimulatable — which are equivalent — as has been done in the case of classical computers. It has been found that there is a kind of universal computer that can do anything, and it doesn't make much difference specifically how it's designed. The same way we should try to find out what kinds of quantum mechanical systems are mutually intersimulatable, and try to find a specific class, or a character of that class which will simulate everything. What, in other words, is the universal quantum simulator (assuming this discretization of space and time)? If you had discrete quantum systems, what other discrete quantum systems are exact imitators of it, and is there a class against which everything can be matched? I believe it's rather simple to answer that question and to find the class, but I just haven't done it.

Suppose that we try the following guess: that every finite quantum mechanical system can be described *exactly*, imitated exactly, by supposing that we have another system such that at each point in space-time this system has only two possible base states. Either that point is occupied, or unoccupied — those are the two states. The mathematics of the quantum mechanical operators associated with that point would be very simple.

$$a = ANNIHILATE = \begin{array}{c|cc} & \text{OCC} & \text{UN} \\ \hline \text{OCC} & 0 & 0 \\ \text{UN} & 1 & 0 \end{array} = \frac{1}{2}(\sigma_x - i\sigma_y)$$

$$a^* = CREATE = \begin{array}{|cc|} \hline 0 & 1 \\ 0 & 0 \\ \hline \end{array} = \frac{1}{2}(\sigma_x + i\sigma_y)$$

$$n = NUMBER = \begin{array}{|cc|} \hline 1 & 0 \\ 0 & 0 \\ \hline \end{array} = a^*a = \frac{1}{2}(1 + \sigma_z)$$

$$\mathcal{I} = IDENTITY = \begin{array}{|cc|} \hline 1 & 0 \\ 0 & 1 \\ \hline \end{array}$$

There would be an operator a which *annihilates* if the point is occupied — it changes it to unoccupied. There is a conjugate operator a^* which does the opposite: If it's unoccupied, it occupies it. There's another operator n called the *number* to ask, "Is something there?" The little matrices tell you what they do. If it's there, n gets a one and leaves it alone; if it's not there, nothing happens. That's mathematically equivalent to the product of the other two, as a matter of fact. And then there's the identity, 1, which we always have to put in there to complete our mathematics — it doesn't do a damn thing! By the way, on the right-hand side of the above formulas the same operators are written in terms of matrices that most physicists find more convenient, because they are Hermitian, and that seems to make it easier for them. They have invented another set of matrices, the Pauli σ matrices:

$$\sigma_z = \begin{pmatrix} 1 & 0 \\ 0 & -1 \end{pmatrix}, \quad \sigma_x = \begin{pmatrix} 0 & 1 \\ 1 & 0 \end{pmatrix}, \quad \sigma_y = \begin{pmatrix} 0 & -i \\ i & 0 \end{pmatrix}, \quad \mathcal{I} = \begin{pmatrix} 1 & 0 \\ 0 & 1 \end{pmatrix}$$

And these are called *spin* — spin one-half — so sometimes people say you're talking about a spin-one-half lattice.

The question is, "If we wrote a Hamiltonian which involved only these operators, locally coupled to corresponding operators on the other space-time points, could we imitate every quantum mechanical system which is discrete and has a finite number of degrees of freedom?" I know, almost certainly, that we could do that for any quantum mechanical system which involves Bose particles. I'm not sure whether Fermi particles could be described by such a system. So I leave that open. Well, that's an example of what I meant by a general quantum mechanical simulator. I'm not sure that it's sufficient, because I'm not sure that it takes care of Fermi particles.

11.5   Can Quantum Systems be Probabilistically Simulated by a Classical Computer?

Now the next question that I would like to bring up is, of course, the interesting one, i.e., "Can a quantum system be probabilistically simulated by a classical (probabilistic, I'd assume) universal computer?" In other words, a computer which will give the same probabilities as the quantum system does. If you take the computer to be the classical kind I've described so far, (not the quantum kind described in the last section) and there're no changes in any laws, and there's no hocus-pocus, the answer is certainly, "No!" This is called the hidden-variable problem: It is impossible to represent the results of quantum mechanics with a classical universal device. To learn a little bit about it, I say let us try to put the quantum equations in a form as close as possible to classical equations so that we can see what the difficulty is and what happens. Well, first of all we can't simulate ψ in the normal way. As I've explained already, there're too many variables. Our only hope is that we're going to simulate probabilities, that we're going to have our computer do things with the same probability as we observe in nature, as calculated by the quantum mechanical system. Can you make a cellular automaton, or something, imitate with the same probability what Nature does, where I'm going to suppose that quantum mechanics is correct, or at least after I discretize space and time it's correct, and see if I can do it. I must point out that you must directly generate the probabilities, the results, with the correct quantum probability. Directly, because we have no way to store all the numbers, we have to just imitate the phenomenon directly.

It turns out then that another thing, rather than the wave function, a thing called the *density matrix*, is much more useful for this. It's not so useful as far as the mathematical equations are concerned, since it's more complicated than the equations for ψ, but I'm not going to worry about mathematical complications, or which is the easiest way to calculate, because with computers we don't have to be so careful to do it the very easiest way. And so with a slight increase in the complexity of the equations (and not very much increase) I turn to the density matrix, which for a single particle of coordinate x in a pure state of wave function $\psi(x)$ is

$$\rho(x, x') = \psi^*(x)\psi(x')$$

This has a special property that is a function of two coordinates x, x'. The presence of two quantities x and x' associated with each coordinate is analogous to the fact that in classical mechanics you have to have two variables to describe the state, x and $\dot{x}$. States are described by a second-order device, with two informations ("position" and "velocity"). So we have to have two pieces of information associated with a particle, analogous to the classical situation, in order to describe configurations. (I've written the density matrix for one particle, but of course there's the analogous thing for R particles, a function of $2R$ variables).

This quantity has many of the mathematical properties of a probability. For example if a state $\psi(x)$ is not certain but is ψ_α with the probability p_α, then the density matrix is the appropriate weighted sum of the matrix for each state α:

$$\rho(x, x') = \sum_\alpha p_\alpha \psi_\alpha^*(x) \psi_\alpha(x')$$

A quantity which has properties even more similar to classical probabilities is the Wigner function, a simple reexpression of the density matrix; for a single particle

$$W(x, p) = \int \rho(x + \frac{y}{2}, x - \frac{y}{2}) e^{ipy} dy$$

We shall be emphasizing their similarity and shall call it "probability" in quotes instead of Wigner function. Watch these quotes carefully, when they are absent we mean the real probability. If "probability" had all the mathematical properties of a probability we could remove the quotes and simulate it. $W(x, p)$ is the "probability" that the particle has position x and momentum p (per dx and dp). What properties does it have that are analogous to an ordinary probability?

It has the property that if there are many variables and you want to know the "probabilities" associated with a finite region, you simply disregard the other variables (by integration). Furthermore the probability of finding a particle at x is $\int W(x, p) dp$. If you can interpret W as a probability of finding x and p, this would be an expected equation. Likewise the probability of p would be expected to be $\int W(x, p) dx$. These two equations are correct, and therefore you would hope that maybe $W(x, p)$ is the probability of finding x and p. And the question then is can we make a device which simulates this W? Because then it would work fine.

Since the quantum systems I noted were best represented by spin one-half (occupied versus unoccupied or spin one-half is the same thing), I tried to do the same thing for spin one-half objects, and it's rather easy to do. Although before, one object only had two states, occupied and unoccupied, the full description — in order to develop things as a function of time — requires twice as many variables, which mean two slots at each point which are occupied or unoccupied (denoted by + and − in what follows), analogous to the x and $\dot{x}$, or the x and p. So you can find four numbers, four "probabilities" $\{f_{++}, f_{+-}, f_{-+}, f_{--}\}$ which act just like, and I have to explain why they're not exactly like, but they act just like probabilities to find things in the state in which both symbols are up, one's up and one's down, and so on. For example, the sum $f_{++} + f_{+-} + f_{-+} + f_{--}$ of the four "probabilities" is 1. You'll remember that one object now is going to have two indices, two plus/minus indices, or two ones and zeros at each point, although the quantum system had only one. For example, if you would like to know whether the first index is positive, the probability of that would be

Prob(first index is +) = f_{++} + f_{+-} [spin z up]

i.e., you don't care about the second index. The probability that the first index is negative is

Prob(first index is −) = f_{-+} + f_{--} [spin z down]

These two formulas are exactly correct in quantum mechanics. You see I'm hedging on whether or not "probability" f can really be a probability without quotes. But when I write probability without quotes on the left-hand side I'm not hedging; that really is the quantum mechanical probability. It's interpreted perfectly fine here. Likewise the probability that the second index is positive can be obtained by finding

Prob(second index is +) = f_{++} + f_{-+} [spin x up]

and likewise

Prob(second index is −) = f_{+-} + f_{--} [spin x down]

You could also ask other questions about the system. You might like to know, "What is the probability that both indices are positive?" You'll get in trouble. But you could ask other questions that you won't get in trouble with, and that get correct physical answers. You can ask, for example, "What is the probability that the two indices are the same?" That would be

Prob(match) = f_{++} + f_{--} [spin y up]

Or the probability that there's no match between the indices, that they're different,

Prob(no match) = f_{+-} + f_{-+} [spin y down]

All perfectly all right. All these probabilities are correct and make sense, and have a precise meaning in the spin model, shown in the square brackets above. There are other "probability" combinations, other linear combinations of these f's which also make physically sensible probabilities, but I won't go into those now. There are other linear combinations that you can ask questions about, but you don't seem to be able to ask questions about an individual f.

11.6   Negative Probabilities

Now, for many interacting spins on a lattice we can give a "probability" (the quotes remind us that there is still a question about whether it's a probability) for correlated possibilities:

$$F(s_1, s_2, \dots, s_n) \quad (s_i \in \{++, +-, -+, --\})$$

Next, if I look for the quantum mechanical equation which tells me what the changes of F are with time, they are exactly of the form that I wrote above for the classical theory:

$$F_{t+1}(\{s\}) = \sum_{\{s'\}} \left[\prod_i M(s_i|s'_j, s'_k \ldots) \right] F_t(\{s'\})$$

but now we have F instead of P. The $M(s_i|s'_j, s'_k \ldots)$ would appear to be interpreted as the "probability" per unit time, or per time jump, that the state at i turns into s_i when the neighbors are in configuration s'. If you can invent a probability M like that, you write the equations for it according to normal logic, those are the correct equations, the real, correct, quantum mechanical equations for this F, and therefore you'd say, "Okay, so I can imitate it with a probabilistic computer!"

There's only one thing wrong. These equations unfortunately cannot be so interpreted on the basis of the so-called "probability", or this probabilistic computer can't simulate them, because the F is not necessarily positive. Sometimes it's negative! The M, the "probability" (so-called) of moving from one condition to another is itself not positive; if I had gone all the way back to the f for a single object, it again is not necessarily positive. An example of possibilities here are

$$f_{++} = 0.6 \quad f_{+-} = -0.1 \quad f_{-+} = 0.3 \quad f_{--} = 0.2$$

The sum $f_{++} + f_{+-}$ is 0.5, that's 50% chance of finding the first index positive. The probability of finding the first index negative is the sum $f_{-+} + f_{-+}$ which is also 50%. The probability of finding the second index positive is the sum $f_{++} + f_{-+}$ which is nine tenths, the probability of finding it negative is $f_{+-} + f_{--}$ which is one-tenth, perfectly alright, it's either plus or minus. The probability that they match is eight-tenths, the probability that they mismatch is plus two-tenths; every physical probability comes out positive. But the original f's are not positive, and therein lies the great difficulty. The only difference between a probabilistic classical world and the equations of the quantum world is that somehow or other it appears as if the probabilities would have to go negative, and that we do not know, as far as I know, how to simulate. Okay, that's the fundamental problem. I don't know the answer to it, but I wanted to explain that if I try my best to make the equations look as near as possible to what would be imitable by a classical probabilistic computer, I get into trouble.

11.7   Polarization of Photons — Two-State Systems

I would like to show you why such minus signs cannot be avoided, or at least that you have some sort of difficulty. You probably have all heard this example of the Einstein-Podolsky-Rosen paradox, but I will explain this little example of a physical experiment which can be done, and which has been done, which does give

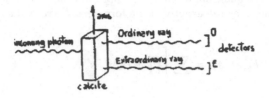

Fig. 11.2.

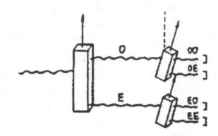

Fig. 11.3.

the answers quantum theory predicts, and the answers are really right, there's no mistake, if you do the experiment, it actually comes out. And I'm going to use the example of polarizations of photons, which is an example of a two-state system. When a photon comes, you can say it's either x polarized or y polarized. You can find that out by putting in a piece of calcite, and the photon goes through the calcite either out in one direction, or out in another — actually slightly separated, and then you put in some mirrors, that's not important. You get two beams, two places out, where the photon can go (See Figure 11.2).

If you put a polarized photon in, then it will go to one beam called the ordinary ray, or another, the extraordinary one. If you put detectors there you find that each photon that you put in, it either comes out in one or the other 100% of the time, and not half and half. You either find a photon in one or the other. The probability of finding it in the ordinary ray plus the probability of finding it in the extraordinary ray is always 1 — you have to have that rule. That works. And further, it's never found at both detectors. (If you might have put two photons in, you could get that, but you cut the intensity down — it's a technical thing, you don't find them in both detectors.)

Now the next experiment: Separation into 4 polarized beams (see Figure 11.3). You put two calcites in a row so that their axes have a relative angle ϕ, I happen to have drawn the second calcite in two positions, but it doesn't make a difference

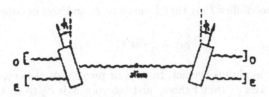

Fig. 11.4.

if you use the same piece or not, as you care. Take the ordinary ray from one and put it through another piece of calcite and look at its ordinary ray, which I'll call the ordinary-ordinary $(O - O)$ ray, or look at its extraordinary ray, I have the ordinary-extraordinary $(O - E)$ ray. And then the extraordinary ray from the first one comes out as the $E - O$ ray, and then there's an $E - E$ ray, alright. Now you can ask what happens.

You'll find the following. *When a photon comes in, you always find that only one of the four counters goes off.*

If the photon is O from the first calcite, then the second calcite gives $O - O$ with probability $\cos^2\phi$ or $O - E$ with the complementary probability $1 - \cos^2\phi = \sin^2\phi$. Likewise an E photon gives a $E - O$ with the probability $\sin^2\phi$ or an $E - E$ with the probability $\cos^2\phi$.

11.8 Two-Photon Correlation Experiment

Let us turn now to the two photon correlation experiment (see Figure 11.4). What can happen is that an atom emits two photons in opposite direction (e.g., the $3s \to 2p \to 1s$ transition in the H atom). They are observed simultaneously (say, by you and by me) through two calcites set at ϕ_1 and ϕ_2 to the vertical. Quantum theory and experiment agree that the probability P_{OO} that both of us detect an ordinary photon is

$$P_{OO} = \frac{1}{2}\cos^2(\phi_2 - \phi_1)$$

The probability P_{EE} that we both observe an extraordinary ray is the same

$$P_{EE} = \frac{1}{2}\cos^2(\phi_2 - \phi_1)$$

The probability P_{OE} that I find O and you find E is

$$P_{OE} = \frac{1}{2}\sin^2(\phi_2 - \phi_1)$$

and finally the probability P_{EO} that I measure E and you measure O is

$$P_{EO} = \frac{1}{2}sin^2(\phi_2 - \phi_1)$$

Notice that you can always predict, from your own measurement, what I shall get, O or E. For any axis ϕ_1 that I chose, just set your axis ϕ_2 to ϕ_1 then

$$P_{OE} = P_{EO} = 0$$

and I must get whatever you get.

Let us see now how it would have to be for a *local* probabilistic computer. Photon 1 must be in some condition α with the probability $f_\alpha(\phi_1)$ that determines it to go through as an ordinary ray [the probability it would pass as E is $1 - f_\alpha(\phi_1)$]. Likewise photon 2 will be in a condition β with probability $g_\beta(\phi_2)$. If $p_{\alpha\beta}$ is the conjoint probability to find the condition pair α, β the probability P_{OO} that both of us observe O rays is

$$P_{OO}(\phi_1, \phi_2) = \sum_{\alpha\beta} p_{\alpha\beta} f_\alpha(\phi_1) g_\beta(\phi_2) \qquad \sum_{\alpha\beta} p_{\alpha\beta} = 1$$

likewise

$$P_{OE}(\phi_1, \phi_2) = \sum_{\alpha\beta} p_{\alpha\beta}(1 - f_\alpha(\phi_1)) g_\beta(\phi_2) \quad \text{etc.}$$

The conditions α determine how the photons go. There's some kind of correlation of the conditions. Such a formula cannot reproduce the quantum results above for any $p_{\alpha\beta}$, $f_\alpha(\phi_1)$, $g_\beta(\phi_2)$ if they are real probabilities — that is all positive, although it is easy if they are "probabilities" — negative for some conditions or angles. We now analyze why that is so.

I don't know what kinds of conditions they are, but for any condition the probability $f_\alpha(\phi)$ of its being extraordinary or ordinary in any direction must be either one or zero. Otherwise you couldn't predict it on the other side. You would be unable to predict with certainty what I was going to get, unless, every time the photon comes here, which way it's going to go is absolutely determined. Therefore, whatever condition the photon is in, there is some hidden inside variable that's going to determine whether it's going to be ordinary or extraordinary. This determination is done deterministically, not probabilistically; otherwise we can't explain the fact that you could predict what I was going to get *exactly*. So let us suppose that something like this happens. Suppose we discuss results just for angles which are multiples of 30°.

On each diagram (Figure 11.5) are the angles 0°, 30°, 60°, 90°, 120°, and 150°. A particle comes out to me, and it's in some sort of state, so what it's going to give

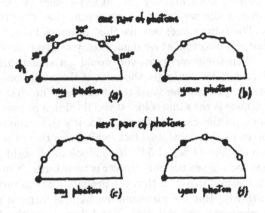

Fig. 11.5.

for 0°, for 30°, etc. are all predicted — determined — by the state. Let us say that in a particular state that is set up the prediction for 0° is that it'll be extraordinary (black dot), for 30° it's also extraordinary, for 60° it's ordinary (white dot), and so on (Figure 11.5a). By the way, the outcomes are complements of each other at right angles, because, remember, it's always either extraordinary or ordinary; so if you turn 90°, what used to be an ordinary ray becomes the extraordinary ray. Therefore, whatever condition it's in, it has some predictive pattern in which you either have a prediction of ordinary or of extraordinary — three and three — because at right angles they're not the same color. Likewise the particle that comes to you when they're separated must have the same pattern because you can determine what I'm going to get by measuring yours. Whatever circumstances come out, the patterns must be the same. So, if I want to know, "Am I going to get white at 60°?" You just measure at 60°, and you'll find white, and therefore you'll predict white, or ordinary, for me. Now each time we do the experiment the pattern may not be the same. Every time we make a pair of photons, repeating this experiment again and again, it doesn't have to be the same as Figure 11.5a. Let's assume that the next time the experiment my photon will be O or E for each angle as in Figure 11.5c. Then your pattern looks like Figure 11.5d. But whatever it is, your pattern has to be my pattern exactly — otherwise you couldn't predict what I was going to get exactly by measuring the corresponding angle. And so on. Each time we do the experiment, we get different patterns; and it's easy: There are just six dots and three of them are white, and you chase them around different way — everything can happen. If we measure at the same angle, we always find that with this kind of arrangement we would get the same result.

Now suppose we measure at $\phi_2 - \phi_1 = 30°$, and ask, "With what probability do we get the same result?" Let's first try this example here (Figure 11.5a, 11.5b). With what probability would we get the same result, that they're both white, or they're both black? The thing comes out like this: Suppose I say, "After they come out, I'm going to choose a direction at random, I tell you to measure 30° to the right of that direction." Then whatever I get, you would get something different if the neighbors were different. (We would get the same if the neighbors were the same.) What is the chance that you get the same result as me? The chance is the number of times that the neighbor is the same color. If you'll think a minute, you'll find that two thirds of the time, in the case of Figure 11.5a, it's the same color. The worst case would be black/white/black/white/black/white, and there the probability of a match would be zero (Figure 11.5c, 11.5d). If you look at all eight possible distinct cases, you'll find that the biggest possible answer is two-thirds. You cannot arrange, in a classical kind of method like this, that the probability of agreement at 30° will be bigger than two-thirds. But the quantum mechanical formula predicts $\cos^2 30°$ (or $3/4$) — and experiments agree with this — and therein lies the difficulty. That's all. That's the difficulty. That's why quantum mechanics can't seem to be imitable by a local classical computer.

I've entertained myself always by squeezing the difficulty of quantum mechanics into a smaller and smaller place, so as to get more and more worried about this particular item. It seems to be almost ridiculous that you can squeeze it to a numerical question that one thing is bigger than another. But there you are — it is bigger than any logical argument can produce, if you have this kind of logic. Now, we say "this kind of logic"; what other possibilities are there? Perhaps there may be no possibilities, but perhaps there are. It's interesting to try to discuss the possibilities. I mentioned something about the possibility of time — of things being affected not just by the past, but also by the future, and therefore that our probabilities are in some sense "illusory." We only have the information from the past, and we try to predict the next step, but in reality it depends upon the near future which we can't get at, or something like that. A very interesting question is the origin of the probabilities in quantum mechanics. Another way of putting things is this: We have an illusion that we can do any experiment that we want. We all, however, come from the same universe, have evolved with it, and don't really have any "real" freedom. For we obey certain laws and have come from a certain past. Is it somehow that we are correlated to the experiments that we do, so that the apparent probabilities don't look like they ought to look if you assume that they are random. There are all kinds of questions like this, and what I'm trying to do is to get you people who think about computer-simulation possibilities to pay a great deal of attention to this, to digest as well as possible the real answers of quantum mechanics, and see if you can't invent a different point of view than the physicists have had to invent to describe this. In fact the physicists have no good point of view. Somebody mumbled something about a many-world picture, and that many-world picture says that the wave function ψ is what's real, and damn the torpedos if there are so many variables, N^R. All these different worlds and every arrangement of configurations are all there just like our arrangement of configurations, we just happen to be sitting in this one. It's possible, but I'm not very happy with it.

So, I would like to see if there's some other way out, and I want to emphasize, or bring the question here, because the discovery of computers and the thinking about computers has turned out to be extremely useful in many branches of human reasoning. For instance, we never really understood how lousy our understanding of languages was, the theory of grammar and all that stuff, until we tried to make a computer which would be able to understand language. We tried to learn a great deal about psychology by trying to understand how computers work. There are interesting philosophical questions about reasoning, and relationship, observation, and measurement and so on, which computers have stimulated us to think about anew, with new types of thinking. And all I was doing was hoping that the computer-type of thinking would give us some new ideas, if any are really needed. I don't know, maybe physics is absolutely okay the way it is. The program that Fredkin is always pushing, about trying to find a computer simulation of physics, seem to me to be an excellent program to follow out. He and I have had wonderful, intense, and interminable arguments, and my argument is always that the real use of it would be with quantum mechanics. and therefore full attention and acceptance of the quantum mechanical phenomena — the challenge of explaining quantum mechanical phenomena — has to be put into the argument, and therefore these phenomena have to be understood very well in analyzing the situation. And I'm not happy with all the analyses that go with just the classical theory, because Nature isn't classical, dammit, and if you want to make a simulation of Nature, you'd better make it quantum mechanical, and by golly it's a wonderful problem, because it doesn't look so easy. Thank you.

11.9   Discussion

Question: Just to interpret, you spoke first of the probability of A given B, versus the probability of A and B jointly — that's the probability of one observer seeing the result, assigning a probability to the other; and then you brought up the paradox of the quantum mechanical result being 3/4, and this being 2/3. Are those really the same probabilities? Isn't one a joint probability, and the other a conditional one?

Answer: No, they are the same. P_{OO} is the *joint probability* that both you and I observe an ordinary ray, and P_{EE} is the *joint* probability for two extraordinary rays. The probability that our observations match is

$$P_{OO} + P_{EE} = cos^2 30° = 3/4$$

Question: Does it in some sense depend upon an assumption as to how much information is accessible from the photon, or from the particle? And second, to take your question of prediction, your comment about predicting, is in some sense reminiscent of the philosophical question, "Is there any meaning to the question of whether there is free will or predestination?", namely, the correlation between the observer and the experiment, and the question there is, "Is it possible to construct a test in which the prediction could be reported to the observer?", or instead, has the ability to represent information already been used up? And I suspect that you may have already used up all the information so that prediction lies outside the range of the theory.

Answer: All these things I don't understand; deep questions, profound questions. However physicists have a kind of a dopy way of avoiding all of these things. They simply say, now look, friend, you take a pair of counters and you put them on the side of your calcite and you count how many times you get this stuff, and it comes out 75% of the time. Then you go and you say, "Now can I imitate that with a device which is going to produce the same results, and which will operate locally", and you try to invent some kind of way of doing that, and if you do it in the ordinary way of thinking, you find that you can't get there with the same probability. Therefore some new kind of thinking is necessary, but physicists, being kind of dull minded, only look at Nature, and don't know how to think in these new ways.

Question: At the beginning of your talk, you talked about discretizing various things in order to go about doing a real computation of physics. And yet it seems to me that there are some differences between things like space and time, and probability that might exist at some place, or energy, or some field value. Do you see any reason to distinguish between quantization or discretizing of space and time, versus discretizing any of the specific parameters or values that might exist?

Answer: I would like to make a few comments. You said quantizing or discretizing. That's very dangerous. Quantum theory and quantizing is a very specific type of theory. Discretizing is the right word. Quantizing is a different kind of mathematics. If we talk about discretizing... of course I pointed out that we're going to have to change the laws of physics. Because the laws of physics as written now have, in the classical limit, a continuous variable everywhere, space and time. If, for example, in your theory you were going to have an electric field, then the electric field could not have (if it's going to be imitable, computable by a finite number of elements) an infinite number of possible values, it'd have to be digitized. You might be able to get away with a theory by redescribing things without an electric field, but supposing for a moment that you've discovered that you can't do that and you want to describe it with an electric field, then you would have to say that, for example, when fields are smaller than a certain amount, they aren't there at all, or something. And those are very interesting problems, but unfortunately they're not good problems for classical physics because if you take the example of a star a hundred light years away, and it makes a wave which comes to us, and it gets weaker, and weaker, and weaker, and weaker, the electric field's going down, down, how low can we measure? You put a counter out there and you find "clunk," and nothing happens for a while, "clunk," and nothing happens for a while. It's not discretized at all, you never can measure such a tiny field, you don't find a tiny field, you don't have to imitate such a tiny field, because the world that you're trying to imitate, the physical world, is not the classical world, and it behaves differently. So the particular example of discretizing the electric field, is a problem which I would not see, as a physicist, as fundamentally difficult, because it will just mean that your field has gotten so small that I had better be using quantum mechanics anyway, and so you've got the wrong equations, and so you did the wrong problem! That's how I would answer that. Because you see, if you would imagine that the electric field is coming out of some 'ones' or something, the lowest you could get would be a full one, but that's what we see, you get a full photon. All these things suggest that it's really true, somehow, that the physical world is representable in a discretized way, because every time you get into a bind like this, you discover that the experiment does just what's necessary to escape the trouble that would come if the electric field went to zero, or you'd never be able to see a star beyond a certain distance, because the field would have gotten below the number of digits that your world can carry.

12

QUANTUM ROBOTS

Paul Benioff

Abstract

Validation of a presumably universal theory, such as quantum mechanics, requires a quantum mechanical description of systems that carry out theoretical calculations and sytems that carry out experiments. The description of quantum computers is under active development. No description of systems to carry out experiments has been given. A very small step in this direction is taken here by giving a description of quantum robots as mobile systems with on board quantum computers that interact with environments of quantum systems. The dynamics of these systems are described in terms of tasks that consist of sequences of computation and action phases. For each task a step operator is defined which can then be used to define finite time interval step dynamics or a task Hamiltonian. A specific task carried out on a very simple environment is used to illustrate the models.

12.1 Introduction

Much of the impetus to study quantum computation, either as networks of quantum gates [1, 2] (See [3] for a review) or as Quantum Turing Machines [4–8], is based on the increased efficiency of quantum computers compared to classical computers for solving some important problems [9, 10]. Realization of this goal or use of quantum computers to simulate other physical systems [6, 11] requires the eventual physical construction of quantum computers. However, as emphasized repeatedly by Landauer [12], there are serious obstacles to such a physical realization.

There is, however, another reason to study quantum computers that is less dependent on whether or not such machines are ever built. It is based on the fact that testing the validity of a physical theory such as quantum mechanics requires the comparison of numerical values calculated from theory with experimental results. If quantum mechanics is universally valid (and there is no reason to assume otherwise), then both the systems that carry out theoretical calculations and the systems that carry out experiments must be described within quantum mechanics. It follows that the systems that test the validity of quantum mechanics must be described by the same theory whose validity they are testing. That is, quantum mechanics must describe its own validation to the maximum extent possible [13].

Because of these self referential aspects, limitations in mathematical systems expressed by the Gödel theorems lead one to expect that there may be interesting questions of self consistency and limitations in such a description. Limitations on self observation by quantum automata [14–16] may also play a role here.

In order to investigate these questions it is necessary to have well defined completely quantum mechanical descriptions of systems that compute theoretical values and of systems that carry out experiments. So far there has been much work on quantum computers. These are systems that can, in principle at least, carry out computation of theoretical values for comparison with experiment. However there has been no comparable development of a quantum mechanical description of robots. These are systems that can, in principle at least, carry out experiments.

Another reason quantum robots are interesting is that it is possible that they might provide a *very small* first step towards a quantum mechanical description of systems that are aware of their environment, make decisions, are intelligent, and create theories such as quantum mechanics [17–19]. If quantum mechanics is universal, then these systems must also be described in quantum mechanics to the maximum extent possible.

The main point of this paper is that quantum robots and their interactions with environments may provide a well defined platform for investigation of many interesting questions generated by the above considerations. To this end some general aspects of quantum robots and their interactions with environments are discussed in the next section. A quantum robot is defined as a mobile system consisting of an on board quantum computer and needed ancillary systems that moves in and interacts with an environment of quantum systems. The concept of tasks, as sequences of computations and actions, carried out by quantum robots is also introduced.

Section 12.3 provides a more detailed description of quantum robots and gives a dynamical model for quantum robots interacting with environments. The on board computer is taken to be a quantum Turing machine consisting of a multistate head moving on a closed lattice of qubits and ancillary output, memory and control systems. The dynamics is defined in terms of step operators that are a sum of computation and action phase step operators. These can be used to describe finite time interval step dynamics or used to construct a Hamiltonian based on Feynman's prescription [20]. Locality and other conditions that the step operators must satisfy are discussed in detail.

A specific example of a task for a quantum robot in a very simple example of an environment is analyzed in detail in Section 12.4. The environment consists of a single spinless particle on a 1-D space lattice and the task is "search to the right for the particle, if found bring back to the initial quantum robot location". Detailed properties of the action and computation phases are given with some mathematical aspects discussed in the Appendix.

The last section contains a discussion of additional aspects. The importance of having a well defined quantum mechanical platform for asking relevant questions is amplified. Also the speculative possibility of a Church-Turing type hypothesis for the class of physical experiments is noted.

It must be emphasized that the language used in this paper to describe quantum robots is carefully chosen to avoid any suggestions that these systems are aware of their environment, make decisions, carry out experiments or make measurements, or have other properties characteristic of intelligent or conscious systems. The quantum robots described here have no awareness of their environment and do not make decisions or measurements. Their description differs in detail only, from that used to describe any other system in quantum mechanics.

Some aspects of the ideas presented here have already occurred in earlier work. Physical operations have been described as instructions for well-defined realizable and reproducible procedures [21], and quantum state preparation and observation procedures have been described as instruction booklets or programs for robots [22]. However these concepts were not described in detail and the possibility of describing these procedures or operations quantum mechanically was not mentioned. Also quantum computers had not yet been described. More recently Helon and Milburn [23] have described the use of the electronic states of ions in a linear ion trap as an apparatus (and a quantum computer register) to measure properties of vibrational states of the ions. In other work quantum mechanical Maxwell's demons have been described [24].

Also there is much work on the interactions between quantum computers and the environment. However, these interactions are considered as a source of noise or errors to be minimized or corrected by use of quantum error correction codes [25]. Here interactions between a quantum robot and the environment are emphasized as an essential part of the overall system dynamics. Other work on environmentally induced superselection rules [26, 27] also emphasizes interactions between the environment and a measurement apparatus that stabilize a selected basis (the pointer basis) of states of the apparatus.

12.2   Quantum Robots

Here quantum robots are considered to be mobile systems that have a quantum computer on board and any other needed ancillary systems. Quantum robots move in and interact (locally) with environments of quantum systems. Since quantum robots are mobile, they are limited to be quantum systems with finite numbers of degrees of freedom.

Environments consist of arbitrary numbers and type of systems moving in 1-, 2-, or 3-dimensional spatial universes. The component systems can have spin or other internal quantum numbers and can interact with one another or be free. Environments can be open or closed. If they are open then there may be systems that remain for all time outside the domain of interaction with the quantum robot that can interact with and establish correlations with other environment systems in the domain on the robot. Quantum field theory may be useful to describe environments containing an infinite number of degrees of freedom. To keep things simple, in this paper environments will be considered to consist of systems in discrete space lattices instead of in continuous space.

The quantum computer that is on board the quantum robot can be described as a quantum Turing machine, a network of quantum gates, or any other suitable model. If it is a quantum Turing machine, it consists of a finite state head moving on a finite lattice of qubits. The lattice can have distinct ends. However it seems preferable if the lattice is closed (i.e. cyclic). If the on board computer is a network of quantum gates then it should be a cyclic network with many closed internal quantum wire loops and a limited number of open input and output quantum wires (narrow bandwidth). Even though acyclic networks are sufficient for the purposes of quantum computation [28] cyclic ones are preferable for quantum robots. One reason is that interactions between these networks and the environment are simpler to describe and understand than those containing a large number of input and output lines. Also the only known examples of *very* complex systems that are aware of their environment and are presumably intelligent, contain large numbers of internal loops and internal memory storage.

The overall dynamics of a quantum robot and its interactions with the environment is described in terms of *tasks*. Tasks can be divided into different types according to their goals. For one type the goals can be described in terms of specified changes in the state of the environmental systems. This type is similar to the association of functions with quantum computers in that a quantum computer, starting with a specified initial state containing the function argument, outputs a final state with the value of the function.

Another type has as a goal the carrying out of a measurement by transfer of information from the environment to the quantum robot. Some tasks may combine both types of goals. Other types of goals may also be possible.

An example of the first type of task is "move each system in region R 3 sites to the right if and only if the destination site is unoccupied." Implementation of such a task requires specification of a path to be taken by the quantum robot in executing the task. Some method of determining when it is inside or outside of the specified region and making appropriate movements must be available. In this case if there are n systems in region R at locations $x_1, x_2, \cdots, x_n$ in region R then the initial state of the regional environment, $|\underline{x}\rangle = \otimes_{j=1}^{n}|x_j\rangle$ becomes $\otimes_{j=1}^{n}|x_j + 3\rangle$ provided all destination sites are unoccupied.

If the initial state of the regional environment is a linear superposition of states $\psi = \sum_{\underline{x}} c_{\underline{x}}|\underline{x}\rangle$ of n-system position states $|\underline{x}\rangle$ in R then the final state of the regional environment is given in general by a density operator even if all destination sites are unoccupied. This is a consequence of the fact that in general the actions of the quantum robot introduce correlations between the states of the robot systems and the different initial environment component states $|\underline{x}\rangle$. When the task is completed on all components $|\underline{x}\rangle$, the overall state of the robot plus environment is given by a linear sum over robot regional environment states of the form $\sum_{\underline{x}} c_{\underline{x}} \theta_{\underline{x}} |\underline{x}\rangle$. Here $\theta_{\underline{x}}$ is the final state of the quantum robot resulting from carrying out the task on the regional environment in state $|\underline{x}\rangle$. Taking the trace over the robot system variables gives the density operator form for the regional environment state.

The above description shows that quantum robots can carry out the same task on many different environments simultaneously. This can be done by use of an initial state of the quantum robot plus environment that is a linear superposition of different environment basis states. For quantum computers the corresponding property of carrying out many computations in parallel has been known for some time [6]. Whether the speedup provided by this parallel tasking ability can be preserved for some tasks, as is the case for Shor's [9] or Grover's algorithms [10] for quantum computers, remains to be seen.

The above described task is an example of a reversible task. There are also many tasks that are irreversible. An example is the task "clean up the region R of the environment" where "clean up" has some specific description such as "move all systems in R to some fixed pattern". This task is irreversible because many initial states of systems in R are taken into the same final state. This task can be made reversible by storing somewhere in the environment outside of R a copy of each component in some basis B of the initial state of the systems in R. For example if $B = \{|\underline{x}\rangle\}$ and $\psi = \sum_{\underline{x}} c_{\underline{x}} |\underline{x}\rangle$ is the initial state, then the copy operation is given by $\sum_{\underline{x}} c_{\underline{x}} |\underline{x}\rangle |\underline{0}\rangle_{cp} \longrightarrow \sum_{\underline{x}} c_{\underline{x}} |\underline{x}\rangle |\underline{x}\rangle_{cp}$.

This operation of copying relative to the states in some basis avoids the limitations imposed by the no-cloning theorem [29] because an unknown state ψ is not being copied. The price paid is that copying relative to some basis introduces branching into the process in that correlations are introduced between the state of systems in the copy region and states of systems in R. This is the quantum mechanical equivalent of the classical case of making a calculation of a many-one function reversible by copying and storing the input [30].

In the above case carrying out the cleanup on the state $\sum_{\underline{x}} c_{\underline{x}} |\underline{x}\rangle |\underline{x}\rangle_{cp}$ corresponds to the operation $\sum_{\underline{x}} c_{\underline{x}} |\underline{x}\rangle |\underline{x}\rangle_{cp} \longrightarrow |\underline{y}\rangle \sum_{\underline{x}} c_{\underline{x}} |\underline{x}\rangle_{cp}$ where $|\underline{y}\rangle$ is the clean up state for the region R. The overall process is reversible as it can be described by the transformation $\sum_{\underline{x}} c_{\underline{x}} |\underline{x}\rangle |\underline{0}\rangle_{cp} \longrightarrow |\underline{y}\rangle \sum_{\underline{x}} c_{\underline{x}} |\underline{x}\rangle_{cp}$. If the final state of the quantum robot depends on the initial state of the systems in region R, then correlations remain and the overall transformation corresponding to carrying out the cleanup task is given by $\sum_{\underline{x}} c_{\underline{x}} |\underline{x}\rangle |\underline{0}\rangle_{cp} \theta_i \longrightarrow |\underline{y}\rangle \sum_{\underline{x}} c_{\underline{x}} |\underline{x}\rangle_{cp} \theta_{\underline{x}}$. Here θ_i and $\theta_{\underline{x}}$ are the initial and final states of the quantum robot.

Each task is considered here to consist of a sequence of computation and action phases. The purpose of each computation phase is to determine what action the quantum robot should take in the following action phase. Information on the state of local environmental systems may also be recorded. The input to the computation includes the local state of the environment and any other pertinent information, such as the output of the previous computation phase. During a computation phase the robot does not move. Changes in the state of the environment are limited to those resulting from observation states of an on board ancillary system, the output system (o) which is also changed. These states determine the action taken following completion of the computation.

During each action phase the the environment state in the neighborhood of the quantum robot is changed or the quantum robot moves on the lattice. Neither or both of these can also occur. Depending on the model used, each action phase can consist of one step or several steps. Here one step consists of the robot moving to at most an adjacent lattice site, or the local environment state changing, or both. During an action phase the state of the (o) system, which determines the action to be carried out, and the state of the on board quantum computer, is not changed. Also the quantum robot may or may not observe the local environment. Examples of actions that do not and do require observations are "rotate the qubit (as a spin system) by an angle ϕ" and "rotate the qubit by an angle ϕ only if it is in state $|0\rangle$. If the qubit is in state $|1\rangle$ move to an adjacent site."

The description of tasks carried out by quantum robots requires the use of completion or halting flags to determine when individual action and computation phases are completed as well as when the overall task is completed. Such flags are necessary if the overall quantum robot plus environment dynamics is described by a Hamiltonian because the unitarity of e^{-iHt} requires that system motion occurs somewhere even after the task is completed.

Note that there are many examples of tasks that never halt. Nonhalting of tasks can arise from several sources. The task may consist of a nonterminating sequence of computation and action phases. Or either a computation of an action phase may never halt. An example of an action that is multistep, does not halt, and requires local environment interactions at each step is "search along a path on a space lattice until a particle is found" where the path contains no particles at all.

As is well known, there are many ways to define local interactions. In the interests of simplicity, a very local delta function interaction will be assumed in the following. That is the local environment for the quantum robot is limited to the environment at the quantum robot location. The same assumption is made for the head moving on the qubit lattice in the model used for the on board quantum computer.

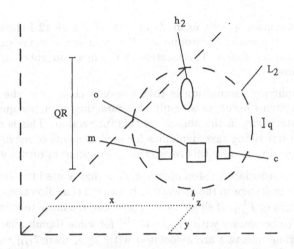

Fig. 12.1. A Schematic Model of a Quantum Robot and its Environment. The environment is a 3-D space lattice containing various types of quantum systems (not shown). The quantum robot QR consists of an on board Quantum Turing machine, finite state memory (m) and output (o) systems, and a control qubit (c). The on board QTM consists of a finite closed lattice $\mathcal{L}_2$ of qubits and a finite state head h_2 that moves on $\mathcal{L}_2$. The location (q) of a marker qubit is shown. The position $\underline{x} = x, y, z$ of the quantum robot h_1 on the environment lattice is shown by an arrow.

12.3   Models of Quantum Robots plus Environments

Here a model of quantum robots plus environments is described that illustrates the above material. To keep things simple the model will be limited to a description of information bearing degrees of freedom only. The relevance of this for the development of quantum computers has been noted by Landauer [31]. Also space will be considered to be discrete rather than continuous.

As noted quantum robots (QR)s consist of an on board quantum computer and ancillary systems. Here the on board quantum computer will be taken to be a quantum Turing machine consisting of a finite state head moving on a closed circular lattice $\mathcal{L}_2$ of $N+1$ qubits. The states of N qubits are used for computation and one qubit, taken to be ternary, is used for a marker. Ancillary systems present are an output system (o), a memory system (m), and a control qubit (c). Both (o) and (m) are described by quantum states in finite dimensional Hilbert spaces.

Simplifying assumptions for the environment include the use of discrete instead of continuous space. As a result environments of quantum robots consist of 1,2,3-D space lattices containing arbitrary numbers of different types of systems. Simple examples of environments consist of a 1-D lattice of qubits (which is a quantum register) and a 1-D lattice containing just one spinless particle. The latter example will be used to discuss a specific example of a task. Figure 12.1 shows a quantum robot in a 3-D space lattice environment. Environment systems external to the quantum robot are not shown. The location of the quantum robot in the lattice is shown by an arrow.

Another simplifying assumption is that the only changes in the states of the environmental systems occur as a result of interacting with the quantum robot. The states are stationary in the absence of this interaction. This is done to avoid complications in describing task dynamics for environments of moving interacting systems. It is hoped to remove this restrictive assumption in future work.

To each task is associated a step operator T_{QR} that is used to describe the task dynamics. Single task steps in the forward or backward time directions are described respectively by T_{QR} or $T_{QR}^{\dagger}$. If single task steps occur in a finite time interval t, then T_{QR} is required to be unitary with $T_{QR} = e^{-iHt}$ for some Hamiltonian H [6, 7, 32]. If infinitesimal time intervals are associated with T_{QR}, then T_{QR} can be used to directly construct a Hamiltonian according to [20]:

$$H = K(2 - T_{QR} - T_{QR}^{\dagger}) \tag{12.1}$$

where K is an arbitrary constant. In this model which has been used elsewhere [5, 8], T_{QR} need not be unitary or even normal ($T_{QR}T_{QR}^{\dagger} \neq T_{QR}^{\dagger}T_{QR}$ is possible).

Since each task consists of a sequence of computation and action phases, T_{QR} can be written as a sum of computation and action phase operators

$$T_{QR} = T_c + T_a \tag{12.2}$$

where T_a and T_c describe respectively single steps of action and computation phases of the quantum robot.

A computation phase (T_c active) accepts as input the states of (o) and (m) and the local environment. The computation, which is in general multistep, determines new states of (o) and (m) as output. These states determine the action to be carried out in the next action phase. During a computation phase there is no change in the environment state (other than that resulting from local observation) or the location of the quantum robot.

An action phase (T_a active) accepts as input the states of (o) and possibly (m). Actions include motion of the quantum robot and local changes of the environment state. They may be single step or multistep and may or may not require local observation of the environment. The states of (o), (m), and the on board quantum computer are not changed. An example of an action that does not require observation is "move one site in the +y direction". An example requiring observation is "if spin 1/2 system is at the QR location rotate spin by θ, else move 1 site in -z direction".

The function of the control qubit (c) is to regulate which phase type is active. In particular T_c or T_a is active if (c) is in state $|0\rangle$ or $|1\rangle$. The last step, or iteration of T_c or T_a, of the computation or action phase includes the respective change $|0\rangle_c \to |1\rangle_c$ or $|1\rangle_c \to |0\rangle_c$.

The conditions that T_c and T_a must satisfy can be expressed in terms of properties of these operators relative to a reference basis for the quantum robot and environment. To this end let the reference basis be given by

$$B_R = \{|m,k,\underline{t}\rangle|\ell_1\rangle_o|\ell_2\rangle_m|i\rangle_c|\underline{x}\rangle_{QR}|E\rangle\}.$$

Here $|m,k,\underline{t}\rangle$ denotes the state of the on board quantum Turing machine with $|m\rangle$ and $|k\rangle$ the respective internal state and $\mathcal{L}_2$ location of the head h_2, $|\underline{t}\rangle = \otimes_{j=1}^{N+1}|\underline{t}_j\rangle$ is the state of the qubits on $\mathcal{L}_2$ with $\underline{t}_j = 0,1$ for N qubits and $\underline{t}_j = 0,1,2$ for the marker qubit. The state $|\underline{x}\rangle_{QR} = |x,y,z\rangle_{QR}$ gives the lattice site location of the quantum robot, denoted by the arrow in Figure 12.1. The state $|E\rangle$ denotes a basis state in some basis B_E of environment states.

In the following all environmental observations carried out by T_c are assumed to commute, with B_E a common eigenbasis for the observations. In this case the action of T_c depends on but does not change the states $|E\rangle$. This, along with the requirement that T_c not change the QR location, gives

$$T_c = \sum_{\underline{x},E} P_{\underline{x},E} T_c P_{\underline{x},E} P_0^c \qquad (12.3)$$

where $P_{\underline{x},E} = |\underline{x},E\rangle\langle\underline{x},E|$ is the projection operator for the QR at site $\underline{x}$ and P_0^c is the projection operator for the control system in state $|0\rangle$. This equation shows that T_c is diagonal in states $|\underline{x},E\rangle$. The action of T_c on states that are linear superpositions of the basis states $|\underline{x},E\rangle$ will in general introduce entanglements between these states and states of the quantum computer. The presence of P_0^c shows that T_c is inactive if the control qubit is in state $|1\rangle$.

Locality conditions for $T_c^{\underline{x},E}$ acting on the component systems of the quantum robot are given by

$$\langle m',k',\underline{t}'|T_c^{\underline{x},E}|m,k,\underline{t}\rangle = 0 \text{ if } \left[\begin{array}{l} \underline{t}' \neq \underline{t} \text{ at sites } \neq k \\ |k'-k| > 1 \end{array} \right. \qquad (12.4)$$

These conditions express requirements that single step changes in the on board quantum computer limit the head h_2 motion to at most one $\mathcal{L}_2$ site in either direction and that changes in the lattice qubit state are limited to the qubit at the head location. Here $\langle \underline{x},E|T_c|\underline{x},E\rangle = T_c^{\underline{x},E}$ is the computation phase operator for the quantum robot at site $\underline{x}$ and the environment in state $|E\rangle$, Eq. 12.3.

An additional condition is that the operator $T_c^{\underline{x},E}$ depends on local environmental conditions only. That is if $|E\rangle$ and $|E'\rangle$ are two environment states that are the same at site $\underline{x}$ but may differ at other locations then

$$T_c^{\underline{x},E'} = T_c^{\underline{x},E} \tag{12.5}$$

To understand this condition better, suppose that the environment consists of n systems each with internal degrees of freedom described by states $|f\rangle$ with f taking on M possible values. For qubits $M = 2$ with $f = 0, 1$. If the systems are distinguishable, a convenient basis representation for the states $|E\rangle$ is $|E\rangle = \otimes_{\ell=1}^{n} |\underline{x_\ell}, f_\ell\rangle$. If the systems are considered as bosons or fermions, then these states must be given the appropriate symmetry.

If one system, the ℓth, is at site $\underline{x}$ and the rest are elsewhere, then $|E\rangle = |E_{\underline{x}}\rangle |E_{\neq \underline{x}}\rangle$ where $|E_{\underline{x}}\rangle = |\underline{x_\ell}, f_\ell\rangle$ and $|E_{\neq \underline{x}}\rangle = |\underline{x_1}, f_1\rangle, \cdots, |\underline{x_{\ell-1}}, f_{\ell-1}\rangle, |\underline{x_{\ell+1}}, f_{\ell+1}\rangle, \cdots, |\underline{x_n}, f_n\rangle$. If systems $\ell_1, \ell_2, \cdots, \ell_m$ with $m < n$ are at site $\underline{x}$ and the rest elsewhere, then $|E_{\underline{x}}\rangle = \otimes_{j=1}^{m} |\underline{x}, f_{\ell_j}\rangle_\ell$, and $E_{\neq \underline{x}}\rangle = \otimes_h |\underline{x_h}, f_h\rangle_h$ where $\otimes_h$ is taken over all $n - m$ systems for which $\underline{x_h} \neq \underline{x}$.

Using this notation the locality condition, Eq. 12.5 can be written as

$$\langle \underline{x}, E|T_c|\underline{x}, E\rangle = \langle E_{\underline{x}}|T_c|E_{\underline{x}}\rangle \tag{12.6}$$

Here $|E\rangle = |E_{\underline{x}}\rangle |E_{\neq \underline{x}}\rangle$ and $\langle \underline{x}|\underline{x}\rangle = \langle E_{\neq \underline{x}}|E_{\neq \underline{x}}\rangle = 1$ have been used. Note that $\langle \underline{x}, E|T_c|\underline{x}, E\rangle = T_c^0$ if no environmental systems are at site $\underline{x}$. This is a valid step operator that defines the computation phase steps if no environmental systems are at the location of the quantum robot. This will be made use of later in discussing a specific example.

Note that $\langle E_{\underline{x}}|T_c|E_{\underline{x}}\rangle$ can be very complex as it can depend on all the m variables $f_{\ell_1}, \cdots, f_{\ell_m}$ as well as on which of the n systems are at site $\underline{x}$. If the n environmental systems are all identical fermions, then the complexity is much reduced because for each value of f at most one system can be in the state $|\underline{x}, f\rangle$. If T_c interacts with at most one environmental system then $\langle E_{\underline{x}}|T_c|E_{\underline{x}}\rangle$ represents at most $M + 1$ distinct computation phase operators, one for each value of f and one if no systems are at the QR location. If T_c can interact with more than one environmental system then more combinations are posible. If the environmental systems are identical bosons, then T_c may have a more complex dependence on the local environment as an arbitrary number of systems in the same internal state can be present at the QR location and T_c may depend on the number of systems present.

Much of the above discussion also applies to the action phase operator T_a. This operator depends on but does not change the states of (o) and (m) relative to some basis. This condition can be expressed by an equation similar to Eq. 12.3:

$$T_a = \sum_{\ell_1, \ell_2} P_{\ell_1 \ell_2}^{o,m} T_a P_{\ell_1 \ell_2}^{o,m} P_1^c. \tag{12.7}$$

where $P_{\ell_1 \ell_2}^{o,m}$ is the projection operator for (o) in state $|\ell_1\rangle$ and (m) in state $|\ell_2\rangle$ and

P_1^c is the projection operator for (c) in state $|1\rangle$. This shows that T_a is diagonal in the states $|\ell_1\rangle_o|\ell_2\rangle_m$ and is inactive when (c) is in state $|0\rangle$.

The operator T_a also satisfies locality conditions similar to those for T_c, Eq. 12.4:

$$\langle \underline{x}'E'|T_a^{\ell_1\ell_2}|\underline{x}E\rangle = 0 \text{ if } \left[\begin{array}{l} |E'\rangle \neq |E\rangle \text{ at sites } \neq \underline{x},\underline{x}' \\ |\underline{x}' - \underline{x}| > 1 \end{array} \right. \tag{12.8}$$

where $T_a^{\ell_1\ell_2} = \langle \ell_1\ell_2|T_a|\ell_1\ell_2\rangle$. These conditions express the facts that during an action phase changes in the environment state are limited to sites $\underline{x}',\underline{x}$ and that the quantum robot moves at most one site in any direction. The states $|m,k,\underline{t}\rangle$ of the on board quantum computer are suppressed in the above as T_a is the identity operator in the supspace spanned by these states.

The requirement that the action is independent of states of environmental systems distant from the quantum robot can be expressed in a fashion similar to that for T_c:

$$\langle \underline{x}'E'|T_a^{\ell_1\ell_2}|\underline{x}E\rangle = \langle E'_{\neq \underline{x}',\underline{x}}|E_{\neq \underline{x}',\underline{x}}\rangle\langle \underline{x}'E'_{\underline{x}',\underline{x}}|T_a^{\ell_1\ell_2}|\underline{x}E_{\underline{x}',\underline{x}}\rangle \tag{12.9}$$

for all $\underline{x}',\underline{x}$ such that $|\underline{x}' - \underline{x}| \leq 1$. The states $|E_{\underline{x}',\underline{x}}\rangle$ and $|E_{\neq \underline{x}',\underline{x}}\rangle$ describe the environment at sites $\underline{x}',\underline{x}$ and elsewhere. The definition of these states is similar to that given earlier for $|E_{\underline{x}}\rangle$ and $|E_{\neq \underline{x}}\rangle$. Also $|E\rangle = |E_{\underline{x}',\underline{x}}\rangle|E_{\neq \underline{x}',\underline{x}}\rangle$ has been used. The matrix element $\langle E'_{\neq \underline{x}',\underline{x}}|E_{\neq \underline{x}',\underline{x}}\rangle = 1$ if and only if $|E'\rangle = |E\rangle$ at sites $\neq \underline{x},\underline{x}'$. Otherwise it equals 0.

The right-hand matrix element expresses the condition that one action phase step can change the environment at most at the initial and final locations of the quantum robot. If environmental changes at other locations are allowed then Eqs. 12.8 and 12.9 must be changed accordingly.

Several additional aspects of the properties of T_a and T_c need to be noted. One is that to avoid complications, the need for history recording has not been discussed. Both the computation and action phases may need to record some history. For example when T_c is active, the change $|\ell_2\rangle_o|\ell_1\rangle_m \longrightarrow |\ell_2'\rangle_o|\ell_1'\rangle_m$ requires history recording if the change is not reversible. Where records are stored (on board the quantum computer or in the environment) depends on the model. Also the task carried out by the quantum robot may not be reversible unless the initial environment is copied or recovered.

Initial and final states for the starting and completion of tasks may be needed. For example at the outset the memory, output, and control systems might be in the state $|0\rangle_m|\ell_i\rangle_o|0\rangle_c$ and the environment would be in some suitable initial state. The process begins with the on board quantum computer active.

Completion of a task could be described by designating one or more states $|\ell_f\rangle$ as final output states and arranging matters so that motion of some type occurs that does not destroy the relevant parts of the final task state. This ballast motion can occur on board the quantum computer or consist of motion of the quantum computer or some other system along a path in the environment without changing the environmental state. If the ballast motion occurs on board the quantum computer and it is described by states in a finite dimensional Hilbert space, the stability of the final task state lasts for a finite time only before the task is undone. Because the evolution operator e^{-iHT} is unitary, continued motion of some type is necessary.

The conditions given above for T_c and T_a are sufficiently general to allow for branching tasks with states describing entangled activities. For example during a computation phase T_c can take (o) and (m) states $|\ell_1, \ell_2\rangle$ into linear superpositions $\sum_{\ell_1 \ell_2} c_{\ell_1 \ell_2} |\ell_1, \ell_2\rangle$. Similarly the action of T_a can take environment and QR position states $|\underline{x}, E\rangle$ into linear superpositions $\sum_{\underline{x}' E'} c_{\underline{x}' E'} |\underline{x}' E'\rangle$. In this case the sum is limited to values of $\underline{x}' E'$ that satisfy Eq. 12.8. Note that Eq. 12.7 is satisfied separately by each branch.

Additional branching is possible if the action of T_c or T_a takes control qubit states into linear sums of $|0\rangle$ and $|1\rangle$. This allows for entanglements of action and computation phases.

12.4   A Specific Task in a Simple Environment

Here a specific task in a very simple environment will be considered to illustrate some aspects of the models discussed above. The environment consists of one spinless particle (p) on a 1-D lattice. In this case a convenient environmental basis is given by the states $|x\rangle$ that denote the position of the particle on the lattice.

The task considered is "search to the right for the particle. If it is found, bring it back to the initial location of the quantum robot." This task consists of stepwise motion of the QR to the right, examining each successive site for the particle. If the particle is found, the QR returns to the initial location with the particle. It is clear that in order to carry out this task reversibly, the on board quantum computer must keep track of the number of sites searched. Reversibility also requires permanent recording of the distance between the QR and the particle, if found.

The overall quantum robot plus environment state transformation resulting from carrying out the task can be represented as $|j\rangle_{QR} \theta(i)|x\rangle_p \longrightarrow |j\rangle_{QR} \theta(x-j)|j\rangle_p$ provided the particle is found. Here $|j\rangle_{QR}|x\rangle_p$ denote the respective lattice positions of the quantum robot and the particle, and $\theta(i)$ denotes the initial state of internal degrees of freedom of the on board quantum computer. The state $\theta(x-j)$ is the final state of the quantum computer with the distance from the quantum robot to the particle recorded in the memory.

If the initial state is a linear superposition of QR and (p) position states the

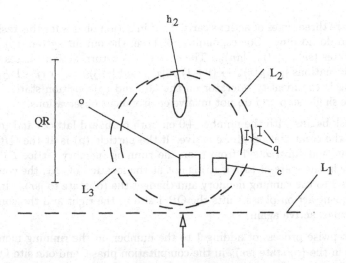

Fig. 12.2. A Schematic Model of a Quantum Robot for the Specific Task on a 1-D Environment Space Lattice. The particle (p) is not shown. The other systems are as in Figure 12.1 except that the (m) systems is expanded into an $N+2$ qubit lattice $\mathcal{L}_3$. The position of the quantum robot on the environment lattice is shown with an arrow.

overall task transformation is given by

$$\Psi_i = \sum_{j,x} c_{j,x}|j\rangle_{QR}|x\rangle_p \theta(i) \Rightarrow\Rightarrow \sum_{j,x}' c_{j,x}|j\rangle_{QR}|j\rangle_p \theta(x-j) + \psi_{nf} \qquad (12.10)$$

The prime on the sum means that it is limited to values of $x - j$ such that $0 \leq x - j \leq 2^N - 1$. For these values the QR will find the particle. What happens if $x-j$ is outside this range (the particle is not found) depends on model assumptions. The state ψ_{nf} represents the the task transformation if the particle is not found. The overall task transformation is reversible provided the states $\theta(d)$ are pairwise orthogonal for different values of d and are orthogonal to the initial state $\theta(i)$.

For carrying out this task the $\mathcal{L}_2$ qubit lattice of the on board quantum Turing machine contains $N + 2$ qubits: N qubits are used for numbers $0, 1, \cdots, 2^N - 1$, one qubit, which is ternary, is a marker, and the remaining qubit adjacent to the marker denotes the sign of the number ($|1\rangle \sim +, |0\rangle \sim -$). This lattice will be used as a short term memory to keep a running count of the number of sites the QR moves at each step.

The memory system (m) is expanded to be another $N + 2$ qubit lattice $\mathcal{L}_3$ like $\mathcal{L}_2$. It is used to record permanently the distance $x - j$ between the initial location of the QR and (p). It corresponds to $\theta(x - j)$ in Eq. 12.10. Figure 12.2 shows the setup on a 1-D lattice environment.

There are three types of actions carried out in action phases for this task: search, return, and do nothing. Corresponding to these, the output system (o) has three internal states $|sr\rangle_o$, $|rt\rangle_o$, $|dn\rangle_o$. The search and return action phases carry out the transformations $|j\rangle_{QR}|x\rangle_p \to |j+1\rangle_{QR}|x\rangle_p$ and $|j\rangle_{QR}|j\rangle_p \to |j-1\rangle_{QR}|j-1\rangle_p$. Do nothing is the identity operator on the QR and (p) position states. All these actions are single step and do not involve environment observations.

The task begins with the number $+0$ on both on board lattices and (o) in state $|dn\rangle_o$ and the computation phase active. If the particle (p) is at the QR location, the computation subtracts 1 from 0 on the running memory lattice $\mathcal{L}_2$ and does not change in the state of (o). If (p) is not at the location of QR, the computation phase adds 1 to the running memory and changes the (o) state to $|sr\rangle_o$. In this case the subsequent action phase shifts the QR 1 site to the right and the computation phase becomes active again.

This stepwise process of adding 1 to the number on the running memory with no change in the (o) state $|sr\rangle_o$ in the computation phase, and one site QR motion in the action phase continues until (p) is located. At this point the computation phase copies the number from running memory to the permanent memory qubit lattice $\mathcal{L}_3$, subtracts 1 from the running memory, and changes the (o) state to $|rt\rangle_o$. The next action phase moves both the QR and (p) back one lattice site.

This process continues until the number 0 appears on the running memory as part of the input to a computation phase. This computation subtracts 1 from the running memory and changes the state of (o) to $|dn\rangle_o$. At this point the task is completed and the ballast phase begins. In the model described here ballast phase motion consists of repeated subtraction of 1 from the running memory with do nothing action phases.

The task dynamics described above is shown schematically in Figure 12.3 as a decision tree. The round circles including "sr", "rt", and "dn" denote action phases. The square boxes between successive action phases, denote memory system states (d = running memory and st = permanent memory), and questions with answers based on local environmental states. The collection of boxes and arrows between successive actions shows what is done during each computation phase. The left hand column shows the dynamics during the search part of the task. The central column, with horizontal arrows only, shows changes made in memory states when the particle (p) is found, and the righthand column shows the dynamics during the return part of the task. The righthand row at the top shows progress during the ballast part of the task. The far righthand zero and that at the end of the search phase are explained below.

The ballast motion continues until the number $-(2^N - 1)$ appears on the running memory. Here the model step operator T_c is defined so that it annihilates the system state when it becomes active provided this number is in the running memory. Since the overall evolution is unitary with the Hamiltonian of Eq. 12.1, the effect of the

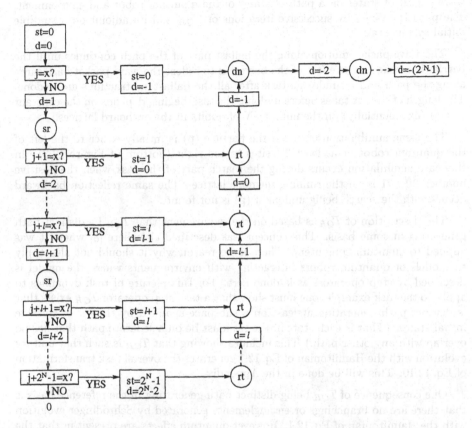

Fig. 12.3. Decision Tree for the Example Task. Task process motion is indicated by the arrows. Circles represent action phases. Square boxes show relevant states of systems. Permanent storage and running memory are shown respectively by "st" and "d". The boxes between adjacent action phase circles show what occurs during a computation phase. The lefthand column shows task progress during the first search part. The center column with horizontal arrows shows what happens in a computation phase when (p) is first located. The righthand column shows task progress during the return part. The ballast activities that occur when the task is complete are shown in the upper right.

annihilation is to provide an infinite reflecting barrier at the quantum robot + environment path state where the annihilation occurs. The progress of the task under the evolution operator e^{-iHt} can be considered as motion of an overall system wave packet of states on a path of states of the quantum robot and environment. The path is defined by successive iterations or T_{QR} and its adjoint on a suitable initial system state.

The wave packet motion along the ballast part of the path continues until the barrier is reached. The effect of the barrier is to reflect the wave packet backwards along the path and to undo the task after all the ballast subtractions are undone. The length of time it takes before undoing the task begins, depends on the constant K and (exponentially) on the number N of qubits in the on board lattices.

The same annihilation occurs if the particle (p) is intially either to the left of the quantum robot or at least 2^N sites to the right (bottom of Figure 12.3). In this case annihilation occurs during the search part of the task when the positive number $2^N - 1$ is on the running memory lattice. The same reflection backward occurs with the search being undone if (p) is not found.

The description of T_{QR} is based on the requirement that T_{QR} be distinct path generating in some basis. This concept was described elsewhere [8] where it was applied to quantum computers. There is no reason why it should not also apply to models of quantum robots interacting with environments where the model is described by step operators as is done here. For this picture of task dynamics to apply to the task example, one must show that a task step operator T_{QR} exists that is distinct path generating at least on a subspace spanned by the set of suitable initial states. (That is each state in the set must be on a separate path that has no overlap with any other path.) This includes showing that T_{QR} is such that unitary evolution with the Hamiltonian of Eq. 12.1 generates the overall task transformation of Eq. 12.10. This will be done in the Appendix.

One consequence of T_{QR} being distinct path generating in the reference basis is that there are no branchings or entanglements generated by Schrödinger evolution with the Hamiltonian of Eq. 12.1. However quantum effects are present in that the motion along the task state paths corresponds to motion of a quantum system on a 1-D lattice on which infinite (or finite) reflecting barriers may be present [8]. Also any linear superposition present in the initial state will be preserved. This is shown by Eq. 12.10.

12.5   Discussion

It must be strongly emphasized that a main reason for studying quantum robots and their interactions with environments of quantum systems is that it provides a well defined platform for investigation of many interesting questions. For example "What properties must a quantum system have so that one can conclude that it is aware of its environment, makes decisions, and has other properties of intelligence?" Answering such a question, even for models of quantum robots plus environments as defined here, seems difficult enough. Without any such model it seems impossible.

The fact that the only known examples of intelligent quantum systems are very complex and contain the order of 10^{23} degrees of freedom only emphasizes this point. A close study of simple systems introduced here may help to show exactly how such systems can be made more complex so that in some well defined limit (if a limit is needed) they become aware of their environment and intelligent. Without some such well defined platform it seems hopeless to try to answer such questions.

It is also worthwhile to consider the following speculative ideas. The close connection between quantum computers and quantum robots interacting with environments suggests that the class of all possible physical experiments may be amenable to characterization just as is done for the computable functions by the Church-Turing hypothesis [6, 33, 34]. That is there may be a similar hypothesis for the class of physical experiments.

The description of tasks carried out by quantum robots (Section 12.2) lends support to this idea in that there may be an equivalent Church-Turing hypothesis for the collection of all tasks that can be carried out. The earlier work that characterizes physical proceedures as collections of instructions [21, 35], or state preparation and observation proceedures as instruction booklets or programs for robots [22] also supports this idea. On the other hand much work needs to be done to give a precise characterization of physical experiments, if such is indeed possible.

Appendix

The requirements that the computation phase step operator T_c must satisfy in order to carry out the example can be given as a set of 9 conditions. The conditions, which follow are based on the task description and on the decision tree.

$$|\underline{0}\rangle_{st}|\underline{0}\rangle_{ru}|dn\rangle_o|j\rangle_{QR}|x\rangle_p \overset{(1)}{\Longrightarrow} |\underline{0}\rangle_{st}|\underline{-1}\rangle_{ru}|dn\rangle_o|j\rangle_{QR}|x\rangle_p \text{ if } j = x$$

$$\overset{(2)}{\Longrightarrow} |\underline{0}\rangle_{st}|\underline{1}\rangle_{ru}|sr\rangle_o|j\rangle_{QR}|x\rangle_p \text{ if } j \neq x$$

$$|\underline{0}\rangle_{st}|\underline{d}\rangle_{ru}|sr\rangle_o|j\rangle_{QR}|x\rangle_p \overset{(3)}{\Longrightarrow} |\underline{0}\rangle_{st}|\underline{d+1}\rangle_{ru}|sr\rangle_o|j\rangle_{QR}|x\rangle_p$$

$$\text{if } j \neq x \text{ and } d < 2^N - 1$$

$$\overset{(4)}{\Longrightarrow} 0 \text{ if } j \neq x \text{ and } d = 2^N - 1$$

$$\overset{(5)}{\Longrightarrow} |\underline{d}\rangle_{st}|\underline{d-1}\rangle_{ru}|rt\rangle_o|j\rangle_{QR}|x\rangle_p \text{ if } j = x$$

$$|\underline{d}\rangle_{st}|\underline{\ell}\rangle_{ru}|rt\rangle_o|j\rangle_{QR}|j\rangle_p \overset{(6)}{\Longrightarrow} |\underline{d}\rangle_{st}|\underline{\ell-1}\rangle_{ru}|rt\rangle_o|j\rangle_{QR}|j\rangle_p \text{ if } \ell > 0$$

$$\overset{(7)}{\Longrightarrow} |\underline{d}\rangle_{st}|\underline{-1}\rangle_{ru}|dn\rangle_o|j\rangle_{QR}|j\rangle_p \text{ if } \ell = 0$$

$$|\underline{d}\rangle_{st}|\underline{-\ell}\rangle_{ru}|dn\rangle_o|j\rangle_{QR}|j\rangle_p \overset{(8)}{\Longrightarrow} |\underline{d}\rangle_{st}|\underline{-\ell-1}\rangle_{ru}|dn\rangle_o|j\rangle_{QR}|j\rangle_p \text{ if } \ell < 2^N - 1$$

$$\overset{(9)}{\Longrightarrow} 0 \text{ if } \ell = 2^N - 1$$

In these conditions only the relevant component system states are listed. Also the control qubit transformation $|0\rangle_c \to |1\rangle_c$ common to all the conditions has been left out. The state subscripts "st" and "ru" refer to the permanent storage and running memory qubit lattices, $\mathcal{L}_3$ and $\mathcal{L}_2$ respectively. The underlined number variables denote the qubit string equivalent of the underlined number. Here for both on board lattices increasing significance of the bits in the strings is in the counterclockwise direction from the marker qubit (Figure 12.2). The state of the most significant $N + 1$st qubit (adjacent to the marker in the clockwise direction) is the sign qubit $|1\rangle \sim +$ and $|0\rangle \sim -$. The condition numbers above the arrows are present for discussion purposes only.

Conditions 2 and 3 show the overall state changes during each computation phase in the search part of the task (the left hand column of Figure 12.3). Condition 4 applies in case the particle is not found. Conditions 1 and 5 apply for a computation phase when the particle is first located at the quantum robot site during the search. Conditions 6 and 7 apply to the return part of the task (righthand column of Figure 12.3), and conditions 8 and 9 describe the computation phases during the ballast motion when the task is completed. Note that in condition 5 the value of d recorded in the storage lattice is the distance from (p) to the initial position of the quantum robot.

The arrow denotes that each computation phase consists of many steps or iterations of T_c. For example the locality conditions for the head h_2 motion on $\mathcal{L}_2$ and $\mathcal{L}_3$ mean that adding or subtracting 1 to the number recorded on the running memory lattice takes several steps. For example the change $211110 + 1 = 200001$ where 2 is the marker qubit and h_2 begins and ends at the marker takes at least 10 steps.

Conditions 3 and 4 show a different computation phase for the same input state depending on whether $d < 2^N - 1$ or $d = 2^N - 1$. This choice does not require a complete search of $\mathcal{L}_2$ each time a search computation is carried out to determine which alternative applies. This follows from the fact that h_2 in moving past a string of 1s to find a 0 will encounter the marker before a 0 if and only if $2^N - 1$ is on the "ru" lattice.

A similar situation applies for the choice between conditions 8 and 9 except that here in carrying out the subtraction h_2 must each time move 1 qubit beyond the first 0 encountered to see if the 0 is or is not adjacent to the marker. If not condition 8 applies, else condition 9 holds. The choice needed for conditions 6 and 7 is similar. Here h_2 moves 1 qubit beyond the first 1 encountered to see if it is adjacent to the marker. If not the 1 is changed to a 0 and all the prior passed 0s change to 1s. If the 1 is adjacent to the marker it is changed to a 0 and the qubit on the other side of the marker is changed to a 1.

The conditions satisfied by the action phase operator T_a are much simpler. They are

$$|sr\rangle_o |j\rangle_{QR} |x\rangle_p \quad \rightarrow \quad |sr\rangle_o |j+1\rangle_{QR} |x\rangle_p$$
$$|rt\rangle_o |j\rangle_{QR} |j\rangle_p \quad \rightarrow \quad |rt\rangle_o |j-1\rangle_{QR} |j-1\rangle_p$$
$$|dn\rangle_o |j\rangle_{QR} |x\rangle_p \quad \rightarrow \quad |dn\rangle_o |j\rangle_{QR} |x\rangle_p$$

As was the case for the conditions for T_c only the relevant changes are included. the control qubit change $|1\rangle_c \rightarrow |0\rangle_c$ common to each condition was not included. Also T_a is diagonal in the states of the on board quantum Turing machine. The conditions show explicitly how T_a depends on but does not change the (o) system states. Each of these transformations corresponds to a single step or iteration of T_a.

The above conditions and discussion are sufficient to see that $T_{QR} = T_c + T_a$ is distinct path generating, at least on the subspace spanned by the basis states that are relevant to the task. The actions of T_{QR} on other basis states is not specified because they are not relevant to the task. The reason is that there will never be any overlap between these states and those that occur during carrying out of the task. Examples of these other states are those in which (o) is in state $|dn\rangle$ and a positive number is in the running memory.

Acknowledgements

This work is supported by the U.S. Department of Energy, Nuclear Physics Division, under contract W-31-109-ENG-38.

References

[1] V. Vedral, A. Barenco, and A. Ekert, *Phys. Rev. A* **54** 147 (1996).

[2] A. Barenco, C. H. Bennett, R. Cleve, D. P. DiVincenzo, N. Margolus, P. Shor, T. Sleator, J. A. Smolin, and H. Weinfurter, *Phys. Rev. A* **52** 3457 (1995).

[3] A. Ekert and R. Jozsa, *Revs. Modern Phys.* **68** 733 (1996).

[4] P. Benioff, *Jour. Stat. Phys.* **22** 563 (1980); Phys. Rev. Letters **48** 1581 (1982).

[5] P. Benioff, *Ann. NY Acad. Sci.* **480** 475 (1986).

[6] D. Deutsch, *Proc. Roy. Soc.* (London) A **400** 997 (1985).

[7] D. Deutsch, *Proc. Roy. Soc.* (London) A **425** 73 (1989).

[8] P. Benioff, *Phys. Rev. A* **54** 1106 (1996); *Phys. Rev. Letters* **78** 590 (1997); *Fortschrifte der Physik*, **46** 423 (1998).

[9] P. Shor, in *Proceedings of the 35th Annual Symposium on the Foundations of Computer Science*, edited by S. Goldwasser, p. 124 (IEEE Computer Society, Los Alamitos, CA 1994); Siam Jour. Comput. **26**, 1481 (1997).

[10] L.K.Grover, in *Proceedings of 28th Annual ACM Symposium on Theory of Computing*, p. 212 (ACM Press New York 1996); Phys. Rev. Letters, **78** 325 (1997); G. Brassard, Science **275** 627 (1997).

[11] R. P. Feynman, *International Jour. of Theoret. Phys.* **21** 467 (1982).

[12] R. Landauer, *Physics Letters A* **217** 188 (1996); *Phil. Trans. R. Soc. Lond. A* **353** 367 (1995); *Physics Today* bf 44 23 (1991) May; *IEEE Transactions on Electron Devices*, **43** 1637 (1996).

[13] A. Peres and W. Zurek, *Amer. Jour. Phys.* **50** 807 (1982).

[14] D. Albert, *Physics Letters* **98A** 249 (1983); *Philosophy of science* **54** 577 (1987); *The Quantum Mechanics of Self-measurement* in *Complexity, Entropy and the Physics of Information*, proceedings of the 1988 workshop in Santa Fe, New Mexico, 1989, W. Zurek, Ed. (Addison Wesely Publishing Co. 1990).

[15] T. Breuer, *Philos. Science* **62** 197 (1995).

[16] A. Peres, *Phys. Letters*, **A101** 249 (1984).

[17] R. Penrose, *The Emperor's New Mind* (Penguin Books, New York, 1991).

[18] H. P. Stapp, *Mind, Matter, and Quantum Mechanics* (Springer Verlag, Berlin 1993).

[19] E. Squires, *Conscious Mind in the Physical World* (IOP Publishing, Bristol England, 1990).

[20] R. P. Feynman, *Optics News* **11** 11 (1985); reprinted in *Foundations of Physics* **16** 507 (1986).

[21] C. H. Randall and D. J. Foulis, *Amer. Math. Monthly*, **77** 363 (1970); D. J. Foulis and C. H. Randall, *Jour. Math. Phys.*, **13** 1667 (1972).

[22] P. Benioff and H. Ekstein, *Phys. Rev. D* **15** 3563, (1977); *Nuovo Cim.* **40 B** 9 (1977).

[23] C. D. Helon and G. J. Milburn, *Quantum Measurements with a Quantum Computer*, Los Alamos Archives preprint, quant-ph/9705014.

[24] S. Lloyd, *Phys. Rev. A* **56** 3374 (1997).

[25] P. W. Shor, *Phys. Rev. A* **52** R2493 (1995); R. LaFlamme, C. Miquel, J.P. Paz, and W. H. Zurek, *Phys. Rev. Letters* **77** 198 (1996); E. Knill and R. Laflamme, *Phys. Rev. A* **55**, 900 (1997); D. P. DiVincenzo and P. W. Shor, *Phys. Rev. Letters* **77** 3260 (1996).

[26] W. H. Zurek, *Phys. Rev. D* **24**, 1516 (1981); **26**, 1862 (1982); J. R. Anglin and W. H. Zurek, *Phys. Rev. D* **53**, 7327 (1996).

[27] E. Joos and H. D. Zeh, *Z. Phys. B* **59**, 223 (1985).

[28] A. Yao, in *Proceedings of the 34th Annual Symposium on Foundations of Computer Science*, pp. 352-361 (IEEE Computer Society, Los Alamitos, CA, 1993).

[29] W. K. Wootters and W. H. Zurek, *Nature* **299**, 802 (1982); H. P. Yuen, *Physics Letters* **113A**, 405 (1986); H. Barnum, C. M. Caves, C. A. Fuchs, R. Jozsa, and B. Schumacher, *Phys. Rev. Letters* **76** 2818 (1996); L. M. Duan and G. C. Duo, Los Alamos Archives, preprint no. quant-ph/9705018.

[30] C. H. Bennett, *IBM Jour. Res. Dev.* bf 17, 525 (1973).

[31] R. Landauer, *Zig-Zag Path to Understanding* in Proceedings of the Workshop on Physics and Computation, PhysComp 94, Los Alamitos (IEEE Computer Society Press, 1994).

[32] E. Bernstein and U. Vazirani, in *Proceedings of the 1993 ACM Symposium on Theory of Computing*, pp 1-20 (ACM, New York, 1993).

[33] A. Church, *Am. Jour. Math.* **58**,345 (1936); A. M. Turing, *Proc. Lond. Math. Soc. 2* **42**, 230 (1936).

[34] M. A. Nielsen, *Phys. Rev. Letters*, **79** 2915 (1997); K Svozil, *The Church-Turing thesis as a Guiding Principle for Physics* Los Alamos Archives preprint quant-ph/9710052.

[35] H. Ekstein, *Phys. Rev.* **153**, 1397 (1967); **184**, 1315 (1969).

13

QUANTUM INFORMATION THEORY

Charles H. Bennett

Abstract

The new theory of coherent transmission and transformation of information in the form of intact quantum states represents a major extension and generalization of classical information and computation theory, and has a number of distinctive features, including dramatically faster algorithms for certain problems, more complex kinds of channel capacity and cryptography, and a second, quantifiable kind of information—entanglement—which interacts with classical information in phenomena such as quantum teleportation.

13.1 Introduction

Quantum mechanics has many paradoxes, but perhaps the greatest is the qualitative disparity between quantum laws and the macroscopic world. For most of the 20th century, physicists and chemists have used quantum mechanics to build an edifice (Fig. 13.1) of quantitative explanation and prediction covering almost all features of our everyday world: The rigidity of stone, the transparency of glass, the luminosity of the sun, among many other things. But if we look at the foundations of this structures we find a set of laws (Figs. 13.2,13.3) that, like the Ten Commandments, are marvelously concise but seem to bear almost no relation the way things work in everyday life. In order to see quantum effects like superposition or entanglement face to face, rather than through a hard-to-understand chain of indirection, one must go to a textbook or a laboratory. Aside from its intrinsic scientific interest, the burgeoning field of quantum information processing may have an important cultural side effect of bringing the formerly arcane foundations of quantum mechanics into popular awareness, much as the computer revolution has done with the formerly arcane foundations of computer science, notions such as computational universality which are in essence understood by anyone who goes into software shop and asks "Do you have a version of this that will run on my Mac?"

The distinctive features of quantum data processing are sketched in Fig. 13.4. In classical computations we use bits to represent data, and if n bits are present we have 2^n possible states. In quantum computation the bits are replaced by *qubits*—quantum systems, such as polarized photons or spin-1/2 particles, with a 2-dimensional Hilbert space and capable of existing in a pair of orthogonal states identified with the Boolean values (for example $\leftrightarrow = |0\rangle$ and $\updownarrow = |1\rangle$) as well as all

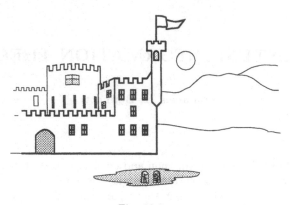

Fig. 13.1.

superpositions. A set of n qubits can exist in 2^n Boolean states and all possible superpositions, so that the state is represented by a vector, or more precisely a ray, in a 2^n dimensional Hilbert space. Just as any transformation of classical data can be expressed as a sequence of simple gates (e.g. NOT and AND) acting on the bits one and two at a time, any transformation of quantum data can be expressed as an array of two-qubit controlled-NOT gates (also called CNOT or XOR) and one-qubit unitary rotations. The central feature of quantum data processing is the superposition principle: If a quantum gate is fed a superposition of inputs, it yields the corresponding superposition of outputs. As in a classical computer logic diagram, gates are interconnected by "wires", representing whatever physical mechanism is use to coherently store or transmit a qubit from one gate operation to the next; but, unlike classical data, the data in a bundle of quantum wires is generally entangled—not expressible by separately specifying the state of each wire. A simple and important example of entanglement—an Einstein-Podolsky-Rosen (EPR) state of two qubits—arises as a necessary consequence of the superposition principle when a CNOT is applied to non-Boolean but unentangled data as in Fig. 13.4b (we often use shading to indicate entanglement between wires). Either of the entangled qubits alone appears completely random, yet together they are in a definite state.

Because quantum mechanics encompasses classical, classical computations may be viewed as a subset of quantum computations, as real numbers are a subset of complex numbers. A classical bit may, without loss of generality, be viewed as a qubit promised to be in one of the states $|0\rangle$ or $|1\rangle$, and a classical wire is a wire that conducts these two states reliably, but introduces a random disturbance if it is asked to conduct a superposition. A classical wire may be viewed as a kind of noisy quantum channel, in which the incoming qubit interacts via a CNOT with an ancillary qubit, which is then dumped into the environment (cf. Fig. 13.4c). The wire's environment, in other words, makes a quantum nondemolition measurement

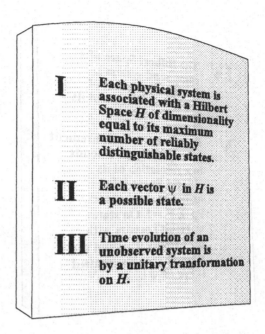

Fig. 13.2.

in the $|0\rangle, |1\rangle$ basis, on the data passing through. In logic diagrams we use thick lines to represent classical wires and thin lines to represent quantum wires.

13.2   Quantum Computing

Interest in quantum information processing has centered on quantum computing—the possibility of using quantum operations in the intermediate stages of a computation to greatly speed up the solution of certain classically hard problems (cf Fig. 13.5). I am tempted to say that, as a topic of fundamental scientific inquiry, quantum computing in this sense is so close to finished as not to be interesting any more. Of course there are some practical details to be worked out, like actually building a quantum computer, but the basic outlines are well understood: Some computations, such as integer factoring, can be sped up exponentially on a quantum computer [1], others, including many NP-type optimization and search problems, can be sped up quadratically [2, 3], while yet other problems cannot be sped up at all [4]. It is actually an exaggeration to declare quantum computation a dead topic, because a number of important theoretical questions remain, notably determining which other classically hard problems can be sped up by quantum computers. Prime candidates are graph isomorphism and computing the permanent of a matrix. Quantum computers have been shown capable of simulating the dynamics

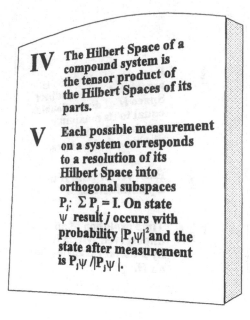

Fig. 13.3.

of a general quantum system given its Hamiltonian; for many systems no efficient classical algorithm for doing so is known.

Practically, we are still far from building a quantum computer that would realize these speedups in a useful way. The goal of doing so looked quite unrealistic until the development, over the last two or three years, of efficient quantum error-correction techniques [5], which in principle allow computers made of unreliable parts to do arbitrarily long quantum computations reliably, if the error and decoherence per elementary gate operation can be made less than some threshold, estimated to be around 10^{-3} to 10^{-6}. Quantum fault-tolerant computing may be viewed as a generalization of the fault-tolerant circuits that were developed in the early days of classical computing, but are scarcely needed nowadays owing to the high intrinsic reliability of today's hardware. In the quantum realm, by contrast, the error and decoherence rates of today's rudimentary quantum hardware is still several orders of magnitude too high for fault-tolerant techniques to take hold. Quantum computing is like controlled fusion—possible in principle, but maybe not practical for a long time to come.

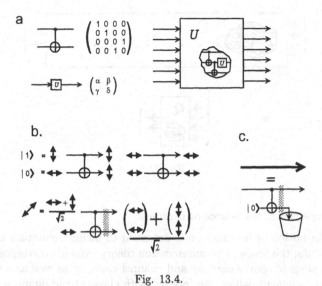

Fig. 13.4.

13.3 Transmission of Quantum Information

Besides quantum computing per se, quantum information processing encompasses a range of other topics which broadly generalize classical communication and information theory. It is here that I think the most exciting theoretical work remains to be done. This is the wild west, or internet, of quantum information science.

Problems in classical information transmission or communication involve several parties, e.g. a sender ("Alice") and a receiver ("Bob"), and in cryptographic scenarios an eavesdropper ("Eve"). The parties have some informational task they wish to accomplish, which may be viewed as a successful transition from an initial state to a corresponding final state of the entire apparatus, for example an initial state in which Alice holds a bit string x chosen from some source distribution X, and a final state in which Bob holds x. The protocol used to achieve this goal should work with certainty, or with high probability, and it should be economical of communication resources, such as number of uses of a noiseless or noisy classical channel. In adversarial settings such as cryptography, the goal of one subset of the parties should be achievable despite the efforts of another subset; for example, Bob should learn x while Eve is prevented from gaining any information about x.

Among the most important notions of classical information theory are source entropy and channel capacity and the related techniques of source and channel coding. These techniques make sources and channels *fungible,* in the sense that the number of channels asymptotically required to reliably communicate the output of a source depends only on the ratio of source entropy to channel capacity, and not

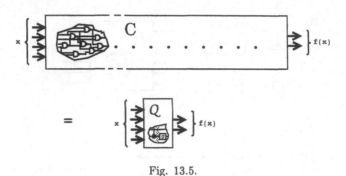

Fig. 13.5.

on other properties of the source or channel.

When the notion of information is extended to include quantum states as well as classical data, the scope of communication theory expands correspondingly, with quantum analogs of source entropy and channel capacity as well as a new fungible resource, entanglement, which can interact with classical and quantum information in a variety of ways.

If a quantum source emits states ψ_i with probability p_i, its von Neumann entropy, $S = -\text{Tr}\rho \log_2 \rho$, where $\rho = \sum_i p_i |\psi_i\rangle\langle\psi_i|$, determines the minimum asymp-

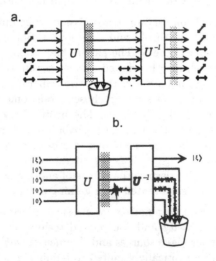

Fig. 13.6.

totic number of qubits to which its signals can be compressed by a quantum encoder and still faithfully recovered by a quantum decoder. This is the analog of classical data compression or source coding, by which redundant classical data is compressed and faithfully regenerated. But quantum data compression [6] differs in that it can be applied to non-orthogonal states (for example equiprobable horizontal and diagonal photons, as shown in Fig. 13.6a) which would be spoiled if one tried to compress them classically. Also, because the states are non-orthogonal, the encoder cannot retain a copy of them, or indeed any memory of them, if they are to be faithfully reconstructed at the receiving end. This difference is a manifestation of the quantum no-cloning principle, according to which cloning a general quantum state, ie the mapping $|\psi\rangle \otimes |0\rangle \rightarrow |\psi\rangle \otimes |\psi\rangle$, where ψ is an unknown state in a 2- or higher-dimensional Hilbert space, is non-physical and cannot be accomplished by any apparatus. By contrast the copying of classical data is easy, much to the consternation of software manufacturers. A quantum encoder is thus like a discreet telegrapher, who transmits other peoples' messages without remembering them.

Source coding removes redundancy so data can be sent more efficiently through a noiseless channel; error-correcting or channel coding, by contrast, introduces redundancy to enable data to withstand transmission through a noisy channel. The simplest classical error-correcting code is the threefold repetition code $0 \rightarrow 000$, $1 \rightarrow 111$, which permits the encoded bit to be faithfully recovered after up to one transmission error in the three-bit codeword. Analogous error-correcting codes exist for quantum data, but they require more redundancy because they need to protect not only Boolean states, but also arbitrary superpositions of them [7, 8, 10]. Thus the simplest single-error-correcting quantum code (Fig. 13.6b) encodes an arbitrary input qubit $|\xi\rangle$ into an entangled state of five qubits, in such a way that if any one is corrupted enroute, the decoder can funnel the effects of the error into the four ancillary qubits, while restoring the first qubit to its original state. By the same token, a noisy quantum channel's capacity Q for faithfully transmitting qubits (determined by the amount of redundancy quantum error-correcting codes require to achieve asymptotically perfect fidelity transmission in the limit of large blocksize), is generally less, and can never be greater, than its capacity C for transmitting classical bits. The inequality $C \geq Q$ holds for all channels because if a channel can faithfully transmit a general qubit, then it can certainly transmit the particular qubits $|0\rangle$ and $|1\rangle$.

Besides C and Q, quantum channels have a third capacity, Q_2, for transmitting qubits faithfully with the help of two-way classical communication between sender and receiver. This classically-assisted quantum capacity will be discussed later, after we describe another uniquely quantum form of communication, quantum teleportation.

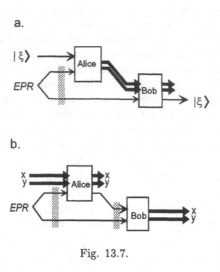

Fig. 13.7.

13.4   Applications of Quantum Entanglement

Two forms of quantum information transmission having no classical counterpart, but closely related to each other, are quantum teleportation [11] (Fig. 13.7a) and quantum superdense coding [12] (Fig. 13.7b). These involve an initial stage in which an EPR pair is shared between two parties, followed by a second stage in which this shared entanglement is used to achieve, respectively, transmission of a qubit via two classical bits, or transmission of two classical bits via one qubit. Quantum teleportation illustrates the fact that transmission of intact quantum states requires two qualitatively different resources, viz. a quantum resource that cannot be cloned, and a directed resource that cannot travel faster than light. In direct transmission of a qubit, these two functions are performed by the same particle. In teleportation the former function is provided by the shared EPR pair, the latter by the two classical bits. This situation may be summarized by saying that classical information theory involves one species of information, and one kind of noiseless communication primitive (transmission of a bit), whereas quantum information theory involves two species (classical information and entanglement), and three primitives (transmitting a bit, transmitting a qubit, and sharing an EPR pair) which are related through superdense coding and teleportation.

Aside from its avoidance of a direct quantum channel from Alice to Bob, teleportation is noteworthy in that it is irreversible: one EPR pair's worth of entanglement is consumed, and two random bits, which are utterly uncorrelated with the state being teleported, are generated and eventually must be discarded into the environment, typically as waste heat, in obedience to Landauer's principle. If a qubit is

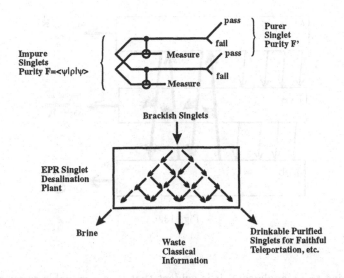

Fig. 13.8.

teleported from Alice to Bob, and then back to Alice, there will be zero net effect on the quantum system, but two EPR pairs of entanglement will have been used up and four bits worth of waste heat generated. Such cyclic but irreversible processes can also be found in classical thermodynamics, for example when an ideal gas freely expands into an evacuated chamber, and then is isothermally compressed back to its original volume.

Through teleportation, remote parties can use classical communication and shared EPR pairs to do any quantum operation they could have done had they been in the same location. For example, if Alice wants to bring one of her qubits into interaction with one of Bob's qubits, she can teleport it him, have him do the interaction, then have him teleport the post-interaction qubit back to her. Conversely, one may ask if there is a single primitive *interaction* to which all communication can be reduced. It is evident that all forms of communication, classical or quantum, can be implemented if we have the ability to do local operations and non-local CNOTs between Alice's and Bob's data. In view of this, it is reasonable to take as our primitive of interaction any Hamiltonian which generates a two-qubit gate equivalent to an Alice-Bob CNOT when integrated over time, for example

$$\mathcal{H} = \frac{\epsilon}{4} \, (1 - \sigma_z)_A \otimes (1 - \sigma_z)_B = \begin{pmatrix} 0 & 0 & 0 & 0 \\ 0 & 0 & 0 & 0 \\ 0 & 0 & 0 & 0 \\ 0 & 0 & 0 & \epsilon \end{pmatrix}, \qquad (13.1)$$

which does so when allowed to act for a time π/ϵ.

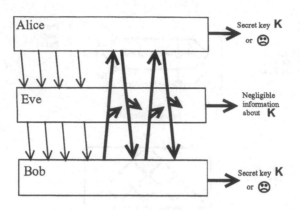

Fig. 13.9.

Returning to entanglement, it is evident that it is a valuable resource complementary in many respects to classical communication. For this reason, it would be good to have a way of measuring it, not only for maximally-entangled EPR states, but also for partly entangled pure states of a bipartite system such as $\cos\theta|00\rangle + \sin\theta|11\rangle$ and for mixed states. For pure states, a good measure is the "entropy of entanglement," defined as the von Neumann entropy of either the Alice or Bob subsystem taken alone. In the example given, this entropy is $H_2(\cos^2\theta) = -\cos^2\theta\log_2\cos^2\theta - \sin^2\theta\log_2\sin^2\theta$. Entropy of entanglement is a good entanglement measure for pure states because for any pure state Ψ of a bipartite system, it is asymptotically equal [13] both to the number of standard EPR states required to prepare one instance of Ψ, and the number of standard EPR pairs that can be prepared from one instance of Ψ, using local operations and classical communication. This fungibility of entanglement justifies the term *ebit* for the amount of entanglement in a maximally-entangled EPR state of two qubits, e.g. a singlet state $(|\uparrow\downarrow\rangle - |\downarrow\uparrow\rangle)/\sqrt{2}$ of two spin-1/2 particles.

For mixed states, the situation is more complicated [14], and it appears likely that for some mixed states, the *entanglement of formation*—the asymptotic number of pure ebits required to prepare an instance of the state by local operations and classical communication—may exceed the *distillable entanglement*—the asymptotic number of pure singlets that can be prepared from an instance of the state, using local operations and classical communication. For pure states both these measures reduce to the entropy of entanglement.

The possibility of distilling pure entanglement from mixed states gives rise to the classically-assisted quantum capacity Q_2 mentioned above. A noisy channel $\mathcal{N}$, if it is not too noisy, can be used for reliable quantum communication by the following indirect procedure [8, 9].

Transformation	Resource Produced		Resources Used
Cloning	$\lvert\psi\rangle \otimes \lvert 0\rangle \Rightarrow \lvert\psi\rangle \otimes \lvert\psi\rangle$		Impossible
Classical Copying	$\lvert x\rangle \otimes \lvert 0\rangle \Rightarrow \lvert x\rangle \otimes \lvert x\rangle$	$\leq$	1 Bit$^{\rightarrow}$
EPR sharing	1 Ebit	$\leq$	1 Qubit$^{\rightarrow}$
Use Qubit to send Bit	1 Bit$^{\rightarrow}$	$\leq$	1 Qubit$^{\rightarrow}$
Teleportation	1 Qubit$^{\rightarrow}$	$\leq$	1 Ebit + 2 Bits$^{\rightarrow}$
Superdense coding	2 Bits$^{\rightarrow}$	$\leq$	1 Ebit + 1 Qubit$^{\rightarrow}$
Quantum Source Coding	1 R-random Qubit$^{\rightarrow}$	$\preceq$	R Qubits$^{\rightarrow}$
Quantum Channel Coding	$Q(\mathcal{N})$ noiseless Qubits$^{\rightarrow}$	$\preceq$	1 use$^{\rightarrow}$ of channel $\mathcal{N}$
Entang. Concentration	E pure singlets	$\preceq$	1 E-entangled pure state
Entanglement Dilution	1 E-entangled pure state	$\preceq$	E singlets +Bits$^{\rightarrow}$
Class. Assisted Q Commun.	$Q_2(\mathcal{N})$ noiseless Qubits$^{\rightarrow}$	$\preceq$	1 use$^{\rightarrow}$ of $\mathcal{N}$ + Bits$^{\leftrightarrow}$
Quantum Key Distrib.	Shared Secret Key Bit,	$\preceq$	Eavesdropped Qubits$^{\rightarrow}$
	or Failure		+Bits$^{\leftrightarrow}$

Table 13.1. In the first line, the cloning of an unknown state ψ is impossible and cannot be achieved by use of any combination of resources. In the other lines, $\leq$ signifies an exact reducibility, ie the resource on the left can be exactly produced by use of the resources on the right. $\preceq$ signifies an asymptotic reducibility, ie m instances of the resource on the left can be approximately produced by use of n instances of the resources on the right, with the ratio m/n and the fidelity of the approximation both approaching unity as $n \to \infty$. The superscript arrows (eg Bit$^{\rightarrow}$) indicate the direction of transmission for resources such as classical bits, qubits, or noisy channel uses; Bit$^{\leftrightarrow}$ indicates that bidirectional classical communication is required. An R-random qubit is a qubit drawn from an ensemble of von Neumann entropy R. An E-entangled pure state is a pure bipartite state having entropy of entanglement E.

1. Alice prepares a supply of EPR pairs and sends half of each pair to Bob through the noisy channel. This leaves Alice and Bob with a supply of shared but noisy EPR pairs, in other words, bipartite mixed states $\rho_{A,B}$.

2. If the channel $\mathcal{N}$ was too noisy, the mixed states will have zero distillable entanglement and nothing more can be done. On the other hand, if the shared pairs have nonzero distillable entanglement, Alice and Bob, by performing local operations and measurements and sacrificing some of the pairs, can distill a smaller number of arbitrarily pure EPR pairs. This purification process (cf. Fig. 13.8) resembles fractional distillation or water desalination, and uses two-way classical communication between Alice and Bob to communicate their measurement results and decide which pairs to keep and which to discard.

3. Alice then uses the good EPR pairs to teleport an unknown input state reliably to Bob, with the help of additional classical communication.

The Q_2 capacity of the noisy channel $\mathcal{N}$ is defined as the asymptotic number of reliable qubits per channel use that can be communicated in this fashion, with the assistance of unrestricted two-way classical communication. Some noisy channels, for example the 50 per cent depolarizing channel (which depolarizes half the photons passing through), have positive Q_2 capacity even though their direct quantum capacity Q is zero. For other channels, the two quantum capacities are equal. It is not known whether there are channels for which Q_2 exceeds the classical capacity C.

13.5 Quantum Cryptography

Quantum cryptographic key distribution [9, 15–19, 22, 23] is a protocol involving both quantum and classical communication among three parties, the legitimate users Alice and Bob and an eavesdropper Eve. Alice and Bob's goal is to use quantum uncertainty to do something that would be impossible by purely classical public communication—agree on a secret random bit string K, called a cryptographic key, that is informationally secure in the sense that Eve has little or no information on it[1]. In the quantum protocol (cf. Fig. 13.9), Eve is allowed to interact with the quantum information carriers (e.g. photons) enroute from Alice to Bob—at the risk of disturbing them—and can also passively listen to all classical communication between Alice and Bob, but she cannot alter or suppress the classical messages. Sometimes (e.g. if the Eve jams or interacts strongly with the quantum signals) Alice and Bob will conclude that the quantum signals have been excessively disturbed, and therefore that no key can safely be agreed upon (designated by a frown in the figure); but, conditionally on Alice and Bob's concluding that it is safe to agree on a key, Eve's expected information on that key should be negligible.

13.6 Summary

Table 13.1 compares some transformations of classical and quantum information in terms of resources produced and consumed.

Acknowledgements

I wish to thank Chris Fuchs, David DiVincezo, Rolf Landauer, Sandu Popescu, John Smolin, and Bill Wootters for discussions. Part of the work was done at the ISI Quantum Computation Workshop with the support of Elsag-Bailey, and part of it was done with the support of the US Army Research Office.

---

[1]Purely classical protocols for key agreement exist and are in widespread use, but these result in a key that is only computationally secure—an adversary with sufficient computing power could infer it from the messages exchanged between Alice and Bob. In particular, the most widely used classical key agreement protocols could be easily broken by a quantum computer, if one were available.

References

[1] P. W. Shor *SIAM J. Computing* **26**, 1484, eprint quant-ph/9508027 (1997).

[2] L. K. Grover, *Phys. Rev. Lett.* **79**, 325-328, eprint quant-ph/9706033 (1997).

[3] C. H. Bennett, E. Bernstein, G. Brassard, and U. Vazirani, "Strengths and Weaknesses of Quantum Computation," eprint quant-ph/9701001.

[4] Y. Ozhigov, "Quantum Computer Can Not Speed Up Iterated Applications of a Black Box," eprint quant-ph/9712051, see also Farhi, et al quant-ph/9802045 and B.M. Terhal (unpublished).

[5] see Preskill, "Reliable Quantum Computers," eprint quant-ph/9705031 (1997) and references cited therein.

[6] R. Jozsa and B. Schumacher, *J. Modern Optics* **41**, 2343-2349 (1994).

[7] P. W. Shor,*Phys. Rev. A* **52**, 2493 (1995), A.R. Calderbank and P.W. Shor, eprint quant-ph/9512032, (1995), A. Steane, "Multiple Particle Interference and Quantum Error Correction," eprint quant-ph/9601029 (1996).

[8] C. H. Bennett, G. Brassard, B. Schumacher, S. Popescu, J. Smolin, and W. K. Wootters, *Phys. Rev. Lett.* (1996); C. H. Bennett, D. P. DiVincenzo, J. Smolin, and W. K. Wootters, "Mixed State Entanglement and Quantum Error Correction" Phys. Rev A **54** 3824-3851, eprint quant-ph/9604024 (1996).

[9] D. Deutsch, A. Ekert, R. Jozsa, C. Macchiavello, S. Popescu, and A. Sanpera, *Phys. Rev. Lett.***77**, 2818 (1996), **80** 2022 (1998).

[10] E. Knill and R. Laflamme, *Phys. Rev. A* **55**, 900-911 (1997) eprint quant-ph/9604034 (1996), D. Gottesman, "A class of quantum error-correcting codes saturating the Hamming bound," eprint quant-ph/9604038 (1996).

[11] C. H. Bennett, G. Brassard, C. Crépeau, R. Jozsa, A. Peres, and W. K. Wootters, *Phys. Rev. Lett.* **70**, 1895 (1993).

[12] C. H. Bennett and S. J. Wiesner, *Phys. Rev. Lett.* **69**, 2881 (1992).

[13] C. H. Bennett, H. J. Bernstein, S. Popescu, and B. Schumacher, "Concentrating Partial Entanglement by Local Operations", *Phys. Rev. A* **53**, 2046 (1996); H.-K. Lo and S. Popescu, "Concentrating entanglement by local actions—beyond mean values," eprint quant-ph/9707038 (1997).

[14] V. Vedral and M. Plenio, "Entanglement Measures and Purification Procedures", quant-ph/9707035; eprint quant-ph/9603022; E. M. Rains, "Entanglement purification via separable superoperators," eprint quant-ph/9707002 (1997); W. K. Wootters, "Entanglement of Formation of an Arbitrary State of two Qubits," eprint quant-ph/9709029 (1997) M. Horodecki, P. Horodecki, and R. Horodecki, "Mixed-state entanglement and distillation," eprint quant-ph/9801069 and references therein.

[15] C. H. Bennett and G. Brassard, *Proceedings of IEEE International Conference on Computers, Systems, and Signal Processing*, Bangalore, India, pp. 175 (IEEE, New York 1984).

[16] C. H. Bennett, *Phys. Rev. Lett.* **68**, 3121-3124 (1992).

[17] A. Ekert, *Phys. Rev. Lett.* **67**, 661 (1991).

[18] C. H. Bennett, F. Bessette, G. Brassard, L. Salvail, and J. Smolin, *J. of Cryptology* **5**, 3 (1992).

[19] C. H. Bennett, G. Brassard, and N. D. Mermin, *Phys. Rev. Lett.* **68**, 557-559 (1992).

[20] C. A. Fuchs, N. Gisin, R. B. Griffiths, C.-S. Niu, A. Peres "Optimal Eavesdropping in Quantum Cryptography. I" quant-ph/9701039; R.B. Griffiths, C.-S. Niu "Optimal...II, Quantum Circuit", eprint quant-ph/9702015.

[21] E. Biham, M. Boyer, G. Brassard, J. van de Graaf, T. Mor "Security of Quantum Key Distribution Against All Collective Attacks", eprint quant-ph/9801022.

[22] D. Mayers and A. Yao, "Unconditional Security of Quantum Cryptography" eprint quant-ph/9802025.

[23] H.-K. Lo and H.-F. Chau, "Security of Quantum Cryptography" eprint quant-ph/9803006.

14

QUANTUM COMPUTATION

Richard J. Hughes

Abstract

The remarkable developments in theoretical and experimental quantum computation that have been inspired by Feynman's seminal papers on the subject are reviewed. Following an introduction to quantum computation, the implications for cryptography of quantum factoring are discussed. The requirements and challenges for practical quantum computational hardware are illustrated with an overview of the ion trap quantum computation project at Los Alamos. The physical limitations to quantum computation with trapped ions are analyzed and an assessment of the computational potential of the technology is made.

"...it seems that the laws of physics present no barrier to reducing the size of computers until bits are the size of atoms, and quantum behavior holds dominant sway." R. P. Feynman, 1985 [1].

"I think I can safely say that nobody understands quantum mechanics." R. P. Feynman, 1965 [2].

14.1 Introduction

A naive extrapolation of computer technology suggests that within 20 years quantum phenomena will become relevant. This observation led Feynman [1] to investigate how a computational device might be implemented with information represented quantum mechanically. Specifically, he considered the representation of a single bit of information by an "atom" that is in one or the other of two possible states, denoted $|0\rangle$ and $|1\rangle$. (A single bit of this form is now known as a qubit.) An L-bit number can then be represented as a state of a "register" of L such two-state systems. Feynman's motivation was to determine whether quantum physics, and the Uncertainty Principle specifically, imposed any limitations on computation. He was able to show that the reversible Boolean operations previously investigated in studies of the thermodynamics of computation [3, 4] could be implemented as unitary transformations on quantum systems. However, in his conclusion he noted that he had "...not really used many of the specific qualities of the differential equations of quantum mechanics," and that what he did was "...only to imitate as closely as possible the digital machine of conventional sequential architecture." But in another paper Feynman [5] hinted that quantum mechanics might offer greater computational power than conventional computers.

Other authors have taken up the leads suggested in Feynman's seminal papers (see also Benioff [6]). Instead of merely attempting to achieve some improvement over conventional computational technologies based on irreversible Boolean logic, in terms of clock speed for instance, the modern concept of a quantum computer is a truly new computational paradigm. For certain problems that are "intractable" on conventional computers the peculiarly quantum phenomena of superposition, "entanglement" [7, 8][1] and interference would allow a dramatic reduction in their computational complexity on a quantum computer[2]. In particular, quantum algorithms with compelling applications to cryptography have been invented. As a result, the field of quantum computation has seen tremendous growth, with realistic hardware proposals and related experiments; and the development of quantum error correcting codes. In this article I shall describe why quantum computation is now such an active field; why it is experimentally difficult; and the prospects for quantum computation with trapped ions.

It was Deutsch [9, 10] who first suggested that by using non-Boolean unitary operations the quantum superposition principle could be exploited to achieve greater computational power than with conventional computation. Bernstein and Vazirani [11] showed that Deutsch's general model of quantum computation was both efficient and "reasonable." Their results were extended by Yao [12] who proved that quantum circuits (introduced in Reference 10) are polynomially equivalent to Deutsch's quantum Turing machine of Reference 9. Early work by Deutsch and Joszsa [13] showed how to exploit computationally the power afforded by the superposition principle. But it was not until the work of Shor [14] in 1994 that this "quantum parallelism" was shown to offer an efficient solution of an interesting computational problem. Building on earlier work of Simon [15], Shor invented polynomial-time quantum algorithms for solving the integer factorization and discrete logarithm problems [14]. The difficulty of solving these two problems with conventional computers underlies the security of much of modern public key cryptography [16]. Shor's algorithms are sufficiently compelling that the daunting scientific and technological challenges involved in practical quantum computation are now worthy of serious experimental study. Since 1994 other interesting quantum algorithms, including a new class typified by Grover's "database search" algorithm [17], have been invented, and there is considerable research directed toward defining a general mathematical framework encompassing all quantum algorithms [18–20].

When Feynman wrote his papers on quantum computation there were no viable hardware schemes. Subsequently, various proposals requiring the invention of new technologies were made, and fundamental problems with them were identi-

---

[1] "...would not call [quantum entanglement] one but rather the characteristic trait of quantum mechanics, the one that enforces its entire departure from classical lines of thought." E. Schrodinger, 1935 [8].

[2] The clock speed of a quantum computer would probably not be particularly high, so for general problems, outside the special class amenable to efficient quantum algorithms, it would not offer any computational advantage.

fied [21, 22]. Since then there have been many relevant experimental developments in the foundations of quantum mechanics, and although experimental quantum computation is still in its infancy, there are now several very promising hardware concepts based on existing technologies that avoid most of these fundamental problems. One of these new schemes, using the quantum states of laser-cooled ions in an electromagnetic trap [23], is particularly promising. In 1994 Cirac and Zoller showed that such systems have the necessary characteristics to perform quantum computation. The relevant coherence times can be adequately long; mechanisms for performing the quantum logic gate operations exist; and a high-probability readout method is possible. (For a detailed description see Reference 24.) Several groups, including our own [25], are now investigating quantum computation with trapped ions. A single logic operation using a trapped beryllium ion has been demonstrated [26]. However, even algorithmically simple computations will require the creation and controlled evolution of quantum states that are far more complex than have so far been achieved experimentally. It is therefore important to quantify the extent to which trapped ions could allow the quantum engineering of the complex states required for quantum computation [27].

Just as with classical computers, quantum computers will have to cope with errors. However, quantum errors are much more challenging than their classical counterparts because they destroy the quantum coherences from which quantum computation derives its power. The phenomenon of "decoherence" [28] arising from interactions between a quantum system and its "environment" is invoked to explain the absence of macroscopic objects in quantum mechanical superposition states [7]. In quantum computers the damaging effects of decoherence will have to be controlled [29]. During the past two years there have been spectacular developments in the theory of quantum error correcting codes that protect quantum information from decoherence by exploiting entanglement [30, 31]. Furthermore, quantum error correction procedures can be implemented in a fault-tolerant fashion [32], holding out the prospect of unlimited quantum computation with imperfect physical implementations, if certain precision thresholds can be attained [33].

The rest of this paper is organized as follows. In Section 14.2 we review the basic principles of quantum computation and indicate why it is more powerful computationally than conventional computation for certain problems. In Section 14.3 we consider the cryptographic implications of quantum factoring. In Section 14.4 we describe the Cirac-Zoller scheme for ion trap quantum computation, and Section 14.5 is devoted to a description of the different qubit schemes possible with trapped ions. Section 14.6 contains estimates of the intrinsic limits to quantum computation with the two classes of qubits. Finally in Section 14.7 we present some conclusions.

14.2　Basic concepts of quantum computation

The essential idea of quantum computation is to represent binary numbers by two-level quantum systems, such as two energy levels of an electron in an atom or ion, or the two possible states of a spin-1/2 particle or photon [34]. We will use the notation $|0\rangle$, $|1\rangle$ to denote the two distinct quantum states of such a system. These states form an orthonormal basis for the Hilbert space of this qubit, with

$$\langle 0|0\rangle = \langle 1|1\rangle = 1$$
$$\langle 0|1\rangle = \langle 1|0\rangle = 0 \tag{14.1}$$

In general, the quantum state of a qubit can be written as a linear combination with complex coefficients of these two states, in the form

$$|\Psi\rangle = cos(\theta/2)|0\rangle - ie^{i\phi}sin(\theta/2)|1\rangle \tag{14.2}$$

whose time evolution is given by the Schrödinger equation

$$i\hbar\frac{\partial}{\partial t}|\Psi\rangle = H|\Psi\rangle \tag{14.3}$$

where $\hbar$ is Planck's constant, and the Hamiltonian, H, is a hermitian 2x2 matrix. Typically, H will be a sum of two terms,

$$H = H_0 + H_I \tag{14.4}$$

where H_0 is the "free" Hamiltonian describing the evolution of the isolated qubit, and H_I is an interaction term, such as the interaction with an external "drive" that effects transitions between qubit levels. The qubit basis states will be eigenstates of H_0, with eigenvalues that may be chosen as

$$H_0\begin{pmatrix}|0\rangle \\ |1\rangle\end{pmatrix} = \begin{pmatrix}0 \\ \epsilon|1\rangle\end{pmatrix} \tag{14.5}$$

where ϵ is some real number.

It is common to work in the interaction representation in which the time dependence of the qubit's state is determined by H_I alone. When the external drive can be switched on and off for specified amounts of time, the time evolution will then be described by

$$\begin{pmatrix}|0\rangle \\ |1\rangle\end{pmatrix} \rightarrow U\begin{pmatrix}|0\rangle \\ |1\rangle\end{pmatrix} \tag{14.6}$$

where U is a unitary 2x2 matrix of the form

$$U = exp(-i\int dtH_I/\hbar) \tag{14.7}$$

With multiple qubits it becomes possible to represent larger numbers in binary notation. For example with an L-qubit quantum register, it is possible to represent numbers, a, between 0 and $2^L - 1$, as the state

$$|a\rangle = \prod_{i=0}^{L-1} |a_i\rangle \qquad (14.8)$$

where

$$a = \sum_{i=0}^{L-1} a_i 2^i \qquad (14.9)$$

and $a_i = 0,1$ is the i-th bit of a. Just as with a single qubit, we can consider the time evolution of a quantum register as the action of a unitary operation. In particular, the logic operations required for conventional computation can be produced in this way. However, because the Schrödinger equation possesses time-reversal symmetry we must use reversible Boolean logic with quantum states. It is known that all conventional logic operations can be accomplished with the following three reversible logic operations [1].

The logical NOT operation,

$$NOT : \begin{pmatrix} |0\rangle \\ |1\rangle \end{pmatrix} \rightarrow U_{NOT} \begin{pmatrix} |0\rangle \\ |1\rangle \end{pmatrix} = \begin{pmatrix} 0 & 1 \\ 1 & 0 \end{pmatrix} \begin{pmatrix} |0\rangle \\ |1\rangle \end{pmatrix} = \begin{pmatrix} |1\rangle \\ |0\rangle \end{pmatrix} = \begin{pmatrix} |\bar{0}\rangle \\ |\bar{1}\rangle \end{pmatrix} \qquad (14.10)$$

which we may also write as,

$$NOT : |a\rangle \rightarrow |\bar{a}\rangle, \quad a = 0, 1 \qquad (14.11)$$

is readily verified to be reversible.

A reversible two-qubit operation is the controlled-NOT gate, in which the state of one qubit (the "target") is flipped if the other qubit (the "control") has the value "1" but unchanged if it has the value "0"

$$CNOT_{c,t} : |a\rangle_c |b\rangle_t \rightarrow |a\rangle_c |a \oplus b\rangle_t, \quad a, b = 0, 1 \qquad (14.12)$$

where the subscripts c, t denote "control" and "target" respectively, and "$\oplus$" denotes the logical exclusive-OR ("XOR") operation or addition mod 2. Note that a second application of this operation returns the state of the qubits to their starting state. Also, note that this operation gives a reversible implementation of the XOR operation, at the expense of carrying forward some additional information (the control bit) that allows for reversibility.

A third reversible logical operation involves three qubits and is known as the controlled-controlled-NOT (or Toffoli) operation:

$$CCNOT_{c1,c2,t} : |a\rangle_{c1} |b\rangle_{c2} |c\rangle_t \rightarrow |a\rangle_{c1} |b\rangle_{c2} |(a \wedge b) \oplus c\rangle_t, \quad a, b, c = 0, 1 \qquad (14.13)$$

where "$\wedge$" is the logical AND operation.

From these three operations it is now straightforward to produce elementary arithmetic operations, such as a simple adder, using a sequence of unitary operations:

$$ADD : |a\rangle_1|b\rangle_2|0\rangle_3 \rightarrow CNOT_{1,2}(CCNOT_{1,2,3}|a\rangle_1|b\rangle_2|0\rangle_3) = |a\rangle_1|a \oplus b\rangle_2|a \wedge b\rangle_3$$
$$(14.14)$$

which produces the sum bit on the second qubit and the carry bit on the third qubit. More complex combinations of the above primitives can be used to produce arbitrary logic operations. But note that input data must typically be carried forward to the output to allow for reversibility. Feynman showed that in general the amount of extra information that must be carried forward is just the input itself. So, to evaluate a function, F, reversibly we will need an input register with the appropriate number of qubits to hold the function's argument, a, and an output register of enough qubits, initially all in the $|0\rangle$ state, to hold the value, $F(a)$:

$$|a\rangle|0\rangle \rightarrow |a\rangle|F(a)\rangle \qquad (14.15)$$

The input register value can be prepared from the $|0\rangle$ state by applying NOT operations to the appropriate bits and then the function value can be produced in the output register using combinations of the above logical operations. The readout of the result of the quantum computation is performed through quantum measurements, which are represented as (non-unitary) projections in Hilbert space. For a single qubit, the projection

$$P = |0\rangle\langle0| \qquad (14.16)$$

projects onto the $|0\rangle$ state, giving the outputs

$$P|0\rangle = |0\rangle$$
$$P|1\rangle = 0 \qquad (14.17)$$

i.e. the $|0\rangle$ state passes the test, whereas the $|1\rangle$ state fails. With bit-wise projections on a register a readout of the result of a computation can be obtained.

Thus far, the discussion has only shown how quantum systems might reproduce the procedures of conventional computation, and has not revealed any particular advantage to quantum mechanical computation, i.e. after each clock cycle the quantum computer is in a basis state. However, the Schrodinger equation defines a continuous time-evolution of a quantum state, so that we are not restricted to the discrete Boolean operations introduced above. In general, the quantum state of a qubit can be written as a coherent superposition state in which it has, in a sense, both values at once (cf. Eq. 14.2.) On applying a readout measurement, P, to a qubit in a state such as Eq. 14.2, quantum mechanics only allows us to predict that we will obtain the result, $|0\rangle$, with probability $\cos^2(\theta/2)$. (Also, note that the effect of the measurement is to project the qubit's state onto either $|0\rangle$ or $|1\rangle$: A phenomenon known as "collapse of the wavefunction.") Such superposition states can be created from the computational basis states using non-Boolean unitary operations such as

$$V(\theta,\phi): \begin{array}{l} |0\rangle \rightarrow \cos(\theta/2)|0\rangle - iexp(i\phi)sin(\theta/2)|1\rangle \\ |1\rangle \rightarrow \cos(\theta/2)|1\rangle - iexp(-i\phi)sin(\theta/2)|0\rangle \end{array} \qquad (14.18)$$

For example, the unitary operation $V(\pi/2,\pi/2)$ creates the equally-weighted superpositions

$$V(\pi/2,\pi/2): \begin{array}{l} |0\rangle \rightarrow 2^{-1/2}(|0\rangle + |1\rangle) \\ |1\rangle \rightarrow 2^{-1/2}(-|0\rangle + |1\rangle) \end{array} \qquad (14.19)$$

from the basis states, $|0\rangle$ and $|1\rangle$. Note, however that these superposition states are not simple statistical mixtures of $|0\rangle$ and $|1\rangle$, because a second application of the same unitary operation produces the result

$$V^2(\pi/2,\pi/2): \begin{array}{l} |0\rangle \rightarrow |1\rangle \\ |1\rangle \rightarrow -|0\rangle \end{array} \qquad (14.20)$$

illustrating the phenomenon of interference of quantum amplitudes. Other non-classical states can be created from the basis states when the $V(\pi/2,\pi/2)$ operation is combined with Boolean operations. For example, the two-gate sequence

$$CNOT_{ij} \cdot V_i(\pi/2,\pi/2)|0\rangle_i|0\rangle_j = 2^{-1/2}(|0\rangle_i|0\rangle_j + |1\rangle_i|1\rangle_j) \qquad (14.21)$$

(where V_i acts on the i-th qubit) creates an "entangled" (non-factorizable) state, in which both qubits have the same value, but neither qubit has a definite value. (Multiparticle entangled states occur throughout quantum computation.) Clearly the operational methods for creating such non-classical states are of interest to the study of quantum mechanics, but they are also at the heart of the potential power of quantum computation to solve computational problems that are intractable on conventional computers.

Applying the $V(\pi/2,\pi/2)$ operation to each of the bits in an L-bit register produces the superposition state

$$|0\rangle_{L-1}|0\rangle_{L-2}\ldots|0\rangle_0 \rightarrow \frac{1}{\sqrt{2^L}} \prod_{i=0}^{L-1} (|0\rangle_i + |1\rangle_i) = \frac{1}{\sqrt{2^L}} \sum_{a=0}^{2^L-1} |a\rangle \qquad (14.22)$$

So using only L quantum operations, a superposition of 2^L states has been produced. Of course, a measurement of this state would produce only one of the 2^L possible values, a, with equal probability, showing that the quantum information in this state is not all accessible at once. But this type of state is the starting point of a chain of argument indicative of the potential power of quantum computation.

Consider now our earlier example of quantum function evaluation, Eq. 14.15. We could replace the input by the above equally-weighted superposition state, and then in one operation of the quantum circuit for the function F we would produce all 2^L outputs in superposition:

$$\frac{1}{2^{L/2}} \sum_{i=0}^{L-1} |a\rangle|0\rangle \rightarrow \frac{1}{2^{L/2}} \sum_{i=0}^{L-1} |a\rangle|F(a)\rangle \tag{14.23}$$

Still, this has only the appearance of exponential work in one operation because, as above, only one function value can be obtained by a measurement of the final state. However, if we were interested in some common property shared by the function's values, such as the function's period, this can be determined efficiently if we now bring into play the notion of interference. Shor showed [14] how to extract the period of a sequence, represented as a superposition of states of an L-qubit quantum register, using the quantum Fourier transform (QFT)

$$|a\rangle \rightarrow 2^{-L/2} \sum_{c=0}^{2^L-1} exp\left(i\frac{a \cdot c}{2^L}\right) |c\rangle \tag{14.24}$$

(Note that this transformation is reversible.) Shor showed how to construct the QFT using only $O(L^2)$ non-Boolean quantum operations, in contrast to the $O(L2^L)$ operations required for a conventional discrete Fourier transform. The QFT can be constructed using a single-qubit unitary (Hadamard) operation, H_j, which acts on the j-th qubit to give:

$$H_j : \begin{array}{l} |0\rangle_j \rightarrow 2^{-1/2}(|0\rangle_j + |1\rangle_j) \\ |1\rangle_j \rightarrow 2^{-1/2}(|0\rangle_j - |1\rangle_j) \end{array} \tag{14.25}$$

and a two-qubit operation, A_{jk}, which acts on the j-th and k-th qubits to give:

$$A_{jk} : \begin{array}{l} |0\rangle_j|0\rangle_k \rightarrow |0\rangle_j|0\rangle_k \\ |0\rangle_j|1\rangle_k \rightarrow |0\rangle_j|1\rangle_k \\ |1\rangle_j|0\rangle_k \rightarrow |1\rangle_j|0\rangle_k \\ |1\rangle_j|1\rangle_k \rightarrow exp(i\pi/2^{k-j})|1\rangle_j|1\rangle_k \end{array} \tag{14.26}$$

(Note the different phases in the H operation and the V operation.) The transformation 14.24 can then performed as the sequence of operations (from left to right):

$$H_{L-1}A_{L-2,L-1}H_{L-2}A_{L-3,L-1}A_{L-3,L-2}H_{L-3} \ldots H_1 A_{0,L-1}A_{0,L-2} \ldots A_{0,2}A_{0,1}H_0 \tag{14.27}$$

producing the state

$$2^{-L/2} \sum_{c=0}^{2^L-1} exp\left(i\frac{a \cdot c}{2^L}\right) |b\rangle \qquad (14.28)$$

where $|b\rangle$ is the bit-reversed state of $|c\rangle$. It is therefore necessary to interpret the bits in reverse order after the transformation in order to obtain the QFT. To see this, note that the phase accumulated in the QFT of $|a\rangle$ to $|c\rangle$ is

$$\sum_{0 \le j < l} \pi a_j c_l + \sum_{0 \le j < k < l} \frac{\pi}{2^{k-j}} a_j c_l \qquad (14.29)$$

which can be rewritten as

$$\sum_{0 \le j+k < l} 2\pi \frac{2^j 2^k}{2^l} a_j b_k \qquad (14.30)$$

where $b_k = c_{l-k-1}$, producing the desired result because adding multiples of 2π does not affect the phase.

On four qubits, for example, the QFT is effected by the sequence of operations (applied from left to right):

$$H_3 A_{2,3} H_2 A_{1,3} A_{1,2} H_1 A_{0,3} A_{0,2} A_{0,1} H_0 \qquad (14.31)$$

To see how the QFT can be used, we now consider two sequences, each of period 2, represented as superpositions of the states of four qubits,

$$|\Psi_e\rangle = \frac{1}{2} \sum_{q=0}^{7} |2q\rangle \qquad (14.32)$$

and,

$$|\Psi_0\rangle = \frac{1}{2} \sum_{q=0}^{7} |2q+1\rangle \qquad (14.33)$$

Applying the QFT 14.24 produces the states:

$$|\Psi_e\rangle \rightarrow 2^{-1/2}(|0\rangle + |8\rangle) \qquad (14.34)$$

and

$$|\Psi_0\rangle \rightarrow 2^{-1/2}(|0\rangle - |8\rangle) \qquad (14.35)$$

In each case a measurement of the final state produces a result of $n \cdot 16/r$, where $r = 2$ is the period of the sequence, and $n = 0, 1$. So that for each state there is a

50% probability that the period, r, which is a common property, can be determined. (Note that this illustrates the probabilistic nature of quantum algorithms. In the cases when the outcome is 0 the computation would have to be repeated.) Despite the shift between the above two sequences, the QFT allows their period to be extracted. This "shift-invariance" of the QFT is important to its usefulness in quantum algorithms. For example, if we consider the function

$$F(i) = \left\{ \begin{array}{ll} 0, & i \ \ even \\ 1, & i \ \ odd \end{array} \right. \quad , i = 0, 1, \ldots 15 \tag{14.36}$$

using a four-qubit "left" register to hold the arguments of F and a single-qubit "right" register to hold the values of F, the state

$$\frac{1}{4} \sum_{i=0}^{15} |i\rangle_L |F(i)\rangle_R = 2^{-1/2} \sum_{q=0}^{7} (|\Psi_e\rangle_L |0\rangle_R + |\Psi_0\rangle_L |1\rangle_r) \tag{14.37}$$

can be produced with appropriate logic operations. Then after applying the QFT to the left register the resulting state is

$$\frac{1}{2} \left[(|0\rangle_L + |8\rangle_L)|0\rangle_R + (|0\rangle_L - |8\rangle_L)|1\rangle_R \right] \tag{14.38}$$

from which the period of F can be extracted with 50% probability on measurement of the left register, independently of the values of F. Generalizations of this method for determining the period of a function are at the heart of the quantum factoring and quantum discrete log algorithms.

Here then is how non-Boolean quantum operations can be used to efficiently solve certain problems on a quantum computer: Information can be created and processed efficiently as superpositions of quantum amplitudes to reveal common features through the phenomenon of multi-particle quantum interference. By operating on only L qubits, it is (in a sense) possible to compute with 2^L quantities in parallel. To obtain the computational advantages of quantum parallelism it is essential that no measurements are made before the computation is complete, otherwise the collapse of the register's wavefunction will destroy the large amount of quantum information whose interference is required for the solution of the problem. Clearly we can avoid making any intentional measurements until the calculation is complete, but it is much more difficult to control decoherence, which is the main challenge for practical quantum computation.

All known quantum algorithms for solving interesting problems use either the QFT or one of its variants such as the quantum Hadamard transformation, to exploit multi-particle quantum interference [35]. Shor's algorithms and Simon's algorithm [15] have inspired a mathematical generalization of this type of quantum algorithm, which may lead to the development of efficient quantum algorithms for solving other problems. In each case, the problem can be phrased as finding a hidden subgroup, H, of an Abelian group, G, given a function on G that takes on different but constant values on the cosets of H. By constructing efficient quantum Fourier transforms over (Abelian) groups, this general class of problem is amenable to solution by quantum computation [20]. An active area of current research is the generalization to non-Abelian groups, such as the symmetric group.

A mathematically distinct class of quantum algorithms are those based on Grover's "database search" algorithm. In this problem one is presented with a table of N elements and asked to find the one element that is "marked" (satisfies some property). Classically, this can be solved by making $O(N)$ queries of the table. Grover's quantum algorithm solves this problem with only $O(N^{1/2})$ quantum operations. One version of this algorithm [36] is provably optimal [37]. Note that this class of quantum algorithm does not achieve the exponential speed-up of factoring, but the "square-root" improvement is nevertheless very significant. However, the most celebrated quantum algorithm is Shor's quantum factoring algorithm, which we turn to next.

14.3   Quantum Computation and Public Key Cryptography

"The problem of distinguishing prime numbers from composite numbers and of resolving the latter into their prime factors is known to be one of the most important and useful in arithmetic." K. F. Gauss, 1801 [38].

Every integer can be uniquely decomposed into a product of prime numbers. Most integers are easy to factor because they are products of small primes, but large integers (hundreds of digits in length) that are products of two, distinct, comparably-sized primes can be very difficult to factor with conventional computers [39]. For example, in 1994 the 129-digit number known as RSA129 [40] required 5,000 MIPS-years of computer time over an 8-month period, using more than 1,000 workstations, to determine its 64-digit and 65-digit prime factors [41]. (By convention one MIPS-year is about 3×10^{13} instructions. Current workstations are rated at 200–800 MIPS.) The perceived difficulty of factoring with conventional computers underlies the security of widely-used public key cryptosystems. But a quantum computer (QC) using Shor's algorithm at a clock speed of 100 MHz would have factored RSA129 in only a few seconds. It is often necessary to ensure that encrypted information remains secure for decades, but when encrypted information is transmitted we must assume that it can be monitored and saved for future analysis by eavesdroppers. If information must be secure for X years, a cryptosystem must no longer be used X years before it is projected to become vulnerable. So the possibility that quantum computers could become feasible is not just a potential challenge to the use of public key cryptography in the future, but is a concern for the use of these cryptosystems today [42].

The RSA cryptosystem [43], is based on the following computationally difficult problem:

RSA problem: Given an integer N that is a product of two distinct primes, p and q, an integer e such that $g.c.d.(e, (p-1)(q-1)) = 1$, and an integer C, find the integer M such that, $C = M^e \bmod N$. (Here "$g.c.d$" denotes "greatest common divisor," and "mod N" indicates that arithmetic is being performed modulo N. Solving this problem is conjectured to be equivalent to factoring.)

To understand the significance of the RSA problem we first introduce Euler's quotient function, which for an integer m is defined as,

$$\phi(m) = \text{number of integers less than } m \text{ that are relatively prime to } m. \quad (14.39)$$

Thus, for a prime, p, $\phi(p) = (p-1)$, and for composite moduli of the form $N = p \cdot q$, introduced above we have,

$$\phi(N) = (p-1)(q-1) \quad (14.40)$$

We will also need the following theorem of Euler, which states that for any integer x, relatively prime to m, i.e. $g.c.d.(x, m) = 1$,

$$x^{\phi(m)} = 1 \bmod m \quad (14.41)$$

Therefore, we can solve the RSA problem if we can find the integer d defined by

$$d = e^{\phi(\phi(N))-1} \bmod \phi(N) \quad (14.42)$$

because then

$$C^d = M^{ed} = M^{k\phi(N)+1} = M \bmod N \quad (14.43)$$

by Euler's theorem, Eq. 14.41, for some integer k [44]. Clearly, if we know $\phi(N)$ we can find the integer d, and we can determine $\phi(N)$ if we know the factors, p and q, of N. Thus the RSA problem can be solved if we can factor the modulus N, which is computationally hard.

In the RSA cryptosystem the above problem is used to provide cryptographic security as follows. Alice wishes to send a (plaintext) message M (a large integer) to Bob, but wants to be sure that the eavesdropper Eve cannot read the message. So:

1. Bob generates two large, distinct (secret) primes, p and q;

2. He computes their product, N, and the integer $\phi(N)$;

3. Bob selects an integer e, such that g.c.d.(e, $\phi(N)$) = 1, and computes the integer d as above;

4. Bob publishes his public key, comprised of the modulus N and encrypting exponent, e, but keeps his private key (decrypting exponent) d (as well as p and q) secret;

5. Alice uses Bob's public key to compute the ciphertext $C = M^e \bmod N$, and sends C to Bob;

6. Bob recovers Alice's message, M, using his secret decrypting exponent, d, as above.

With an ℓ bit modulus, N, Alice and Bob can encrypt and decrypt their messages using only $O(\ell^3)$ operations, but if Eve wants to decrypt Alice's communication she is faced with the computationally hard problem of factoring Bob's modulus, N. One way for Eve to factor N would be to perform trial divisions by all primes less than $N^{1/2}$. However, this would require $O(\exp[\ell])$ divisions, and so the amount of computational work required would grow exponentially with the size of the modulus.

Modern factoring algorithms (including Shor's quantum factoring algorithm) use a different strategy [45]: They search for non-trivial solutions, y, of Legendre's congruence,

$$y^2 = 1 \bmod N \tag{14.44}$$

from which we have,

$$(y+1)(y-1) = 0 \bmod N \tag{14.45}$$

Then the factors of N will be distributed between the two parentheses, and can be found using

$$g.c.d.(y \pm 1, N) = \text{factor of } N \tag{14.46}$$

The General Number Field Sieve (GNFS) algorithm [46] is the best algorithm is use today for factoring large integers. It is much more efficient than trial division, with a sub-exponential run-time growth, $O[\exp(1.923\ell^{1/3}(\ln(\ell))^{2/3})]$, and is well-adapted to distributed processing. This algorithm was recently used to factor the 130-digit number known as RSA130 [47] in 500 MIPS-years of computer time.

To factor an ℓ-bit integer, N, Shor's quantum factoring algorithm requires a classical integer, x, that is relatively prime to N. This integer is obtained by classical computation, and then the function [48]

$$f(a) = x^a \bmod N, \quad a = 0, 1, ...N^2 - 1 \tag{14.47}$$

is computed quantum mechanically. This function is periodic, with period, r,

$$f(a+r) = f(a) \tag{14.48}$$

where, r, (known as the order of x) is the smallest integer, r, for which

$$x^r = 1 \bmod N \qquad (14.49)$$

After applying the QFT and measuring the result as illustrated in Section 14.2, there is some probability, determined by both number theory and quantum physics, that the order of x can be found. If the order, r, is even, the congruence

$$(x^{r/2} - 1)(x^{r/2} + 1) = 0 \bmod N \qquad (14.50)$$

can be used to factor N using conventional computation, as described above.

Shor's algorithm therefore requires one 2ℓ-bit register to hold the argument of the function, f ; an ℓ-bit register to hold the function values, and some additional memory to allow reversible computation of the function. The amount of additional memory and number of quantum logic gates is somewhat dependent on the specific implementation of the algorithm [49], but in our recent improved version [50] a QC would need L qubits of memory and n_g quantum logic operations, with

$$\begin{aligned} L &= 5\ell + 4 \\ n_g &= 25\ell^3 + O(\ell^2) \end{aligned} \qquad (14.51)$$

to factor an ℓ-bit modulus. The number of logic operations is dominated by the computation of Shor's function, f. In contrast to the (sub)exponential run-time growth of the classical GNFS factoring algorithm, the quantum algorithm has a dramatically slower, polynomial, $O(\ell^3)$, growth. (The ℓ^3-dependence can be understood as arising from the (conditional) multiplication of 2ℓ classical ℓ-bit integers to build the function, f. Each of the multiplications requires $O(\ell^2)$ bit-additions, using "elementary school" multiplication, that can be reduced to elementary quantum logic operations.) If we assume a nominal clock speed of 100 MHz for a QC we find that a 512-bit integer could be factored in about 30 seconds. Furthermore, the quantum factoring algorithm used for this estimate is not optimized and significant improvements are possible.

Of course, improvements in conventional computers are also anticipated, but because of the (sub-)exponential run-time growth of conventional factoring algorithms the security of public key cryptosystems against conventional attacks can be maintained for many years by using integers with only 1,024 or 2,048 bits. In contrast, because of the slow polynomial growth of the quantum factoring algorithm's run-time, it would not be possible to easily ensure security against possible future quantum attacks by only making such modest increases in the size of the modulus. Integers with exponentially larger numbers of bits would be required for security, and integers of such a size would render the encryption and decryption procedures prohibitively long.

From this simple analysis we see that because of the very large numbers of qubits and the long coherence times required we cannot yet state that quantum factoring

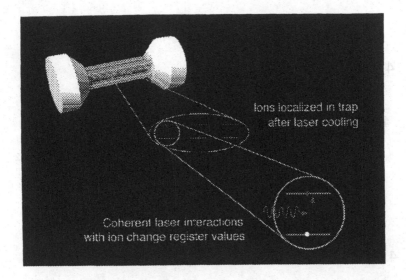

Fig. 14.1. Linear radio frequency quadrupole ion trap for quantum computation.

of cryptographically significant integers will become possible. But the possibility that quantum factoring might become feasible in 20 years time (say) should be a serious concern for public-key cryptography today.

Shor also invented a polynomial-time quantum algorithm for solving the discrete logarithm problem (DLP) [14].

Discrete logarithm problem: Given a prime number, p, an integer $g < p$ and another integer $y < p$, find the integer x, such that $g^x = y \bmod p$.

This problem, which like factoring is computationally intractable, is also widely used in public key cryptography.

14.4 Quantum Computational Hardware

From the foregoing we conclude that there are three essential requirements for quantum computation hardware. Firstly, it must be possible to prepare multiple qubits, adequately isolated from interactions with their environment for the duration of computation, in an addressable form. Secondly, there must be an external drive mechanism for performing the requisite quantum logic operations, with the requisite careful and precise control of the qubits' phases. And thirdly, there must be a readout mechanism for measuring the state of each qubit at the end of the computation. It is clear that it is much easier to write down a sequence of quantum logic operations than it is to perform them in the laboratory. Nevertheless, the above conditions can be satisfied with laser-cooled trapped ions [23], nuclear mag-

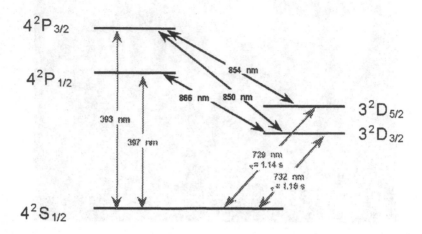

Fig. 14.2. Relevant transitions in calcium ions, showing wavelengths and lifetimes of metastable levels.

netic resonance [51] and cavity quantum electrodynamics [52]. Other technologies that may be suitable for quantum computation include quantum dots and superconducting circuits. We will illustrate the experimental challenges and prospects for practical quantum computation with the example of the trapped ion experiment that is being developed at Los Alamos [25]. In an ion trap quantum computer [23] a qubit would comprise two long-lived internal states, which we shall denote $|0\rangle$ and $|1\rangle$, of an ion isolated from the environment by the electromagnetic fields of a linear radio-frequency quadrupole (RFQ) ion trap. Many different ion species are suitable for quantum computation, and several different qubit schemes are possible, as we shall see below. For example, at Los Alamos we are developing an ion-trap quantum-computer experiment using calcium ions, with the ultimate objective of performing multiple gate operations on a register of several qubits (and possibly small computations) in order to determine the potential and physical limitations of this technology [25]. We have chosen calcium ions for the convenience of the wavelengths required. The heart of our experiment is a linear radio-frequency quadrupole (RFQ) ion trap with cylindrical geometry in which strong radial confinement is provided by radio-frequency potentials applied to four "rod" electrodes and axial confinement is produced by a harmonic electrostatic potential applied by two "end caps." Our ion trap is about 1 cm long and 1.7 mm wide. (See Figure 14.1.)

After Doppler cooling on their 397-nm S–P transition, calcium ions will become localized along the ion trap's axis [53] because their recoil energy (from photon emission) is less than the spacing of the ions' quantum vibrational energy levels in

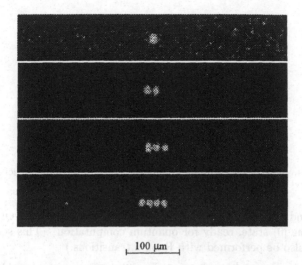

|⎯ 100 μm ⎯|

Fig. 14.3. Strings of calcium ions laser-cooled to rest in the Los Alamos quantum computation ion trap. The image is formed by ion fluorescence in the cooling laser beam at 397nm.

the axial confining potential [54]. (See Figures 14.2 and 14.3.) Although localized to distances much smaller than the wavelength of the cooling radiation, the ions nevertheless undergo small amplitude oscillations. Their lowest frequency mode is the axial center of mass (CM) motion in which all the ions oscillate in phase along the trap axis. The frequency of this mode, whose quantum states will provide a computational "bus," is set by the axial potential. The inter-ion spacing is determined by the equilibrium between this axial potential, which tends to push the ions together, and the ions' mutual Coulomb repulsion.

For example, with a 200-kHz axial CM frequency, the inter-ion spacing is on the order of 30 μm. After this first stage of cooling, the ions form a "quantum register" in which one qubit can be addressed (with a suitable laser beam) in isolation from its neighbors. We have determined that more than 20 ions can be held in an optically addressable configuration. However, before quantum computation can take place, the quantum state of the ions' CM mode must be prepared in its quantum ground state.

Because of the long radiative lifetime of the metastable 3D-states ($\sim$ 1s), the S–D electric quadrupole transition in calcium ions has such a narrow width that it displays upper and lower sidebands separated from the central frequency by the CM frequency. With a laser that has a suitably narrow linewidth, tuned to the lower sideband, an additional stage of laser cooling (beyond Doppler cooling) can be used to prepare the "bus" qubit (CM vibrational mode) in its lowest quantum

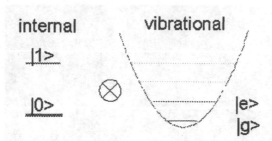

Fig. 14.4. Internal and motional quantum states of a trapped calcium ion.

state ("sideband cooling") [55]. On completion of this stage, the QC would have all qubits in the $|0\rangle$ state, ready for quantum computation. (This second stage of cooling could also be performed with Raman transitions.)

The quantum state of the register of ions will then be manipulated by performing quantum logical gate operations that will be effected by directing a laser beam at individual ions for prescribed times. The laser-ion interaction will coherently change the state of the qubit through the phenomenon of Rabi oscillations. (Several different types of transition are possible.) As we will see below, the $CNOT$ operation can be effected with the help of the quantum states of the ions' CM motion to convey quantum information from one ion to the other.

On completion of the quantum logic operations the result of the quantum computation can be read out using the phenomenon of quantum jumps [56], by turning on a laser that drives the transition between the $|0\rangle$ state and another ionic level that decays rapidly back to $|0\rangle$). An ion in the $|0\rangle$ state will then fluoresce, whereas an ion in the $|1\rangle$ state will remain dark. So, by observing which ions fluoresce and which are dark, a bit value can be obtained.

14.5 Trapped ion qubits

In addition to the two states, $|0\rangle$ and $|1\rangle$ comprising each ionic qubit, in an ion trap QC there is also a computational "bus" qubit formed by the ground, $|g\rangle$, and first excited state, $|e\rangle$, of the ions' CM axial vibrational motion, which is used to perform logic operations between qubits (see Figure 14.4).

By virtue of energy conservation (and possibly other selection rules) it is possible to perform two types of coherent operations on a qubit, using laser pulses directed at an ion: On-resonance transitions that change only an ion's internal state ("V" pulses); and red-sideband transitions (detuned from resonance by the CM frequency) that change both the qubit's internal state and the CM quantum state ("U" pulses). (See Figure 14.5.) The V-pulse Hamiltonian for a particular

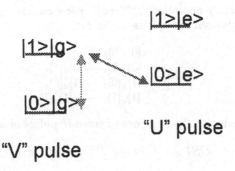

Fig. 14.5. Quantum computational transitions with trapped ions.

ion is,

$$H_V = \frac{\hbar\Omega}{2} \left[e^{-i\phi} |1\rangle\langle 0| + e^{i\phi} |0\rangle\langle 1| \right] \qquad (14.52)$$

and the U-pulse Hamiltonian is,

$$H_U = \frac{\hbar\eta\Omega}{2\sqrt{L}} \left[e^{-i\phi} |1\rangle\langle 0|a + e^{i\phi} |0\rangle\langle 1|a^\dagger \right] \qquad (14.53)$$

Here Ω is the Rabi frequency (proportional to the square root of the laser intensity, I), ϕ is the phase of laser drive, η is the Lamb-Dicke parameter (characterizing the strength of the interaction between the laser and the ions' oscillations), L is the number of ions, and a ($a^\dagger$) is the destruction (creation) operator for quanta of the CM motion, satisfying

$$a|g\rangle = 0, \quad a^\dagger|g\rangle = |e\rangle, \quad [a, a^\dagger] = 1 \qquad (14.54)$$

The unitary operations effected by applying these Hamiltonians to the m-th qubit for a duration given by a parameter θ and phase ϕ are:

$$V_m(\theta, \phi) : \begin{array}{l} |0\rangle_m \to cos(\theta/2)|0\rangle_m - ie^{i\phi}sin(\theta/2)|1\rangle_m \\ |1\rangle_m \to cos(\theta/2)|1\rangle_m - ie^{-i\phi}sin(\theta/2)|0\rangle_m \end{array} \qquad (14.55)$$

and

$$U_m(\theta, \phi) : \begin{array}{l} |0\rangle_m|e\rangle \to cos(\theta/2)|0\rangle_m|e\rangle - ie^{i\phi}sin(\theta/2)|1\rangle_m|g\rangle \\ |1\rangle_m|g\rangle \to cos(\theta/2)|1\rangle_m|g\rangle - ie^{-i\phi}sin(\theta/2)|0\rangle_m|e\rangle \end{array} \qquad (14.56)$$

To perform logic operations on the qubits an additional red-detuned operation involving the transition from $|0\rangle$ to an auxiliary level, $|aux\rangle$, in each qubit is required. with Hamiltonian

$$H_U^{aux} = \frac{\hbar\eta\Omega}{2\sqrt{L}} \left[e^{-i\phi} |aux\rangle\langle 0|a + e^{i\phi} |0\rangle\langle aux|a^\dagger \right] \qquad (14.57)$$

with associated unitary operation $U_m^{aux}(\theta\phi)$. For example, the controlled-sign-flip (CSF) operation between two qubits, c and t

$$CSF_{ct}: \begin{array}{l} |0\rangle_c|0\rangle_t \to |0\rangle_c|0\rangle_t \\ |0\rangle_c|1\rangle_t \to |0\rangle_c|1\rangle_t \\ |1\rangle_c|0\rangle_t \to |1\rangle_c|0\rangle_t \\ |1\rangle_c|1\rangle_t \to -|1\rangle_c|1\rangle_t \end{array} \qquad (14.58)$$

can be accomplished with the sequence of three U-pulses of appropriate duration:

$$CSF_{ct} = U_c(\pi,0)U_t^{aux}(2\pi,0)U_c(\pi,0) \qquad (14.59)$$

From this operation a $CNOT$ gate can be produced as

$$CNOT_{ct} = V_t(\pi/2,\pi/2)CSF_{ct}V_t(\pi/2,\pi/2) \qquad (14.60)$$

The speed of U- and V-pulse transitions is determined by the Rabi frequency, Ω, which is proportional to the square root of the laser intensity. But the U-pulses are slower than the V-pulses because they must put the ions' center of mass into motion, which is a slower process with more ions, and moreover the Lamb-Dicke parameter, η, is less than one. Because of their slowness (smallness of the coupling) the U-operations are the rate-limiting quantities to quantum logic operations. It is therefore desirable to drive these transitions as quickly as possible. However, the laser intensity cannot be made arbitrarily large, in order to avoid driving a V-transition, for instance. In the following we shall only count the duration of the U-pulses to the computational time.

There are two classes of candidates for the qubit levels, which we shall refer to as "metastable state" and "Raman" qubits, respectively. The first category occurs in ions such as Hg^+, Sr^+, Ca^+, Ba^+ and Yb^+ with first excited states that are metastable, with lifetimes ranging from 0.1 s (Hg^+), 0.4s (Sr^+), 1s (Ca^+) (see Figure 14.2); 1 min (Ba^+) and even 10 years (Yb^+). A qubit is comprised of an ion's electronic ground (S) state ($|0\rangle$), and a sublevel ($|1\rangle$) of the metastable excited state (a D-state in Hg, Ca or Ba; an F-state in Yb). The advantage of this scheme is that it requires only a single laser beam to drive the qubit transitions, which greatly simplifies the optics of ion addressing. However, the disadvantage of this scheme is that it requires optical frequency stability of the laser drive that effects coherent transitions between the qubit levels.

Raman qubit schemes use hyperfine sublevels of an ion's ground state, or even Zeeman sublevels in a small magnetic field for ions with zero nuclear spin, with transitions between the qubit levels driven by Raman transitions. The advantages of this type of scheme are that the qubit states can be much longer-lived than the metastable state qubits; only radio frequency stability is required (corresponding to the frequency difference between the sublevels); and there are many more possible choices of ion (Be^+, Ca^+, Ba^+ and Mg^+ for example). Disadvantages are that addressing of the qubits is more complex owing to the requirement for two laser beams; and the readout is more involved than with metastable state qubits.

During quantum computation it is essential that a QC evolves through a sequence of pure quantum states, prescribed by some quantum algorithm. In general there will be some time scale required for a particular computation, and other time scales characterizing the processes that lead to the loss of quantum coherence. By estimating these time scales we can determine if ion trap QCs have the necessary preconditions to allow quantum computation to be performed, and which systems are most favorable. Furthermore, certain decoherence mechanisms become more pronounced with larger numbers of qubits, and there are technological limits to the number of qubits that can be held and addressed. Therefore, there are also memory (space) limitations to quantum computation, as well as time limitations, and it will be important to determine how to optimize quantum algorithms to make best use of the available resources. In our experiment we have determined that more than 20 ions can be held in a linear configuration and optically addressed with minimal cross-talk, using available technology [25].

The various decoherence mechanisms can be separated into two classes: Fundamental or technical. The former are limitations imposed by laws of Nature, such as the spontaneous emission of a photon from a qubit level, or the breakdown of the two-level approximation if a qubit transition is driven with excessive laser power. The technical limits are those imposed by existing experimental techniques, such as the "heating" of the ions' CM vibrational mode, or the phase stability of the laser driving the qubit transitions. One might expect that these limitations would become less restrictive as technology advances.

14.6   Computational limits with trapped ion qubits

We shall consider a quantum algorithm that requires L qubits (ions), and n laser pulses (we count only the slow, U-pulses), each of duration t (a π-pulse, $\theta = \pi$, for definiteness). With metastable state qubits, spontaneous emission of just one photon from one of the qubits' $|1\rangle$ states will destroy the quantum coherence required to complete this computation, so we may set an upper limit on the computational time, nt, in terms of the spontaneous emission lifetime of this level, τ_0. The specific form of the bound depends on the "average" number of qubits that will occupy the $|1\rangle$ state during the computation: we choose this proportion to be $2/3$; giving a bound:

$$nt < 6\tau_0/L \qquad (14.61)$$

So we see that "more" computation can be performed if the logic gate time, t, can be reduced. The duration, t, of a π-pulse (proportional to Ω^{-1}) is determined by the intensity, I, of the laser field: $t \sim I^{-1/2}$. However, t cannot be made arbitrarily small. In an earlier paper we showed that t cannot be smaller than the period of the

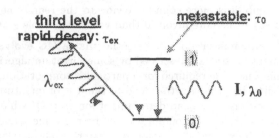

Fig. 14.6. Decoherence induced by breakdown of the two-level approximation.

CM motion, and shorter periods require stronger axial potentials that push the ions closer together [27]. The shortest possible gate time then corresponds to a minimum ion spacing of one wavelength of the interrogating laser light. In this paper we will consider a different mechanism that gives comparable limits: The breakdown of the two level approximation in intense laser fields, first considered in Reference 57.

In addition to the two states comprising each qubit, there are other, "extraneous" ionic levels with higher energies than the $|1\rangle$ state that have rapid electric dipole transitions (lifetime τ_{ex}) to the ground state, and so if some population is transferred to such states during computation their rapid decay will destroy quantum coherence (see Figure 14.6). Although the driving laser frequency is far off-resonance (detuning Δ) from the transition frequency between $|0\rangle$ and a higher lying extraneous level, in intense laser fields there will be some probability, P_{ex}, of occupying this level, given by

$$P_{ex} \sim \frac{\Omega_{ex}^2}{8\Delta^2} \tag{14.62}$$

where Ω_{ex} is the Rabi frequency for the transition from the ground state, $|0\rangle$, to the higher lying, extraneous level.

Therefore, the probability of decoherence through this two-level breakdown is proportional to the laser intensity, I. By requiring that the probability of photon emission from a third level should be less than one during the computation, we obtain the following inequality

$$nt\frac{\Omega_{ex}^2}{8\Delta^2\tau_{ex}} < 1 \tag{14.63}$$

This inequality sets an upper bound on the laser intensity. From the two inequalities 14.61 and 14.63 we obtain the bound

$$nL < \eta \left(\frac{20}{\pi}\right)^{1/2} \left(\frac{\lambda_0}{\lambda_{ex}}\right)^{3/2} \tau_{ex}\Delta \tag{14.64}$$

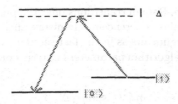

Fig. 14.7.

between an algorithmic quantity (left-hand side) and a physics parameter (right-hand side), where λ_0 is the wavelength of the $|0\rangle$ - $|1\rangle$ transition, and λ_{ex} is the wavelength of the transition from the extraneous level to the $|0\rangle$ state. Using "typical" values of $\tau_{ex} \sim 10^{-8}$s and $\Delta \sim 10^{15}$Hz we see that the value of the right-hand-side of this inequality is $\sim \eta \cdot 10^7$, translating into enough time to perform a very large number ($n \sim 10^5 - -10^6$) of logic operations on tens of qubits. (The Lamb-Dicke parameter, η, for these ions will be ~ 0.01-0.1.)

The inequality (14.64) suggests that longer wavelength qubit transitions allow more computation. Indeed, for specific ions we obtain the bounds:

$$Hg^+ : nL < \eta \cdot 3 \times 10^7$$
$$Sr^+ : nL < \eta \cdot 7 \times 10^7$$
$$Ca^+ : nL < \eta \cdot 1 \times 10^8$$
$$Ba^+ : nL < \eta \cdot 5 \times 10^8$$

suggesting that Ba^+ ions may offer greater computational potential than Hg^+ or Ca^+. However, with $L \sim 60$ qubits the bound (64) in Ba^+ corresponds to a computational time $6\tau_0/L \sim 6$ s, whereas technical sources of decoherence such as ion heating and laser phase stability are likely to limit the computation before this limit is reached. Therefore, Ba^+ ions are not likely to offer any significant computational advantage over Ca+ at present.

When qubits are represented by Zeeman or hyperfine sublevels of an ion's ground state, Raman transitions would be used to drive the computational operations, detuned by an amount Δ below some third level (lifetime τ_1) (see Figure 14.7). The Rabi frequency for Raman transitions is proportional to the laser field intensity,

$$\Omega \sim I/\Delta \tag{14.65}$$

as is the decoherence process of spontaneous emission from the third level,

$$P \sim I/\Delta^2 \tag{14.66}$$

Hence, the probability of a successful computational result is independent of how quickly the computation is performed (at least from the perspective of this decoherence mechanism). Therefore, Raman transitions offer the possibility of completing a computation before technical decoherence mechanisms, such as ion heating, become significant. Using similar arguments as in the last section, we can derive the following inequality for quantum algorithm parameters in terms of the physics parameters for Raman qubits:

$$nL^{1/2} < 8\eta\tau_1\Delta \qquad (14.67)$$

The right-hand side of this inequality has a typical value $\sim \eta.5\mathrm{x}10^5$ which is adequate for a large number of gate operations ($n \sim 10^6$) on tens of qubits. With equal numbers of qubits, the error probability per gate is lower for the Raman transitions than for metastable qubits.

It is also possible to express the computational bounds on trapped ions in terms of typical atomic values of lifetimes, wavelengths etc [58]. However, the bounds obtained this way are considerably more pessimistic than the ones above because real ions have much longer lived metastable levels than is suggested by the atomic unit of electric quadrupole moment, for instance. Therefore, although indicative of the amount of computation possible, this approach does not provide an absolute upper bound on computational capacity in terms of fundamental constants [59].

To translate the above physics bounds on algorithmic quantities into limits on the size of integer that could be factored, it is necessary to determine the computational space and time requirements of quantum factoring. Using the values [50]

$$L = 5\ell + 4$$
$$n = 96\ell^3 + O(\ell^2)$$

(where n is the number of U pulses) in the decoherence bounds above, we obtain the (algorithm-dependent) factoring limits ($\eta = 0.01$):

$$Hg^+ : \ell < 5\text{bits}$$
$$Sr^+ : \ell < 6\text{bits}$$
$$Ca^+ : \ell < 6\text{bits}$$
$$Ba^+ : \ell < 10\text{bits}$$
$$Yb^+ : \ell < 5\text{bits}$$

with metastable qubits. (Larger values may be possible with Raman qubits provided a careful optimization of the parameters is made.) These limits correspond roughly to the size of computation at which the probability of success has fallen to $1/e$. Larger integers could be factored but with a lower success probability. Certainly, these projections of the intrinsic factoring capacity of ion trap QCs are insignificant in comparison with the size of integers that are used in cryptography. Nevertheless, the 6-bit factoring limit with Ca^+ ions (for instance) represents $\sim 20,000$ U-pulses applied to 34 qubits, taking ~ 0.2 s, representing ample opportunity for studying practical small-scale quantum computation. Also, the above limits do not take into account any possible gains from the use of quantum error correction. We note that the total computational time with metastable qubits is $\sim 6\tau_0/L$, so that it might be possible to reduce the computation time by using an algorithm with additional qubits but less gate operations.

These estimates show that decoherence will inevitably limit the size of quantum computation that can be performed with trapped ions. This raises the question of whether the errors caused by decoherence can be corrected. At first sight this does not appear to be promising because we know that we cannot perform measurements during computation without destroying the quantum information we are trying to protect. Also, we know that it is not possible to faithfully duplicate an arbitrary quantum state [60]. Nevertheless it is now known that quantum error correction is possible by encoding a single logical qubit as an entangled state of multiple physical qubits [30, 31]. Errors can be detected by performing quantum logical operations involving the physical qubits and various ancilla qubits, whose state is subsequently measured and the result used to apply corrective unitary operations. Furthermore, it is also known that quantum error correction itself can be performed in a fault tolerant fashion [32], and that if certain thresholds in the precision of gate operations can be attained, these techniques allow indefinite computation in principle [33]. To do full justice to these remarkable results would require an entire article, but we note that they are of the form of an existence proof and that it remains a challenge to determine how to best implement quantum error correction methods for the current hardware schemes.

There are distinct contrasts between the computational bounds for ion trap quantum computation and the recently proposed NMR quantum computation model [61]. Ion trap qubit coherence is limited by spontaneous emission processes whereas NMR qubit decoherence is thermally dominated ($kT >> h\nu$). Ion trap quantum information is consequently more robust. Furthermore, gate times in an ion trap QC could be as short as 1 μs (set by achievable laser intensities, and two-level breakdown), whereas NMR gate times will typically be ~ 0.1-1 s (set by the strength of spin-spin interactions and the need to avoid crosstalk with unintended qubits). Readout in an NMR QC is problematic, with an exponential reduction in magnetization signal with additional qubits, whereas ion trap QC readout is a robust process independent of the number of qubits involved. Moreover, an ion trap QC has the advantage that logic operations can be performed between arbitrary qubits in the register, whereas in NMR only nearest-neighbor operations are possible. Therefore, computation in an NMR QC would use much of the available coherence time in moving qubits around the register until they are adjacent to each other. We estimate that a realistic bound to the computation possible in an NMR QC is about 10 qubits and 100 logic operations. Of these 100 operations many would be used in a typical computation to move separated qubits until they are adjacent. Nevertheless, NMR quantum computation is interesting in its own right, and especially because it allows investigations of small quantum computations with existing hardware.

14.7   Summary and Conclusions

In this paper we have surveyed the remarkable developments in quantum computation that have taken place since Feynman's pioneering work in the field. For certain problems quantum entanglement and multiparticle interference can be exploited to allow much more efficient solutions by quantum parallelism than by conventional computation. We have discussed the cryptographic significance of quantum computation, and we have seen that to factor a 512-bit integer (for example) would require a quantum computer in which the quantum state of a 2,564-qubit register could be controlled in its 2^{2564}-dimensional Hilbert space, while of the order of 10^9 quantum logic operations are performed. We have also seen that with trapped ions we can envision controlling the quantum state of ~ 50 qubits in their 2^{50}-dimensional Hilbert space for $\sim 10^5$ quantum logic operations. Current experiments are only now proceeding beyond a single logic operation on a pair of qubits [62]. Given the enormous disparity between the current state of the art of quantum computation experiments and the requirements for quantum factoring of interesting numbers, it would therefore be easy to dismiss quantum computation as irrelevant for cryptography. However, as we have seen in Section 14.3, it is the possibility that quantum computation might become possible in 20 or 30 years time (say) that must be seriously considered today. Cryptography will therefore be a compelling motivation for the development of quantum computation research. But quantum computation will also open up a wide variety of important fundamental experiments in the foundations of quantum mechanics because a quantum computer allows arbitrary quantum states to be created from a small set of primitive operations.

We have surveyed the prospects for and limitations to quantum computation with trapped ions. It is apparent that with existing technology, adequate time scales and capacity to hold multiple qubits are available to explore quantum computation well beyond the current state-of-the-art. These intrinsic limits (without quantum error correction) only correspond to the factoring of small integers. However, the numbers of qubits and logic operations involved are large enough that this technology is well worth pursuing. Ion traps will therefore be a potent method for exploring whether superpositions and entangled states of large numbers of qubits can be created. Investigations of the type studied here identify the relevant physics issues that must be addressed to achieve computational gains. In particular, we note that there has yet to be a demonstration that more than one ion can be sideband cooled to the vibrational ground state. Furthermore, the heating mechanisms for this vibrational mode are poorly understood [63, 64]. Studies of sideband cooling and reheating of multiple ions will therefore be crucial to the development of ion trap QCs. Once entangled states of three or more qubits can be constructed it will also be possible to determine whether multiparticle decoherence mechanisms are consistent with the model that we have used.

We are probably now entering the "vacuum tube" era of quantum computation. We should expect that many of the technologies now being pursued for quantum computation will be superceded by even more promising ideas. As experiments on larger and larger quantum registers are developed we will have to face the fundamental dichotomy between classical and quantum physics. Perhaps we will encounter some failure of conventional quantum mechanics [65, 66], but in my opinion it is more likely that experimental progress will be limited by the daunting technical challenges. In any event the future will be exciting for both quantum physics and computation. As in a number of other cases, Richard Feynman's unique physical insights have opened up an entirely new field of research.

Postscript

As a postdoc at Caltech from 1980 to 1982 I was fortunate to have an office close to Feynman's, and so I was privileged to experience his unique approach to physics at first hand. I discovered that even the topics he did not consider worthy of publication in the physics literature were often more fascinating than many papers in print. Subsequently I derived immense pleasure from explaining one of his scientific "tricks," known as "Feynman's proof of Maxwell's equations," which remained unpublished until the story was recounted by Dyson [67]. Feynman had shown Dyson that two of the four Maxwell's equations could be "derived" starting only from the Heisenberg equations of motion of a *non-relativistic* quantum mechanical particle, subject to a force of a very general form. That so much could be derived from so little seemed miraculous, especially considering the relativistic nature of Maxwell's equations. 1 simply had to get to the bottom of the problem [68]. I found that Feynman's argument was essentially a rediscovery of some little-known consistency conditions on generalized forces that can be accommodated in classical Lagrangian mechanics, first discovered by Helmholtz in the 19th Century [69]. By understanding Feynman's "trick" I found I had developed new insights into features of classical mechanics that we often take for granted.

Acknowledgements

It is a pleasure to thank R. Cleve, A. K. Ekert, J. M. Ettinger, M. H. Holzscheiter, P. Hoyer, D. F. V. James, R. Josza, P. G. Kwiat, S. K. Lamoreaux, M. Mosca, M. S. Neergaard, V. Sandberg, M. M. Schauer, D. Tupa and W. Warren for helpful discussions. RJH thanks the ISI Foundation, Torino, Italy for hospitality. This research was funded by the National Security Agency.

References

[1] R. P. Feynman, *The Feynman Lectures on Computation*, A. J. G. Hey and R. W. Allen eds., (Addison-Wesley, Reading, MA, 1996); *Found. Phys.* **16**, 507 (1986).

[2] R. P. Feynman, *The Character of Physical Law* (MIT Press, Cambridge, 1965).

[3] C. H. Bennett, *IBM J. Res. Dev.* **17**, 525 (1973).

[4] E. Fredkin and T. Toffoli, *Int. J. Theor. Phys.* **21**, 219 (1982).

[5] R. P. Feynman, *Int. J. Theor. Phys.* **21**, 467 (1982).

[6] P. Benioff, *Phys. Rev. Lett.* **48**, 1581 (1982).

[7] E. Schrodinger, *Naturwiss.* **23**, 807, 823, 844 (1935).

[8] E. Schrodinger, *Proc. Camb. Phil. Soc.* **31**, 555 (1935); 32, 446 (1936).

[9] D. Deutsch, *Proc. Roy. Soc.*, **A400**, 97 (1985).

[10] D. Deutsch, *Proc. Roy. Soc.* **A425**, 73 (1989).

[11] E. Bernstein and U. Vazirani, *SIAM J. Comput.* **26**, 1411 (1997).

[12] A. Yao, "Quantum circuit complexity" in *Proc. 34th IEEE Symposium on the Foundations of Computer Science*, pp. 352 (IEEE Press, Piscataway, NJ, 1993).

[13] D. Deutsch and R. Josza, *Proc. Roy. Soc.* **A439**, 553 (1992).

[14] P. W. Shor, "Algorithms for Quantum Computation: Discrete Logarithm and Factoring," in *Proceedings of the 35th Annual Symposium on the Foundations of Computer Science*, S. Goldwasser ed., pp.124 (IEEE, Los Alamitos, CA, 1994); *SIAM J. Comp.* **26**, 1484 (1997).

[15] D. Simon, in *Proceedings of the 35th Annual Symposium on the Foundations of Computer Science*, S. Goldwasser ed., pp.124 (IEEE, Los Alamitos, CA, 1994).

[16] B. Schneier, *Applied Cryptography* (Wiley, New York, 1996); A. J. Menzenes, P. C. van Oorschot and S. A. Vanstone, *Handbook of Applied Cryptography* (CRC Press, New York, 1997).

[17] L. K. Grover, *Proceedings of the 28th Annual ACM Symposium on the Theory of Computing*, pp. 212 (ACM Press, New York, 1996).

[18] D. Boneh and R. Lipton, "Quantum cryptanalysis of hidden linear functions" *Advances in Cryptology: CRYPTO'95* (Springer, New York, 1995).

[19] A. Kitaev, "Quantum measurements and the Abelian stabilizer problem," quant-ph 9511026.

[20] P. Hoyer, "Efficient quantum transforms," quant-ph/9702028.

[21] R. Landauer, *Phil. Trans. Roy. Soc.* **A353**, 367 (1995).

[22] W. Unruh, *Phys. Rev.* **A51**, 992 (1995).

[23] J. I. Cirac and P. Zoller, *Phys. Rev. Lett.* **74**, 4094 (1995).

[24] D. F. V. James, "Quantum dynamics of cold trapped ions, with application to quantum computation," Los Alamos report LA-UR-97-745, quant-ph 9702053 (1977), *Appl. Phys. B*, **66**, 181 (1998).

[25] R. J. Hughes et al., "The Los Alamos Trapped Ion Quantum Computer Experiment," Los Alamos report LA-UR-97-3301, *Fortschritte der Physik*, **46**, 4, (1998), http://xxx.lanl.gov/ps/quant-ph/9708050; see also http://p23.lanl.gov/Quantum/quantum.html.

[26] C. Monroe et al., *Phys. Rev. Lett.* **75**, 4714 (1995).

[27] R. J. Hughes et al., *Phys. Rev. Lett.* **77**, 3240 (1996); D. F. V. James et al., *Proc. SPIE* **3076**, 42 (1997).

[28] W. H. Zurek, *Physics Today* **44**, 36 (1991).

[29] G. Palma et al., *Proc. Roy. Soc.* **A452**, 567 (1996).

[30] P. Shor, *Phys. Rev.* **A52**, 2493 (1995).

[31] A. M. Steane, *Phys. Rev. Lett.* **77**, 793 (1996).

[32] P. Shor, "Fault tolerant quantum computation," in *Proceedings of the 35th Annual Symposium on the Foundations of Computer Science*, Los Alamitos, CA (1996).

[33] R. Laflamme, E. Knill and W. Zurek, *Science* (in press); J. Preskill, "Reliable quantum computers," quant-ph/9705031.

[34] For a review see, A. K. Ekert, "Quantum Computation," in "Atomic Physics 14" D. J. Wineland et al., eds., pp.450 (AIP, New York, 1995).

[35] R. Cleve et al., "Quantum Algorithms Revisited," quant-ph/9708016.

[36] M. Boyer et al., "Tight bounds on quantum searching," in *Proc. 4th. Workshop on Phys. Comput.*, pp.36, New England Complex Systems Institute, Boston, MA (1996).

[37] C. H. Bennett et al., *SIAM J. Comput.* **26**, 1510 (1997).

[38] K. F. Gauss, *Disquisitiones Arithmeticae (1801)*, translated by A. A. Clarke (Springer, New York, 1986).

[39] C. Pomerance, *Notices of the AMS* **43**, 1473 (1996); L. M. Adelman, "Algorithmic number theory — the complexity connection" in *Proc. 35th Annual Symposium on the Foundations of Computer Science*, pp.88 (IEEE, Los Alamitos, 1994).

[40] M. Gardner, *Sci. Am.* **237** (August), 120 (1977).

[41] D. Atkins et al., *Advances in Cryptology - ASIACRYPT'94*, pp 263 (Springer-Verlag, New York, 1994).

[42] R. J. Hughes, "Cryptography, quantum computation and trapped ions," Los Alamos report LA-UR-97-4986, (quant-ph/9712054) to be published in *Philosophical Transactions of the Royal Society A*.

[43] R. L. Rivest, A. Shamir and L. M. Adelman, *Comm. ACM* **21**, 120 (1978).

[44] See: M. R. Schroeder, *Number theory in science and communication*, (Springer, New York, 1992) for instance.

[45] H. Riesel, *Prime Numbers and Computer Methods for Factorization*, (Birkhauser, Boston, 1994).

[46] A. K. Lenstra and H. W. Lenstra, *Lecture Notes in Mathematics 1554: The development of the Number Field Sieve* (Springer, New York, 1993).

[47] J. Cowie et al., *Advances in Cryptology - ASIACRYPT'96*, pp 382, K. Kim and T. Matsumoto eds. (Springer, New York, 1996).

[48] A. K. Ekert and R. Josza, *Rev. Mod. Phys.* **68**, 733-753 (1996).

[49] D. Beckman et al., *Phys. Rev.* **A54**, 1034 (1996).

[50] R. J. Hughes and. M. S. Neergaard, in preparation.

[51] D. G. Cory, A. F. Fahmy and T. F. Havel, *Proc. Natl. Acad. Sci. USA* **94**, 1634 (1997); N. A. Gershenfeld and I. L. Chuang, *Science* **275**, 350 (1997).

[52] Q. A. Turchette et al., *Phys. Rev. Lett.* **75**, 4710 (1995).

[53] M. G. Raizen et al., *Phys. Rev.* **A45**, 6493 (1992).

[54] R. H. Dicke, *Phys. Rev.* **89**, 472 (1953); C. Cohen-Tannoudji et al., "Atom -Photon Interactions", pg. 518 (Wiley, New York, 1992).

[55] F. Diedrich et al., *Phys. Rev. Lett.* **62**, 403 (1989).

[56] W. Nagourney et al., *Phys. Rev. Lett.* **56**, 2797 (1986); Th. Sauter et al., Phys. Rev. Lett. 57, 1696 (1986); J. C. Bergquist et al., *Phys. Rev. Lett.* **57**, 1699 (1986).

[57] M. B. Plenio and P. L. Knight, *Phys. Rev.* **A53**, 2986 (1996); "Decoherence limits to quantum computation using trapped ions," quant-ph 9610015.

[58] S. Haroche and J.-M. Raimond, *Physics Today* (August 1996) 51.

[59] R. J. Hughes and D. F. V. James, "Prospects for quantum computation with trapped ions," Los Alamos report LA-UR-97-3967, to be published in the proceedings of the University of Maryland Baltimore County conference on "Fundamental problems in quantum theory," August 1997.

[60] W. K. Wooters and W. H. Zurek, *Nature* **299**, 802 (1982); D. Dieks, *Phys. Lett.* **A92**, 271 (1982).

[61] W. S. Warren, *Science* **277**, 1688 (1997).

[62] D. Bouwmeester et al., *Nature* **390**, 575 (1997).

[63] D. J. Wineland et al., *Phys. Rev.* **A50**, 67 (1994); D. J. Wineland and H. G. Dehmelt, *J. Appl. Phys.* **46**, 919 (1975).

[64] S. K. Lamoreaux, *Phys. Rev* **A56**, 4970 (1988).

[65] G. C. Ghirardi et al., *Phys. Rev.* **D34**, 470 (1986).

[66] R. Penrose, *Gen. Rel. Grav.* **28**, 581 (1996).

[67] F. J. Dyson, *Physics Today*, February 1989, 32; *Am. J. Phys.* **58**, 209 (1990).

[68] R. J. Hughes, *Am. J. Phys.* **60**, 301 (1992).

[69] H. v Helmholtz, *J. Reine Angewandte Math.* **100**, 137 (1887).

Part IV

Parallel Computation

15

COMPUTING MACHINES IN THE FUTURE

Richard P. Feynman

Nishina Memorial Lecture, August 9, 1985

15.1 Introduction

It's a great pleasure and an honor to be here as a speaker in memorial for a scientist that I have respected and admired as much as Professor Nishina. To come to Japan and talk about computers is like giving a sermon to Buddha. But I have been thinking about computers and this is the only subject I could think of when invited to talk.

The first thing I would like to say is what I am not going to talk about. I want to talk about the future computing machines. But the most important possible developments in the future are things that I will not speak about. For example, there is a great deal of work to try to develop smarter machines, machines which have a better relationship with humans so that input and output can be made with less effort than the complex programming that's necessary today. This often goes under the name of artificial intelligence but I don't like that name. Perhaps the unintelligent machines can do even better than the intelligent ones. Another problem is the standardization of programming languages. There are too many languages today, and it would be a good idea to choose just one. (I hesitate to mention that in Japan, for what will happen will be that there will simply be more standard languages — you already have four ways of writing now and attempts to standardize anything here result apparently in more standards and not fewer!) Another interesting future problem that is worth working on but I will not talk about, is automatic debugging programs. Debugging means fixing errors in a program or in a machine and it is surprisingly difficult to debug programs as they get more complicated. Another direction of improvement is to make physical machines three dimensional instead of all on a surface of a chip. That can be done in stages instead of all at once — you can have several layers and then add many more layers as the time goes on. Another important device would be one that could automatically detect defective elements on a chip and then the chip automatically rewire itself so as to avoid the defective elements. At the present time when we try to make big chips there are often flaws or bad spots in the chips, and we throw the whole chip away. If we could make it so that we could use the part of the chip that was effective, it would be much more efficient. I mention these things to try to tell you that I am aware of what the real problems are for future machines. But what I want to talk about is simple, just some small technical, physically good things that can be done in principle according to the physical laws. In other words, I would like to discuss the machinery and not the way we use the machines.

I will talk about some technical possibilities for making machines. There will be three topics. One is parallel processing machines which is something of the very near future, almost present, that is being developed now. Further in the future is the question of the energy consumption of machines which seems at the moment to be a limitation, but really isn't. Finally I will talk about the size. It is always better to make the machines smaller, and the question is how much smaller is it still possible, in principle, to make machines according to the laws of Nature. I will not discuss which and what of these things will actually appear in the future. That depends on economic problems and social problems and I am not going to try to guess at those.

15.2   Parallel Computers

The first topic concerns parallel computers. Almost all the present computers, conventional computers, work on a layout or an architecture invented by von Neumann, in which there is a very large memory that stores all the information, and one central location that does simple calculations. We take a number from this place in the memory and a number from that place in the memory, send the two to the central arithmetical unit to add them and then send the answer to some other place in the memory. There is, therefore, effectively one central processor which is working very very fast and very hard while the whole memory sits out there like a vast filing cabinet of cards which are very rarely used. It is obvious that if there were more processors working at the same time we ought to be able to do calculations faster. But the problem is that someone who might be using one processor may be using some information from the memory that another one needs, and it gets very confusing. For such reasons it has been said that it is very difficult to get many processors to work in parallel.

Some steps in that direction have been taken in the larger conventional machines called "vector processors". When sometimes you want to do exactly the same step on many different items you can perhaps do that at the same time. The hope is that regular programs can be written in the ordinary way, and then an interpreter program will discover automatically when it is useful to use this vector possibility. That idea is used in the Cray and in "supercomputers" in Japan. Another plan is to take what is effectively a large number of relatively simple (but not very simple) computers, and connect them all together in some pattern. Then they can all work on a part of the problem. Each one is really an independent computer, and they will transfer information to each other as one or another needs it. This kind of a scheme is realised in the Caltech Cosmic Cube, for example, and represents only one of many possibilities. Many people are now making such machines. Another plan is to distribute very large numbers of very simple central processors all over the memory. Each one deals with just a small part of the memory and there is an elaborate system of interconnections between them. An example of such a machine is the Connection Machine made at MIT. It has 64,000 processors and a system of routing in which every 16 can talk to any other 16 and thus has 4000 routing connection possibilities.

It would appear that scientific problems such as the propagation of waves in some material might be very easily handled by parallel processing. This is because what happens in any given part of space at any moment can be worked out locally and only the pressures and the stresses from the neighboring volumes need to be known. These can be worked out at the same time for each volume and these boundary conditions communicated across between the different volumes. That's why this type of design works for such problems. It has turned out that a very large number of problems of all kinds can be dealt with in parallel. As long as the problem is big enough so that a lot of calculating has to be done, it turns out that a parallel computation can speed up time to solution enormously and this principle applies not just to scientific problems.

What happened to the prejudice of two years ago, which was that the parallel programming is difficult? It turns out that what was difficult, and almost impossible, is to take an ordinary program and automatically figure out how to use the parallel computation effectively on that program. Instead, one must start all over again with the problem, appreciating that we have parallel possibility of calculation, and rewrite the program completely with a new attitude to what is inside the machine. It is not possible to effectively use the old programs. They must be rewritten. That is a great disadvantage to most industrial applications and has met with considerable resistance. But the big programs usually belong to scientists or other, unofficial, intelligent programmers who love computer science and are willing to start all over again and rewrite the program if they can make it more efficient. So what's going to happen is that the hard programs, vast big ones, will be the first to be re-programmed by experts in the new way, and then gradually everybody will have to come around, and more and more programs will be programmed that way, and programmers will just have to learn how to do it.

15.3   Reducing the Energy Loss

The second topic I want to talk about is energy loss in computers. The fact that they must be cooled is an apparent limitation for the largest computers — a good deal of effort is spent in cooling the machine. I would like to explain that this is simply a result of very poor engineering and is nothing fundamental at all. Inside the computer a bit of information is controlled by a wire which either has a voltage of one value or another value. It is called "one bit", and we have to change the voltage of the wire from one value to the other and have to put charge on or take charge off. I make an analogy with water: We have to fill a vessel with water to

ENERGY USE:

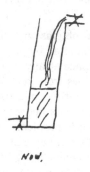

NOW,

Fig. 15.1.

get one level or empty it to get to the other level. This is just an analogy — if you like electricity better you can think more accurately electrically. What we do now is analogous, in the water case, to filling the vessel by pouring water in from a top level (Fig. 15.1), and lowering the level by opening the valve at the bottom and letting it all run out. In both cases there is a loss of energy because of the sudden drop in level of the water, through a height from the top level where it comes in, to the low bottom level, and also when you start pouring water in to fill it up again. In the cases of voltage and charge, the same thing occurs.

It's like, as Mr. Bennett has explained, operating an automobile which has to start by turning on the engine and stop by putting on the brakes. By turning on the engine and then putting on the brakes, each time you lose power. Another way to arrange things for a car would be to connect the wheels to flywheels. Now when the car stops, the flywheel speeds up thus saving the energy — which can then be reconnected to start the car again. The water analog of this would be to have a U-shaped tube with a valve in the center at the bottom, connecting the two arms of the U (Fig. 15.2). We start with it full on the right but empty on the left with the valve closed. If we now open the valve, the water will slip over to the other side, and we can close the valve again, just in time to catch the water in the left arm. Now when we want to go the other way, we open the valve again and the water slips back to the other side and we catch it again. There is some loss and the water doesn't climb as high as it did before, but all we have to do is to put a little water in to correct the loss — a much smaller energy loss than the direct fill method. This trick uses the inertia of the water and the analogue for electricity is inductance. However, it is very difficult with the silicon transistors that we use

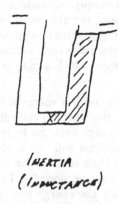

INERTIA

(INDUCTANCE)

Fig. 15.2.

today to make up inductance on the chips. So this technique is not particularly practical with present technology.

Another way would be to fill the tank by a supply which stays only a little bit above the level of the water, lifting the water supply in time as we fill up the tank (Fig. 15.3), so that the dropping of water is always small during the entire effort. In the same way, we could use an outlet to lower the level in the tank, but just taking water off near the top and lowering the tube so that the heat loss would not appear at the position of the transistor, or would be small. The actual amount of loss will depend on how high the distance is between the supply and the surface as we fill it up. This method corresponds to changing the voltage supply with time. So if we

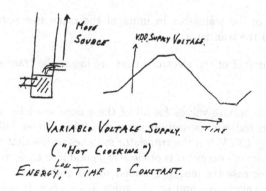

VARIABLE VOLTAGE SUPPLY.

("HOT CLOCKING")

ENERGY · TIME = CONSTANT.

Fig. 15.3.

could use a time varying voltage supply, we could use this method. Of course, there is energy loss in the voltage supply, but that is all located in one place and there it is simple to make one big inductance. This scheme is called "hot clocking", because the voltage supply operates at the same time as the clock which times everything. In addition we don't need an extra clock signal to time the circuits as we do in conventional designs.

Both of these last two devices use less energy if they go slower. If I try to move the water supply level too fast, the water in the tube doesn't keep up with it and there ends being a big drop in water level. So to make the device work I must go slowly. Similarly, the U-tube scheme will not work unless that central valve can open and close faster than the time it takes for the water in the U-tube to slip back and forth. So my devices must be slower — I've saved an energy loss but I've made the devices slower. In fact the energy loss multiplied by the time it takes for the circuit to operate is constant. But nevertheless, this turns out to be very practical because the clock time is usually much larger than the circuit time for the transistors, and we can use that to decrease the energy. Also if we went, let us say, three times slower with our calculations, we could use one third the energy over three times the time, which is nine times less power that has to be dissipated. Maybe this is worth it. Maybe by redesigning using parallel computations or other devices, we can spend a little longer than we could do at maximum circuit speed, in order to make a larger machine that is practical and from which we could still reduce the energy loss.

For a transistor, the energy loss multiplied by the time it takes to operate is a product of several factors (Fig. 15.4):

1. the thermal energy proportional to temperature, kT;

2. the length of the transistor between source and drain, divided by the velocity of the electrons inside (the thermal velocity $\sqrt{3kT/m}$);

3. the length of the transistor in units of the mean free path for collisions of electrons in the transistor;

4. the total number of the electrons that are inside the transistor when it operates.

Putting in appropriate values for all of these numbers tells us that the energy used in transistors today is somewhere between a billion to ten billion or more times the thermal energy kT. When the transistor switches we use that much energy. This is very large amount of energy. It is obviously a good idea to decrease the size of the transistor. We decrease the length between source and drain and we can decrease the number of the electrons, and so use much less energy. It also turns out that a smaller transistor is much faster, because the electrons can cross it faster and make

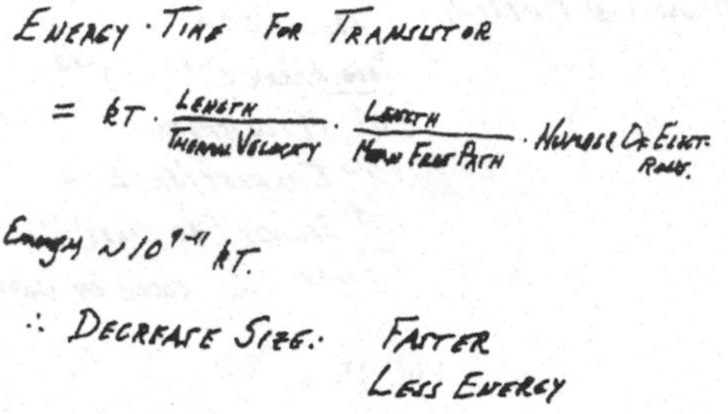

$$\text{ENERGY} \cdot \text{TIME FOR TRANSISTOR}$$

$$= kT \cdot \frac{\text{LENGTH}}{\text{THERMAL VELOCITY}} \cdot \frac{\text{LENGTH}}{\text{MEAN FREE PATH}} \cdot \text{NUMBER OF ELECT.}$$

$$\text{ENERGY} \sim 10^{9-11} \, kT.$$

$$\therefore \text{ DECREASE SIZE.} \quad \text{FASTER} \\ \text{LESS ENERGY}$$

Fig. 15.4.

their decisions to switch faster. For every reason, it is a good idea to make the transistor smaller, and everybody is always trying to do that.

But suppose we come to a circumstance in which the mean free path is longer than the size of the transistor, then we discover that the transistor doesn't work properly any more. It does not behave the way we expected. This reminds me, years ago there was something called the sound barrier. Airplanes were supposed not to be able to go faster than the speed of sound because, if you designed them normally and then tried to put the speed of sound in the equations, the propeller wouldn't work and the wings don't lift and nothing works correctly. Nevertheless, airplanes can go faster than the speed of sound. You just have to know what the right laws are under the right circumstances, and design the device with the correct laws. You cannot expect old designs to work in new circumstances. But *new* designs can work in *new* circumstances, and I assert that it is perfectly possible to make transistor systems, or more correctly, switching systems and computing devices in which the dimensions are smaller than the mean free path. I speak of course, 'in principle', and I am not speaking about the actual manufacture of such devices. Let us therefore discuss what happens if we try to make the devices as small as possible.

15.4 Reducing the Size

So my third topic is the size of computing elements and now I speak entirely theoretically. The first thing that you would worry about when things get very small, is Brownian motion — everything is shaking about and nothing stays in place. How can you control the circuits then? Furthermore, if a circuit does work, doesn't it now have a chance of accidentally jumping back? If we use two volts for the energy of this electric system, which is what we ordinarily use (Fig. 15.5) that is eighty times the thermal energy at room temperature ($kT=1/40$ volt) and the chance that something jumps backward against 80 times thermal energy is e, the base of the

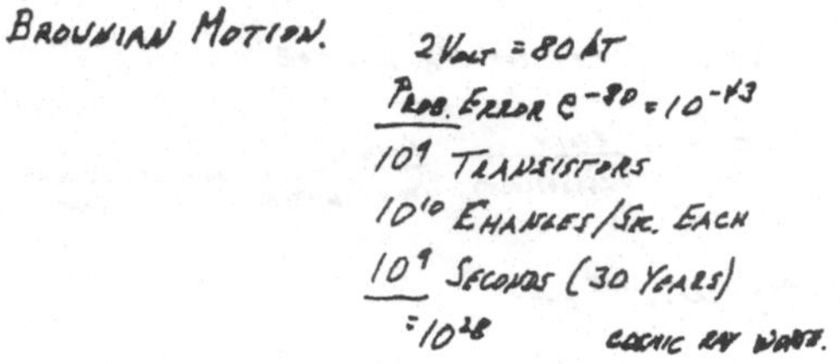

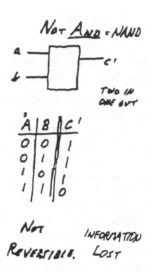

Fig. 15.5.

Fig. 15.6.

natural logarithm, to the power minus eighty, or 10^{-43}. What does that mean? If we had a billion transistors in a computer (which we don't yet have), all of them switching 10^{10} times a second (a switching time of a tenth of a nanosecond), switching perpetually, operating for 10^9 seconds, which is 30 years, the total number of switching operations in such a machine is 10^{28}. The chance of one of the transistors going backward is only 10^{-43}, so there will be no error produced by thermal oscillations whatsoever in 30 years. If you don't like that, use 2.5 volts and then the probability gets even smaller. Long before that, real failures will come when a cosmic ray accidentally goes through the transistor, and we don't have to be more perfect than that.

However, much more is in fact possible and I would like to refer you to an article in a most recent Scientific American by C. H. Bennett and R. Landauer ["The

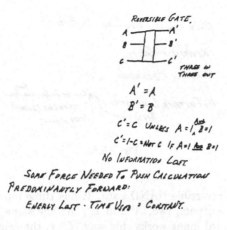

Fig. 15.7.

Fundamental Physical Limits of Computation", Sci. Am. July 1985; Japanese Transl.— SAIENSU, Sept. 1985]. It is possible to make a computer in which each element, each transistor, can go forward and accidentally reverse and still the computer will operate. All the operations in the computer can go forward or backward. The computation proceeds for a while one way and then it undoes itself, 'uncalculates', and then goes forward again and so on. If we just pull it along a little, we can make this computer go through and finish the calculation by making it just a little bit more likely that it goes forward than backward.

It is known that all possible computations can be made by putting together some simple elements like transistors; or, if we want to be more logically abstract, something called a NAND gate, for example (NAND means NOT-AND). A NAND gate has two "wires" in and one out (Fig. 15.6). Forget the NOT for the moment. What is an AND gate? An AND gate is a device whose output is 1 only if both input wires are 1, otherwise its output is 0. NOT-AND means the opposite, thus the output wire reads 1 (i.e. has the voltage level corresponding to 1) unless both input wires read 1; if both input wires read 1, then the output wire reads 0 (i.e. has the voltage level corresponding to 0). Figure 15.6 shows a little table of inputs and outputs for such a NAND gate. A and B are inputs and C is the output. If A and B are both 1, the output is 0, otherwise 1. But such a device is irreversible: Information is lost. If I only know the output, I cannot recover the input. The device can't be expected to flip forward and then come back and compute correctly anymore. For instance, if we know that the output is now 1, we don't know whether it came from $A=0$, $B=1$ or $A=1$, $B=0$ or $A=0$, $B=0$ and it cannot go back. Such a device is an irreversible gate. The great discovery of Bennett and, independently, of Fredkin, is that it is possible to do computation with a different kind of fundamental gate unit, namely, a reversible gate unit. I have illustrated their idea — with a unit

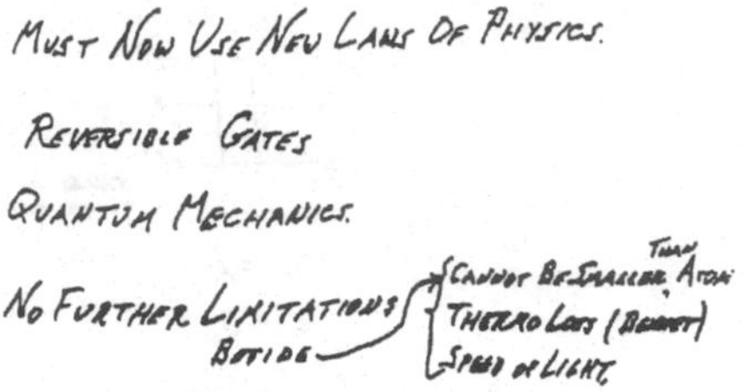

MUST Now USE NEW LAWS OF PHYSICS.

REVERSIBLE GATES

QUANTUM MECHANICS.

No FURTHER LIMITATIONS ⎰CANNOT BE SMALLER THAN ATOM
                     BESIDES ⎰THERMO LOSS (BENNETT)
                              ⎱SPEED OF LIGHT

Fig. 15.8.

which I could call a reversible NAND gate. It has three inputs and three outputs (Fig. 15.7). Of the outputs, two, A' and B', are the same as two of the inputs, A and B, but the third input works this way. C' is the same as C unless A and B are both 1 in which case it changes whatever C is. For instance, if C is 1 it is changed to 0, if C is 0 it is changed to 1 — but these changes only happen if both A and B are 1. If you put two of these gates in succession, you see that A and B will go through, and if C is not changed in both it stays the same. If C is changed, it is changed twice so that it stays the same. So this gate can reverses itself and no information has been lost. It is possible to discover what went in if you know what came out.

A device made entirely with such gates will make calculations if everything moves forward, but if things go back and forth for a while, but then eventually goes forward enough it still operates correctly. If the things flip back and then go forward later it is still all right. It's very much the same as a particle in a gas which is bombarded by the atoms around it. Such a particle usually goes nowhere, but with just a little pull, a little prejudice that makes a chance to move one way a little higher than the other way, the thing will slowly drift forward and travel from one end to the other, in spite of the Brownian motion that is has made. So our computer will compute provided we apply a drift force to pull the thing across the calculation. Although it is not doing the calculation in a smooth way, nevertheless, calculating like this, forward and backward, it eventually finishes the job. As with the particle in the gas, if we pull it very slightly, we lose very little energy, but it takes a long time to get to one side from the other. If we are in a hurry, and we pull hard, then we lose a lot of energy. It is the same with this computer. If we are patient and go slowly, we can make the computer operate with practically no energy loss, even less than kT per step, any amount as small as you like if you have enough time. But if you are in a hurry, you must dissipate energy, and again it's true that the energy lost to pull the calculation forward to complete it multiplied by the time you are allowed to make the calculation is a constant.

With these possibilities in mind, let's see how small can we make a computer. How big must a number be? We all know we can write numbers in base 2 as strings

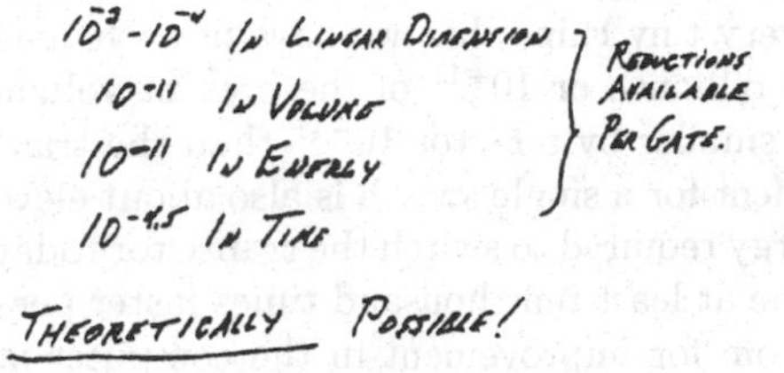

10^3-10^{-4} In Linear Dimension ⎤ Reductions

10^{-11} In Volume   ⎬ Available

10^{-11} In Energy   ⎭ Per Gate.

10^{-15} In Time

THEORETICALLY _Possible!_

Fig. 15.9.

of "bits" each a one or a zero. But how small can I write? Surely only one atom is needed to be in one state or another to determine if it represents a one or a zero. And the next atom could be a one or a zero, so a little string of atoms are enough to hold a number, one atom for each bit. (Actually, since an atom can have more than just two states, we could use even fewer atoms, but one per bit is little enough! So, for intellectual entertainment, we consider whether we could make a computer in which the writing of bits is of atomic size, in which a bit is, for example, whether the spin in the atom is up for 1 or down for 0. And then our 'transistor', which changes the bits in different places, would correspond to some interaction between atoms which will change their states. The simplest example would be a kind of 3-atom interaction to be the fundamental element or gate in such a computer. But again, the device won't work right if we design it with the laws appropriate for large objects. We must use the new laws of physics, quantum mechanical laws, the laws that are appropriate to atomic motion (Fig. 15.8). We therefore have to ask whether the principles of quantum mechanics permit an arrangement of atoms so small in number as a few times the number of gates in a computer that could operate as a computer. This has been studied in principle, and such an arrangement has been found. Since the laws of quantum mechanics are reversible, we must use the invention by Bennett and Fredkin of reversible logic gates. When this quantum mechanical situation is studied, it is found that quantum mechanics adds no further limitations to anything that Mr. Bennett has said from thermodynamic considerations. Of course there is a limitation, the practical limitation anyway, that the bits must be of the size of an atom and a transistor 3 or 4 atoms. The quantum mechanical gate I used has 3 atoms. (I would not try to write my bits onto nuclei, I'll wait till the technological development reaches atoms before I need to go any further!) That leaves us just with: (a) the limitations in size to the size of atoms; (b) the energy requirements depending on the time as worked out by Bennett; and (c) the feature that I did not mention concerning the speed of light — we can't send the signals any faster than the speed of light. These are the only physical limitations on computers that I know of.

If we somehow manage to make an atomic size computer, it would mean (Fig.

15.9) that the dimension, the *linear* dimension, is a thousand to ten thousand times smaller than those very tiny chips that we have now. It means that the volume of the computer is 100 billionth or 10^{-11} of the present volume, because the volume of the 'transistor' is smaller by a factor 10^{-11} than the transistors we make today. The energy requirement for a single switch is also about eleven orders of magnitude smaller than the energy required to switch the transistor today, and the time to make the transitions will be at least ten thousand times faster per step of calculation. So there is plenty of room for improvement in the computer and I leave this to you, practical people who work on computers, as an aim to get to. I underestimated how long it would take for Mr. Ezawa to translate what I said, and I have no more to say that I have prepared for today. Thank you! I will answer questions if you'd like.

Questions and Answers

Q: You mentioned that one bit of information can be stored in one atom, and I wonder if you can store the same amount of information in one quark.

A: Yes. But we don't have control of the quarks and that becomes a really impractical way to deal with things. You might think that what I am talking about is impractical, but I don't believe so. When I am talking about atoms, I believe that some day we will be able to handle and control them individually. There would be so much energy involved in the quark interactions that they would be very dangerous to handle because of the radioactivity and so on. But the atomic energies that I am talking about are very familiar to us in chemical energies, electrical energies, and those are numbers that are within the realm of reality, I believe, however absurd it may seem at the moment.

Q: You said that the smaller the computing element is the better. But, I think equipment has to be larger, because...

A: You mean that your finger is too big to push the buttons? Is that what you mean?

Q: Yes, it is.

A: Of course, you are right. I am talking about internal computers perhaps for robots or other devices. The input and output is something that I didn't discuss, whether the input comes from looking at pictures, hearing voices, or buttons being pushed. I am discussing how the computation is done in principle and not what form the output should take. It is certainly true that the input and the output cannot be reduced in most cases effectively beyond human dimensions. It is already too difficult to push the buttons on some of the computers with our big fingers. But with elaborate computing problems that take hours and hours, they could be done very rapidly on the very small machines with low energy consumption. That's the kind of machine I was thinking of. Not the simple applications of adding two numbers but elaborate calculations.

Q: I would like to know your method to transform the information from one atomic scale element to another atomic scale element. If you will use a quantum mechanical or natural interaction between the two elements then such a device will become very close to Nature itself. For example, if we make a computer simulation, a Monte Carlo simulation of a magnet to study critical phenomena, then your atomic scale computer will be very close to the magnet itself. What are your thoughts about that?

A: Yes. All things that we make are Nature. We arrange it in a way to suit our purpose, to make a calculation for a purpose. In a magnet there is some kind of relation, if you wish, there are some kinds of computations going on, just like there are in the solar system, in a way of thinking. But that might not be the calculation we want to make at the moment. What we need to make is a device for which we can change the programs and let it compute the problem that we want to solve, not just its own magnet problem that it likes to solve for itself. I can't use the solar system for a computer unless it just happens that the problem that someone gave me was to find the motion of the planets, in which case all I have to do is to watch. There was an amusing article written as a joke. Far in the future, the "article" appears discussing a new method of making aerodynamical calculations: Instead of using the elaborate computers of the day, the author invents a simple device to blow air past the wing. (He reinvents the wind tunnel!)

Q: I have recently read in a newspaper article that operations of the nerve system in a brain are much slower than present day computers and the unit in the nerve system is much smaller. Do you think that the computers you have talked about today have something in common with the nerve system in the brain?

A: There is an analogy between the brain and the computer in that there are apparently elements that can switch under the control of others. Nerve impulses controlling or exciting other nerves, in a way that often depends upon whether more than one impulse comes in — something like an AND or its generalization. What is the amount of energy used in the brain cell for one of these transitions? I don't know the number. The time it takes to make a switching in the brain is very much longer than it is in our computers even today, never mind the fancy business of some future atomic computer, but the brain's interconnection system is much more elaborate. Each nerve is connected to thousands of other nerves, whereas we connect transistors only to two or three others. Some people look at the activity of the brain in action and see that in many respects it surpasses the computer of today, and in many other respects the computer surpasses ourselves. This inspires people to design machines that can do more. What often happens is that an engineer has an idea of how the brain works (in his opinion) and then designs a machine that behaves that way. This new machine may in fact work very well. But, I must warn you that that does not tell us anything about how the brain actually works, nor is it necessary to ever really know that, in order to make a computer very capable. It is not necessary to understand the way birds flap their wings and how the feathers are designed in order to make a flying machine. It is not necessary to understand the lever system in the legs of a cheetah — an animal that runs fast — in order to make an automobile with wheels that goes very fast. It is therefore not necessary to imitate the behavior of Nature in detail in order to engineer a device which can in many respects surpass Nature's abilities. It is an interesting subject and I like to talk about it. Your brain is very weak compared to a computer. I will give you a series of numbers, one, three, seven... Or rather, ichi, san, shichi, san, ni, go, ni, go, ichi, hachi, ichi, ni, ku, san, go. Now I want you to repeat them back to me. A computer can take tens of thousands of numbers and give me them back in reverse, or sum them or do lots of things that we cannot do. On the other hand, if I look at a face, in a glance I can tell you who it is if I know that person, or that I don't know that person. We do not yet know how to make a computer system so that if we give it a pattern of a face it can tell us such information, even if it has seen many faces and you have tried to teach it. Another interesting example is chess playing machines. It is quite a surprise that we can make machines that play chess better than almost everybody in the room. But they do it by trying many many possibilities. If he moves here, then I could move here, and he can move there, and so forth. They look at each alternative and choose the best. Computers look at millions of alternatives, but a master chess player, a human, does it differently. He recognizes patterns. He looks at only thirty or forty positions before deciding what move to make. Therefore, although the rules are simpler in Go, machines that play Go are not very good, because in each position there are too many possibilities to move and there are too many things to check and the machines cannot look deeply. Therefore the problem of recognizing patterns and what to do under these circumstances is the thing that the computer engineers (they like to call themselves computer scientists) still find very difficult. It is certainly one of the important things for future computers, perhaps more important than the things I spoke about. Make a machine to play Go effectively!

Q: I think that any method of computation would not be fruitful unless it would give a kind of provision on how to compose such devices or programs. I thought the Fredkin paper on conservative logic was very intriguing, but once I came to think of making a simple program using such devices I came to a halt because thinking out such a program is far more complex than the program itself. I think we could easily get into a kind of infinite regression because the process of making out a certain program would be more complex than the program itself and in trying to automate the process, the automating program would be much more complex and so on. Especially in this case where the program is hard wired rather than being separated as a software. I think it is fundamental to think of the ways of composition.

A: We have some different experiences. There is no infinite regression: it stops at a certain level of complexity. The machine that Fredkin ultimately is talking about and the one that I was talking about in the quantum mechanical case are both universal computers in the sense that they can be programmed to do various jobs. This is not a hard-wired program. They are no more hard-wired than an ordinary computer that you can put information in — the program is a part of the input — and the machine does the problem that it is assigned to do. It is hard-wired but it is universal like an ordinary computer. These things are very uncertain but I found an algorithm. If you have a program written for an irreversible machine, the ordinary program, then I can convert it to a reversible machine program by a direct translation scheme, which is very inefficient and uses many more steps. Then, in real situations, the number of steps can be much less. But at least I know that I can take a program with a 2n steps where it is irreversible, convert it to 3n steps of a reversible machine. That is many more steps. I did it very inefficiently since I did not try to find the minimum — just one way of doing it. I don't really think that we'll find this regression that you speak of, but you might be right. I am uncertain.

Q: Won't we be sacrificing many of the merits we were expecting of such devices, because those reversible machines run so slow? I am very pessimistic about this point.

A: They run slower, but they are very much smaller. I don't make it reversible unless I need to. There is no point in making the machine reversible unless you are trying very hard to decrease the energy enormously, rather ridiculously, because with only 80 times kT the irreversible machine functions perfectly. That 80 is much less than the present day 10^9 or 10^{10} kT, so I have at least 10^7 improvement in energy to make, and can still do it with irreversible machines! That's true. That's the right way to go, for the present. I entertain myself intellectually for fun, to ask how far could we go in principle, not in practice, and then I discover that I can go to a fraction of a kT of energy and make the machines microscopic, atomically microscopic. But to do so, I must use the reversible physical laws. Irreversibility comes because the heat is spread over a large number of atoms and can't be gathered back again. When I make the machine very small, unless I allow a cooling element which is lots of atoms, I have to work reversibly. In practice there probably will never come a time when we will be unwilling to tie a little computer to a big piece of lead which contains 10^{10} atoms (which is still very small indeed) making it effectively irreversible. Therefore I agree with you that in practice, for a very long time and perhaps forever, we will use irreversible gates. On the other hand it is a part of the adventure of science to try to find a limitation in all directions and to stretch a human imagination as far as possible everywhere. Although at every stage it has looked as if such an activity was absurd and useless, it often turns out at least not to be useless.

Q: Are there any limitations from the uncertainty principle? Are there any fundamental limitations on the energy and the clock time in your reversible machine scheme?

A: That was my exact point. There is no further limitation due to quantum mechanics. One must distinguish carefully between the energy lost or consumed irreversibly, the heat generated in the operation of the machine, and the energy content of the moving parts which might be extracted again. There is a relationship between the time and the energy which might be extracted again. But that energy which can be extracted again is not of any importance or concern. It would be like asking whether we should add the mc^2, the rest energy, of all the atoms which are in the device. I only speak of the energy lost times the time, and then there is no limitation. However it is true that if you want to make a calculation at a certain extremely high speed, you have to supply to the machine parts which move fast and have energy, but that energy is not necessarily lost at each step of the calculation; it coasts through by inertia.

A (to no Q): Could I just say with regard to the question of useless ideas, I'd like to add one more. I waited, if you would ask me, but you didn't. So I will answer it anyway. How would we make a machine of such small dimensions where we have to put the atoms in special places? Today we have no machinery with moving parts whose dimension is extremely small, at the scale of atoms or hundreds of atoms even, but there is no physical limitation in that direction either. There is no reason why, when we lay down the silicon even today, the pieces cannot be made into little islands so that they are movable. We could also arrange small jets so we could squirt the different chemicals on certain locations. We can make machinery which is extremely small. Such machinery will be easy to control by the same kind of computer circuits that we make. Ultimately, for fun again and intellectual pleasure, we could imagine machines as tiny as a few microns across, with wheels and cables all interconnected by wires, silicon connections, so that the thing as a whole, a very large device, moves not like the awkward motions of our present stiff machines but in a smooth way of the neck of a swan, which after all is a lot of little machines, the cells all interconnected and all controlled in a smooth way. Why can't we do that ourselves?

Acknowledgement

This lecture was originally published in 1985 as a Nishina Memorial Lecture. It is reprinted here with kind permission of Professor K.Nishijima on behalf of the Nishina Memorial Foundation.

16

INTERNETICS: TECHNOLOGIES, APPLICATIONS AND ACADEMIC FIELDS

or

Parallel Computing
and Computational Science
Do Not Quite Work

Geoffrey C. Fox

Dedicated to Richard Feynman

Abstract

Ten years ago, we were all sure that parallel computing technology and the inter-disciplinary academic field of computational science would be center pieces of both academic and economic growth. We show that this insight was, in principle, correct but was an incomplete vision for large-scale computation implies both increased computer power and increasing numbers of users and applications. Parallel computing undoubtedly works on essentially all problems, but we were unable to produce deployable software systems. Further, few industries could achieve adequate return to justify investment in parallel computers, except in a few areas such as databases. Computational science is the academic field on the interface of computer science with fields such as physics, chemistry, and applied mathematics. This expertise allows you to be very useful and, in principle, is an excellent area of study, but is not a wise field for many students as employers and universities prefer traditional fields.

We show how parallel computing and computational science has evolved into Internetics, which is a vibrant growing and much larger field that surely does work both in principle and in practice. Internetics embodies the technologies and expertise used in building large-scale distributed systems and linking fields like physics not just with parallel computers, but with the Web of complex heterogeneous computers. This is CORBA and Java, and not just MPI and HPF. It is Internetics that is the emerging academic field, and not computational science, and internetics is of growing attraction to students and employers. Using an Internetics base, we will produce much better software environments for parallel systems, but the commercial and academic fields associated with parallelism will not grow in the near future.

We argue that we almost "got it right" and the essential features of the original vision were correct and are part of current broader thrust.

16.1   Introduction

In our first book on parallel computing, we joyfully used the well-known fairy tale centered on a mirror that could be asked the question,

"Mirror mirror on the wall — which is the most powerful computer of them all?"

We thought, in 1987, that the choice was between a microprocessor-based parallel array, and a traditional vector supercomputer. However, the mirror distorted our vision, and what we should have seen was a distributed array, and not just a closely coupled parallel simulation, but a complex metaproblem with multiple concurrent asynchronous components. There should not be one power user using a single large machine, but communities linked by a geographically distributed ensemble that, incidentally, could include one or more large parallel systems.

In our current vision, applications of interest extend from those in science and engineering to the information area; computing with a hint of communications — the original parallel computing thrust — becomes communications with some large-scale computing; compilers become interpreters; Fortran becomes Java, and many changes like these. We term this overall concept as Internetics, which is the field centered on technologies, applications, and services enabled by worldwide computing and communications. This is defined to be interdisciplinary, as both base technologies and applications are included. It includes issues of large scale from all points of view, not just large individual parallel or distributed compute engines or networks, but also one web client talking to one server. This is large scale for a different reason — it is pervasive.

As described in [1, 2], we found in our work at Caltech that interdisciplinary research at the interface of computer science and areas such as physics and chemistry was very rewarding [3]. Indeed, individuals who knew both areas, seemed well placed to lead the expected surge of interest in large-scale simulations using parallel systems. Several other groups came to similar conclusions, and the academic field of computational science was set up in several universities, and studied in many conferences [4]. However, this initiative seems to have stalled as student interest has shifted in both the computer science and application areas. In the latter case, in fact, fields like physics are seeing a general drop of enrollment as this is perceived as a difficult area to get jobs. Classic computational science (computational physics) is not a large enough field to change this. We argue that internetics combined with physics could, however, offer significant growth opportunities for this and similar traditional fields.

16.2   Does Parallel Computing Work?

It has been clear for some ten years or more than one can parallelize the majority of large-scale applications. Further, this parallelization scales and can be implemented on machines with very many nodes. The essential point is that one only needs to parallelize large problems, and these can be usually thought of as an algorithm applied iteratively to a dataset. The computation is large because the dataset is large. Then parallelism is achieved by breaking the dataset up into parts and placing one part in each processor. In Figures 16.1 and 16.2, we illustrate this for the examples we studied at Caltech

- Seismic wave propagation

- Astrophysics

- Computer Chess

- Hadrian's Palace (adapted from an earlier example using his wall)

The datasets are respectively the terrain in which the waves are propagated; the universe in which the galaxies are simulated; the set of moves in a computer generated decision tree; and the set of bricks that the masons must lay to build the wall and tile the floor. The latter analogy (where the "computation" is performed by humans and not digital systems) shows that domain decomposition and parallelism is well known and has established success in all aspects of the human experience. While at Caltech, I use to remark that NASA, when it needed to build a shuttle, did or rather could not hire Superman to address the task; rather some 50,000 workers were hired. These built the shuttles using a dynamic complex heterogeneous decomposition of this single problem. The workers had to be instructed (programmed); be arranged in a hierarchical set of teams (architecture); and the process was designed to ensure workers could proceed effectively and not spend too much time interfering with other team members (minimize inter-processor communication). This was accompanied by dynamic planning and assignment of tasks (adaptive dynamic resource management). The parallel computing terminology is placed in brackets and it is clear that the fundamental computer science issues are familiar concepts in society and that in principal they should be "naturally" soluble. Further, one noted that Nature's parallel systems all communicate via message passing whether they be swarms of bees, colonies of ants, collections of neurons, or teams of human minds.

So we were very confident that parallel computing was possible, and set up the "Caltech Concurrent Computation Program" to demonstrate this. The parameters of hypercubes built first by Seitz and then JPL were chosen to support this type of parallelism. As rather tardily demonstrated [3], we successfully implemented over 50 distinct significant parallel applications.

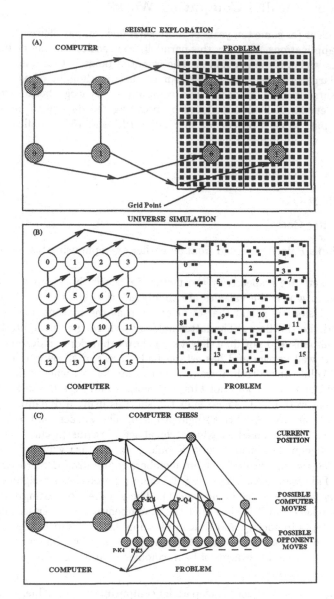

Fig. 16.1. Three examples of parallel computing using domain decomposition to map problem onto computer.

Fig. 16.2. Parallel Construction of Hadrian's Palace — tilers laying dance floor in two-dimensional decomposition with masons building the wall with a one-dimensional decomposition.

Now the above arguments have compelling generality but are, of course, superficial. There are important cases where parallelism is not trivial, including cases where *time* and not *dataset size* is the "large" parameter. Here, we looked at studies of the motion of the solar system, with a few-way parallelism, over long time periods. Solar system studies (using parallelism over planets) cannot use massive parallelism directly but typically planetary evolution is sensitive to poorly known initial conditions. These would be studied by multiple runs with different parameter values. This exploratory work is, as they say today, pleasingly parallel (in Feynman's day, it was "embarrassingly" parallel) and recovers our ability to use parallel systems. Gerry Sussman implemented this parallelism with a specialized digital orrery while on sabbatical at Caltech. This, like the hypercube, was discussed in Feynman's class. A second and more important case is event-driven simulations, which are commonly used in the modeling and simulation of macroscopic systems. The military is a major user of this technology and in the U.S.A., this work is coordinated by the DMSO office [5]. Event-driven simulations must execute entities in the time ordering of event occurrences and this essentially sequentializes the myriad of components. This is in principle insoluble, but in practice parallelism can be found as events in a large simulation are geographically distributed and do not effect each other for long time periods. This feature, combined with various ingenious variations of the time warp rollback mechanism, actually allows large-scale event-driven simulations to run effectively even on relatively loosely coupled distributed systems. In fact, while at Caltech investigations of such problems was a major activity of our collaboration with the Jet Propulsion Laboratory.

However, these nifty technical issues are not the reason why parallel computing is or is not successful. Rather, the critical point was explained one day in a wonderful

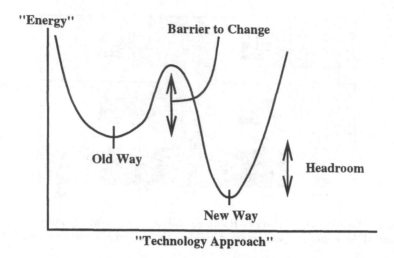

Fig. 16.3. Phase Transitions in Complex Systems

public lecture by Carver Mead in Caltech's Beckman Auditorium. He explained how the computing industry faced and would see many technology transitions. However, any new approach needed enough "headroom" to replace the old way. Changes in deployed technology are like the phase transitions in physics to which Feynman made so many contributions.

Systems can live for a long time in a "false minimum" that is the older technology if there is a substantial energy barrier to change, as shown in Figure 16.3.

We can use a complex systems language as advocated by Feynman's colleague, Gell-Mann, and the Santa Fe Institute. A complex system is a set of interconnected entities that, although governed at a low level by standard laws of physics, have interactions that are best described by a "macroscopic coarse graining" As most enterprises involve some sort of optimization, one can usually associate a phenomenological energy function that is minimized by our complex system. For instance, the "no-arbitrage opportunity" used in economic modeling implies that one can view the stock market as a complex system. Trading is the heat bath providing a myriad of microscopic interaction that equilibrates this system, and financial instruments are priced by maximizing value. In the computer industry, market forces equilibrate the system while innovation causes the complex system to evolve while always maximizing some unclear energy function representing customer satisfaction per unit dollar.

After this digression, we return to our discussion of parallel computing. In the framework of Figure 16.3, taking an over simplified view that captures the essence, there are indeed two minima: the "current sequential computing" and the "large-scale parallel systems" minima. The technical computer science studies show that "Parallel Computing Works" so this second minima is distinct, well defined, and lower than the "sequential minimum" However, sequential computing technology has advanced dramatically over the last 15 years and the "headroom" shown in Figure 16.3 is not so great. We argued that parallel computing was inevitable, as the feature size reduction in chips implied one would "have to spend" one's computer budget on parallel systems as technology reduced the unit sequential system cost. This argument is fallacious for two reasons. Firstly, the industry made the sequential chip architecture "better" as we increased from a few hundred thousand to the current several million transistors in each chip. This did involve parallelism, but only that which could be implemented without user intervention. Current chips are much faster than those of five years ago but, in fact, use transistors less efficiently, as "automatic parallelism" is not as efficient as (user directed) data parallelism. However, this approach stays in the "same minimum" of Figure 16.3 and requires no "phase transitions" Thus, it is chosen by our market forces. A second, and perhaps more important, development is that effectively users are spending less money each on computing. The dominant thrust in the computing industry is not on a few very powerful systems but on wide spread deployment of very many small (PC's) systems. This is the critical point we missed — large-scale computing was as we always said inevitable, but the scaling included not just the number of processors (as we foretold), but also the number of users. Thus, the dominant system today is not a central closely coupled parallel system linking many individuals and their machine together. There are important problems that still need all the computing they can get. These include large- scale academic computations such as astrophysics and quantum chemistry, and many areas of importance to national security, such as the well-known U.S. Department of Energy ASCI program to model nuclear stockpiles. However, more generally, the anticipated growth has not been in this area, but rather in the distributed systems area where the user base has increased at the same rate as the deployment of computer power.

Returning to Figure 16.3, the "barrier to change" is very large and this is central to understanding why the use of large-scale parallel systems has not expanded. We know good parallel algorithms for almost every important problem and can express these in efficient parallel software. Unfortunately, there are three major difficulties with this process. Firstly, there is usually no easy way to "port (migrate)" existing sequential code to parallel systems. Secondly, the current clean parallel languages are all low level — and, in fact, not much better than what we used in Feynman's courses 10-15 years ago. They are based around explicit user-specified message passing, as in the current PVM and MPI systems today, or the CrOS system we used at Caltech. There are better higher-level systems, but these are not universal "silver bullets" and do not provide a clearly excellent broad-base programming environment, so we have no compelling high-level language that expresses the majority of problems. Even for what we know how to do, there is a further difficulty. Namely, the high-performance parallel-computing field is of order 1% in dollar volume of the commodity computer market. However, it has all the software problems that PCs have plus all the additional parallel computing issues. Thus, the field does not have the capital investment or market size to be able to develop quality software. We have argued recently that this implies that the parallel-computing field should rethink its software strategy [6, 7]. It should build wherever possible on top of software produced for commodity markets, such as the web and business enterprise systems. Here we view parallel computers as a special case of distributed systems with especially tight synchronization constraints.

Note that distributed computing assumes problems are already decomposed and designs software to access, store, and integrate decomposed parts together. Parallel computing's central difficulty is different — it is finding a way of expressing tightly integrated problems in a way that they can be efficiently decomposed. We need to focus on this problem, and integrate it with tools taken where possible from the much larger commodity market. Previously, the high-performance parallel-computing community has tried to solve an essentially impossible problem — develop a complete programming environment from scratch with much less available resources than the existing sequential "false minimum" of Figure 16.3.

Thus, we see that parallel computing established its possibilities but did not, in Carver Mead's terminology, have enough headroom to effect transition from the current sequential "meta-stable equilibrium state"

We can ask if this will change? Firstly, we have oversimplified, as always in such broad-based discussions. There are important areas — especially parallel databases where the transition has been successful. This is understandable because once a single tool (the database) was parallelized, all applications of it could take advantage of parallel machines. Secondly, there is one compelling argument in favor of the inevitable adoption of parallel techniques. One notes that personal computer chips will "soon" have so many transistors that designers must use them to implement parallelism. It is argued that this will force the commodity market to take parallelism seriously, and drive the pervasive deployment of this technology. This argument has some truth to it, but it is not clear that the degree of parallelism involved is enough. If one "just" needs to use up a "few-way" parallelism, then the functional approach (as used by (Java) threads in simultaneously processing different components of a web page) may be sufficient. Such parallelism avoids the critical difficulties of large-scale data parallelism, and will stay in current minimum and not drive the phase transition of Figure 16.3. Thus, we expect, over the next five years, that the level of activity and importance of parallel computing will remain roughly constant. It will not become the dominant force we thought in the early 1990s. We do expect that parallel-programming environments will improve significantly, but not enough to make it possible to easily leap across the boundary in Figure 16.3.

16.3   Is Computational Science a New Academic Field?

So while at Caltech, my studies of both high energy physics and parallel computing were helped immeasurably by an excellent group of students who I diligently trained in an interdisciplinary fashion. I have no doubt that this training was highly effective and was essential for the generally accepted success of our activities. However, these students were typically not so successful on their graduation. Their training was a "jack of all trades" (or at least two trades) and getting a good job — especially in universities — requires excellence in one recognized field. I have, for the last 10 years, recommended students not to perform interdisciplinary work until "they get tenure" This advice flies somewhat in the face of the growing interest in interdisciplinary activities by funding agencies — especially the National Science Foundation. However, not entirely, as one can perform interdisciplinary activities in two ways; firstly, using one or more individuals — each of whose expertise spans multiple fields; secondly, one can build a team of specialized individuals whose combined knowledge spans multiple fields. The latter has, in my opinion, been the mode adopted successfully in most recent projects whereas at Caltech, I largely used the first model. Good universities find it hard to hire interdisciplinary faculty. The tenure review system, in spite of some flaws, has successfully built the quality American research universities. This assumes there is a peer group inside and outside the university that can provide reliable information on which to judge the merits of faculty promotions. This is almost impossible to do in interdisciplinary fields where a candidate, whose work falls into multiple areas, will get less than perfect reviews in any of the component areas of his or her expertise.

There is another more mundane problem with implementing interdisciplinary fields. Suppose we have N basic fields — physics, chemistry, biology, medicine, computer science, electrical engineering, environmental studies, and so on. Then we can design 2^N-1 interdisciplinary areas by choosing any combination of these basic fields. This leads to a plethora of subjects with probably limited life times and no good way to choose where to focus. Thus, it seems best to set up academic institutions with a few core subjects (a basic liberal arts education perhaps?) and building around this an evolving web of interdisciplinary studies. Interdisciplinary work can be recognized by certificates, minors, masters or other "lesser degree" forms. Even this modest goal requires changes in university structure as currently the core subjects have too many requirements and a broader educational experience is hard. However, I believe these changes can and, in fact, probably will be made. In particular, one core field physics is seeing a major reduction in enrollment as it is correctly perceived that there are few jobs in the "pure field" of physics. However, I have found physics is, in fact, an excellent training for general interdisciplinary research. Physics teaches problem solving based on fundamental principles — a good approach in most areas.

My attempts to understand the academic role of computational science revealed

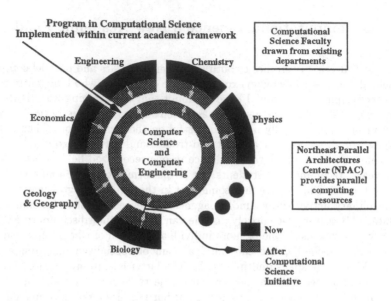

Fig. 16.4. A classical view of computational science as implemented at Syracuse in the early 1990s.

another bothering recurring theme. Namely, nobody could agree as to what it was. Individuals at Caltech insisted it was the same as what I call (Σ_i computational i) i.e., an amalgam of computational physics, chemistry, etc. However, NSF in its recent "partners for advanced computational infrastructure" solicitation, I think views it, as I do, more broadly as shown in Figure 16.4. This includes as well "applied computer science" or those aspects of computer science involved with hardware, software, and algorithms of scientific and engineering computation. Most academic implementations have, in fact, given computational science a central home in some sort of computer science or applied mathematics department and so emphasizing the last component. The academic computing tower of Babel is further confused by the fields of computer engineering, computational science and engineering, and scientific computing. These take the fields we have discussed and given them different emphases, but leading to, again, good educational opportunities with an unclear national recognition. The major problems for students in computational science is that most employees generally have no idea what the word "computational science" means, and computer science is the best academic degree with which to hunt for jobs. This being said, it is also true that it is the applied and not the theoretical computer science skills that most employers demand. Thus, the strategy explained at the beginning of this section is sound. Get a degree labeled by a basic well understood field such as computer science, but arrange one's studies to obtain "lesser degrees" (master, certificates) demonstrating proficiency in fields like computational science, which teach good practical skills.

After these general remarks on interdisciplinary research, let us discuss computational science defined as the academic field lying in between computer science (and applied mathematics/computer engineering) and the various fields of science and engineering that use high-performance (parallel) computing. We show in Figure 16.4, the particular view of computational science we developed at Syracuse University. This was designed in accordance with earlier remarks to be implemented within the existing academic framework and not require a new academic unit. This is consistent with approaches at other universities as described in John Rice's fine summary [4].

Our vision for the success of this field is well captured in these words from Daedalus [2] in 1992.

"This essay is constructed around a single premise: the inexorable increase in the performance of computers can open up new vistas in essentially all fields. We need skilled people to explore and exploit these possibilities, however, and our educational system is behind the times. Current curricula at grade schools and colleges will not educate students to exploit the possibilities opened up by parallel computers and the emergence of the computational methodology. Furthermore, the young but relatively traditional field of computer science will only give us a small fraction of the scientists in the computational wave that will lead the revolution. Computer scientists will develop the wonderful machines — a critical enabling technology. However, what we need most are computational scientists — individuals trained to use computers. High-performance computing is critical to the nation's needs. The Gulf War illustrated this in our military, but the future battles will increasingly be economic. Thus, high-performance computers can assure the industrial competitiveness of the nation, but this can only be true if we educate those who can use parallel computers in new ways for industry."

These were sound arguments, but they embody the flaw exposed in Section 2. Computation will open up new vistas in essentially all fields, but parallel computing *on its own* will not. I moved from permanent summer to winter (Caltech to Syracuse) because I wanted to implement the original computational science vision broadly in a university closely linked to the real world (industry). However, I soon realized the flaw as when I surveyed New York State industry [8–12], I found little interest in parallel computing. In fact, at Syracuse, I developed some reasonable core courses in computational science [CPS615,713 see http://www.npac.syr.edu/Education] but after peaking with some 50 students and two sections in the early 90s, student interest has waned and the current enrollment is down by an order of magnitude. Syracuse students are pretty bright, but they tend to be pragmatic and go to courses where they think there are jobs to be found. Anticipating the discussion of the next section, over the period 1995-98, enrollment in web technology classes has grown by an order of magnitude.

So we do need to change university curricula as computing is impacting every field and this must be reflected in the material students learn. However, I now believe that our original vision of computational science was too limited, and not broad enough to survive the inevitable slings and arrows of uncertain technological progress. We do need some key characteristics — in particular, strong flexible core subject curricula with enough latitude that there is room for students to take interdisciplinary studies. This is a sound lesson we have learnt from the experiments in computational science.

16.4 Internetics — The Correct Vision?

In last two sections, we described parallel computing as the expected driving technology that would increase the role of computation in all fields and so drive a new interdisciplinary academic field of computational science. This vision was not quite right as it did not anticipate the rapid improvement in sequential architectures. Further, it missed a critical feature of "large-scale" namely, it was more important to scale the number of people involved than to scale the power of individual computers.

The new technology vision is the wide spread deployment of computational devices and communication links with the essentially identical architecture whether in a central massively parallel server or a distributed set of digital set-top boxes in a suburban community. All are linked commodity processors exchanging messages between themselves. This scenario will not be trapped in a niche market that is 1% of the total but rather will overtake all computer systems. In this world, the operating system seen by the users is WebWindows [13] as currently illustrated by the integration of Microsoft's Internet Explorer with the PC Windows Operating System. According to the basic market principles, web technology will lead to the best available software as it addresses the largest possible market, and so can amortize software development costs over the largest possible volume. Further, the web has a particularly good creative model as its modular distributed software design is designed and built by a loosely knit world wide software team. Note we have had previous pervasive software models such as those of IBM mainframes or of the PC itself. The web is different from these, as it is a *pervasive complete* software model that addresses distributed heterogeneous computing. We can regard any other computing system as a special case of the web. Studying parallel computing will start with its distributed computing base, and as mentioned earlier, add support for tight synchronization. Studying military or electronic commerce applications will add security to the mix, while CORBA is naturally linked to the web to satisfy the need for managed distributed objects.

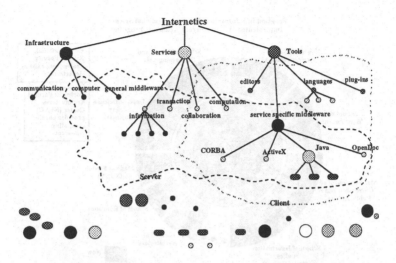

Fig. 16.5. Professor Xiamong Li's view of Internetics

We can follow Xiaoming Li and call the resultant field *Internetics*. This is the study of technologies enabling, and applications enabled by, the world wide, large-scale, object web hardware and software infrastructure. As shown in Figure 16.5, this field includes computing, but also a rich collection of networking and information infrastructure, services, and tools. We had realized the importance of this area from our sad survey at Syracuse, which had shown that industry in New York State was not so interested in parallel computing. We did identify that although large-scale simulation was of "tertiary importance" (as told to me by a now defunct military aircraft company), information processing was of general interest. I also like to recount the tale of a large appliance company that could only find a small, beleaguered audience of six engineers for my talk on the value of simulation and high-performance computing. However, a few years later, they were ecstatic to learn from us how to link their product database to the web.

Even while we were setting up a "classical" computational science program at Syracuse, we realized early on from the feedback from industry that it was incomplete. So, in late 1995, we started an "information track" of the computational science program at Syracuse. The idea is illustrated in Figure 16.6 and generalizes the concept explained in Section 3 that computational science was at the interface of "applied computer science" and a set of applications. In the information track, we replace the science and engineering applications of Figure 16.4 with areas shown in Figure 16.6 where information processing is the key computation task. This includes areas such as education, health care, crisis management, journalism (perhaps using the web for dissemination), and marketing.

Combining Figure 16.4 and 16.6, we find the academic implementation of *Inter-*

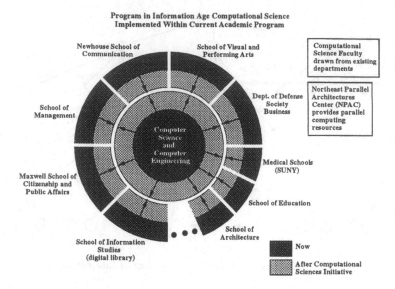

Fig. 16.6. The 1995 extension of Figure 16.4 at Syracuse to include all applications into computational science. This is a forerunner of an academic implementation of Internetics.

netics as the field that lies between modern applied computer science and application areas. We have designed a tentative Internetics curricula running from the K-12 (school children) to graduate level. It starts by teaching school children the essence of the web and how to program in Java. Java is a particularly good language for the K-12 age group, as it has good graphics and obvious utility in improving web pages. Thus, we can easily motivate the language and our beginning programmers get the gratification of better personal web pages to share with their peers. The technologies of Internetics are more social than that of the original computational science. At the graduate level, we designed a six semester course certificate covering technologies (such as the basic Web, VRML, multimedia, collaboration, distributed objects) and a choice of application specializations, such as those in Figures 16.4 and 16.6.

We see a general trend towards Internetics (although, of course, typically not with this name) but so far there is not the necessary consensus to expect widespread adoption. For instance, as a subset of Internetics, an interesting field is called by some just "multimedia" and Syracuse scoped this out, but did not adopt a "masters in multimedia" program. However, we see that Internetics embodies the essential vision of computational science that the use of modern computers and communications systems will revolutionize many fields. Thus, it is essential for both academic and economic reasons to train a generation of students to be familiar with both computing and particular applications. In fact, there is substantial interest from industry in retraining existing workers in the techniques of Internetics.

Thus, although details of our original vision were flawed, it is included in the new broader picture, which will succeed.

Acknowledgement

These ideas owe a lot to my interactions with many people. Feynman taught me much about how one should conduct scientific research maintaining both practical relevance and theoretical depth. The members of the Caltech Concurrent Computation Program (C^3P) taught me interdisciplinary research while Professor Xiaoming Li from Peking has been a wonderful collaborator in many ways during the last few years.

References

[1] G. C. Fox, "Achievements and Prospects for Parallel Computing," *Concurrency: Practice and Experience*, **3**(6), 725-739 (1991). Special Issue: *Practical Parallel Computing: Status and Prospects*. Guest Editors: Paul Messina and Almerico Murli. SCCS-29b, C3P-297b, CRPC-TR90083.

[2] G. C. Fox, "Parallel Computing and Education," *Daedalus Journal of the American Academy of Arts and Sciences*, **121**(1), 111-118, (1992). CRPC-TR91123, SCCS-83. Caltech Report C3P-958.

[3] G. C. Fox, P. C. Messina and R. D. Williams, editors. *Parallel Computing Works!* (Morgan Kaufmann Publishers, San Francisco, CA, 1994). http://www.npac.syr.edu/projects/cpsedu/CSEmaterials/copywrite/pcw/index.html

[4] J. R. Rice, "Computational Science and the Future of Computing Research," in *IEEE Computational Science and Engineering*, **2**(4), 35-41 (1995).

[5] High Level Architecture and Run-Time Infrastructure by DoD Modeling and Simulation Office (DMSO), http://www.dmso.mil/hla

[6] G. C. Fox and W. Furmanski, "Parallel and Distributed Computing using Pervasive Web and Object Technologies," in *Advances in Parallel Computing, Vol.12, Proceedings of the International Conference on Parallel Computing: Fundamentals, Applications and New Directions*, E. H. D'Hollander, G. R. Joubert, F. J. Peters, U. Trottenberg eds. (Elsevier Publishers, 1998). Syracuse University, NPAC Technical Report SCCS-807, February 1998.

[7] G. C. Fox and W. Furmanski, "High Performance Commodity Computing," in *Computational Grids: The Future in High Performance Distributed Computing*, C. Kesselman and I. Foster, editors (Morgan Kaufman Publishers, San Francisco, CA, 1998). Syracuse University, NPAC Technical Report SCCS-808, February 1998.

[8] G. C. Fox, "Parallel Computing in Industry — An Initial Survey," in *Proceedings of 6th Australian Supercomputing Conference* (supplement), pages 1-10. Communications Services, Melbourne, December 1992. Held at World Congress Centre, Melbourne, Australia. Syracuse University, NPAC Technical Report SCCS-302b. CRPC-TR92219.

[9] G. Fox, K. Hawick, M. Podgorny and K. Mills, "The Electronic InfoMall — HPCN Enabling Industry and Commerce," Volume **919** of *Lecture Notes in Computer Science*, pp 360-365 (Springer-Verlag, 1994). Syracuse University, NPAC Technical Report SCCS-665.

[10] G. C. Fox, "Involvement of Industry in the National High Performance Computing and Communication Enterprise," in *Developing a Computer Science Agenda for High Performance Computing*, edited by U. Vishkin (ACM Press). Syracuse University, NPAC Technical Report SCCS-716, Syracuse, NY, May 1994..

[11] G. C. Fox, and W. Furmanski, "The Use of the National Information Infrastructure and High Performance Computers in Industry," in *Proceedings of the Second International Conference on Massively Parallel Processing using Optical Interconnections*, pages 298-312, Los Alamitos, CA, October 1995 (IEEE Computer Society Press). Syracuse University, NPAC Technical Report SCCS-732.

[12] G. C. Fox, "An Application Perspective on High-Performance Computing and Communications,", Research Study Fourth Quarter 1997 (RCI, Ltd). Syracuse University, NPAC Technical Report SCCS-757, April 1996.

[13] G. C. Fox, and W. Furmanski, "SNAP, Crackle, WebWindows!," RCI, Ltd., Management White Paper, 29 (1996). Syracuse University, NPAC Technical Report SCCS-758.

RICHARD FEYNMAN AND THE
CONNECTION MACHINE

W. Daniel Hillis *

17.1 Introduction

One day in the spring of 1983, when I was having lunch with Richard Feynman, I mentioned to him that I was planning to start a company to build a parallel computer with a million processors. (I was at the time a graduate student at the MIT Artificial Intelligence Lab.) His reaction was unequivocal: "That is positively the dopiest idea I ever heard." For Richard, a crazy idea was an opportunity to prove it wrong or prove it right. Either way, he was interested. By the end of lunch he had agreed to spend the summer working at the company.

Richard had as much fun with computers as anyone I ever knew. His interest in computing went back to his days at Los Alamos, where he supervised the "Computers" — that is, the people who operated the mechanical calculators there. He was instrumental in setting up some of the first plug-programable tabulating machines for physical simulation. His interest in the field was heightened in the late 1970s when his son Carl began studying computers at MIT.

I got to know Richard through his son. Carl was one of the undergraduates helping me with my thesis project. I was trying to design a computer fast enough to solve commonsense reasoning problems. The machine, as we envisioned it, would include a million tiny computers, all connected by a communications network. We called it the Connection Machine. Richard, always interested in his son's activities, followed the project closely. He was sceptical about the idea, but whenever we met at a conference or during my visits to Caltech, we would stay up until the early hours of the morning discussing details of the planned machine. Our lunchtime meeting on that spring day in 1983 was the first time he ever seemed to believe we were really going to try build it.

Richard arrived in Boston the day after the company was incorporated. We had been busy raising the money, finding a place to rent, issuing stock and so on. We had found an old mansion just outside the city, and when Richard showed up we were still recovering from the shock of having the first few million dollars in the bank. No one had thought about anything technical for months. We were arguing about what the name of the company should be when Richard walked in, saluted and said "Richard Feynman reporting for duty. OK, boss, what's my assignment?"

---

*This article was originally published in the February 1989 issue of Physics Today.

The assembled group of not-quite-graduated MIT students was astounded. After a hurried private discussion ("I don't know, you hired him...") we informed Richard that his assignment would be to advise the application of parallel processing to scientific problems. "That sounds like a bunch of baloney," he said. "Give me something real to do." So we sent him out to buy some office supplies. While he was gone, we decided that the part of the machine we were most worried about was the router that delivered messages from one processor to another. We were not entirely sure that our planned design would work. When Richard returned from buying pencils, we gave him the assignment of analyzing the router.

17.2   The Machine

The router of the Connection Machine was the part of the hardware that allowed the processors to communicate. It was a complicated object; by comparison, the processors themselves were straightforward. Connecting a separate wire between every pair of processors was totally impractical; a million processors would require 10^{12} wires. Instead, we planned to connect the processors in the pattern of a 20-dimensional hypercube, so that each processor would only need to talk directly to 20 others. Because many processors had to communicate simultaneously, many messages would contend for the same wire. The router's job was to find a free path through this 20-dimensional traffic jam or, if it couldn't, to hold the message in a buffer until a path became free. Our question to Feynman was: "Had we allowed enough buffers for the router to operate efficiently?"

In those first few months Richard began studying the router circuit diagrams as if they were objects of Nature. He was willing to listen to explanations of how and why things worked a certain way, but fundamentally he preferred to figure everything out himself. He would sit in the woods behind the mansion and simulate the action of each circuit with pencil and paper. Meanwhile, the rest of us, happy to have found something to keep Richard occupied, went about the business of ordering the furniture and computers, hiring the first engineers and arranging for the Defense Advanced Research Projects Agency to pay for the development of the first prototype. Richard did a remarkable job of focusing on his "assignment," stopping only occasionally to help wire the computer room, set up the machine shop, shake hands with the investors, install the telephones and cheerfully remind us of how crazy we all were. When we finally picked the name of the company, Thinking Machines Corporation, Richard was delighted. "That's good. Now I don't have to explain to people that I work with a bunch of loonies. I can just tell them the name of the company."

The technical side of the project was definitely stretching our capacities. We had decided to simplify things by starting with only 64,000 processors, but even then the amount of work to be done was overwhelming. We had to design our own silicon integrated circuits, with processors and a router. We also had to invent packaging and cooling mechanisms, write compilers and assemblers, devise ways of testing processors simultaneously and so on. Even simple problems like wiring the boards together took on a whole new meaning when you were working with tens of thousands of processors. In retrospect, if we had had any understanding of how complicated the project was going to be, we would never have started.

17.3 'Get These Guys Organized'

I had never managed a large group before, and I was clearly in over my head. Richard volunteered to help out. "We've got to get these guys organized" he told me. "Let me tell you how we did it at Los Alamos." It seems that every great man has a certain time and place in his life that he takes as a reference point ever after: a time when things worked as they were supposed to and great deeds were accomplished. For Richard, that time was at Los Alamos during the Manhattan project. Whenever things got "cockeyed," Richard would look back and try to understand how now was different from then. Using this formula, Richard decided we should pick an expert in each area of importance to the machine — software, packaging, electronics and so on — to become the "group leader" of that area, just as it had been at Los Alamos.

Part two of Feynman's "Let's Get Organized" campaign was a regular seminar series of invited speakers who might suggest interesting uses for our machine. Richard's idea was that we should concentrate on people with new applications, because they would be less conservative about what kind of computer they would use. For our first seminar he invited John Hopfield, a friend of his from Caltech, to give us a talk on his scheme for building neural networks. In 1983, studying neural networks was about as fashionable as studying ESP, so some people considered Hopfield a little crazy. Richard was certain he would fit right in at Thinking Machines.

What Hopfield had invented was a way of constructing an associative memory," a device for remembering patterns. To use an associative memory, one trains it on a series of patterns — for example, pictures of letters of the alphabet. Later, when the memory is shown a new pattern, it is able to recall a similar pattern it has seen in the past. A new picture of the letter A will "remind" the memory of another A it has seen before. Hopfield figured out how such a memory could be built from devices functionally similar to biological neurons.

Not only did Hopfield's method seem to work; it seemed to work particularly well on the Connection Machine. Feynman figured out the details of how to use one processor to simulate each of Hopfield's neurons, with the strength of each connection represented as a number in the processor's memory. Because of the parallel nature of Hopfield's algorithm, all the processors could be used concurrently with 100 percent efficiency; the Connection Machine would thus be hundreds of times faster than any conventional computer.

17.4 An Algorithm for Logarithms

Feynman worked out in some detail the program for computing Hopfield's network on the Connection Machine. The part that he was proudest of was the subroutine for computing a logarithm. I mention it here not only because it is a clever algorithm, but also because it is a specific contribution Richard made to the mainstream of computer science. He had invented it at Los Alamos.

Consider the problem of finding the logarithm of a fractional number between 1 and 2. (The algorithm can be generalized without too much difficulty.) Feynman observed that any such number can be uniquely represented as a product of numbers of the form $1+2^{-k}$, where k is an integer. Testing for the presence of each of these factors in a binary representation is simply a matter of a shift and a subtraction. Once the factors are determined, the logarithm can be computed by adding together the precomputed logarithms of the factors. The algorithm fit the Connection Machine especially well because the small table of the logarithms of $1+2^{-k}$ could be shared by all the processors. The entire computation took less time than doing a division.

Concentrating on the algorithm for a basic arithmetic operation was typical of Richard's approach. He loved the details. In studying the router he paid attention to the action of each individual gate, and in writing the program he insisted on understanding the implementation of every instruction. He distrusted abstractions that could not be directly related to the facts. When, several years later, I wrote a general-interest article on the Connection Machine for Scientific American, he was disappointed that it left out too many details. He asked "How is anyone supposed to know that this isn't just a bunch of crap?"

Feynman's insistence on looking at the details helped us discover the potential of the machine for numerical computing and physical simulation. We had thought that the Connection Machine would not be efficient at "number crunching," because the first prototype had no special hardware for vectors or floating-point arithmetic. Both of these were "known" to be requirements for number crunching. Feynman decided to test this assumption on a problem he was familiar with in detail: quantum chromodynamics . Quantum chromodynamics is the presently accepted field theory of the strongly interacting elementary particles in terms of their constitutent quarks and gluons. It can, in principle, be used to compute the mass of the proton (in units of the pion mass). In practice, such a computation might require so much arithmetic that it would keep the fastest computers in the world busy for years. One way to do the calculation is to use a discrete four-dimensional lattice to model a section of space-time. Finding the solution involves adding up the contributions of all the possible configurations of certain matrices at the links of the lattice, or at least some large representative sample. (This is essentially a Feynman path integral.) What makes this so difficult is that calculating the contribution of even a single configuration involves multiplying the matrices around every loop in the lattice, and the number of loops grows as the fourth power of the lattice size. Because all these multiplications can take place concurrently, there is plenty of opportunity to keep all 64,000 processors busy.

To find out how well this would work in practice, Feynman had to write a computer program for quantum chromodynamics. Because BASIC was the only computer language Richard was really familiar with, he made up a parallel-processing version of BASIC in which he wrote the program. He then simulated the operation of the program by hand to estimate how fast it would run on the Connection Machine. He was excited by the results: "Hey Danny, you're not gonna believe this, but that machine of yours can actually do something useful!" According to Feynman's calculations, the Connection Machine, even without any special hardware for floating-point arithmetic, would out-perform a machine that Caltech was building explicitly for quantum chromodynamics calculations. From that point on, Richard pushed us more and more toward looking at numerical applications of the machine.

By the end of that summer of 1983, Richard had completed his analysis of the behavior of the router, and much to our surprise and amusement, he presented his answer in the form of a set of partial differential equations. To a physicist this may seem natural, but to a computer designer it seems a bit strange to treat a set of Boolean circuits as a continuous, differentiable system. Feynman's router equations were written in terms of variables representing continuous quantities such as "the average number of 1 bits in a message address." I was much more accustomed to inductive proof and case analysis than to taking the time derivative of "the number of 1's." Our discrete analysis said we needed seven buffers per chip: Feynman's differential equations suggested we only needed five. We decided to play it safe and ignore Feynman.

The decision to ignore Feynman's analysis was made in September, but by the following spring we were up against a wall. The chips we had designed were slightly too big to manufacture, and the only way to solve the problem was to cut the number of buffers per chip back to five. Because Feynman's equations claimed we could do this safely, his unconventional methods of analysis started looking better and better to us. We decided to go ahead and make the chips with the smaller number of buffers. Fortunately, Feynman was right. When we put together the chips, the machine worked. The first program run on the machine was John Horton Conway's Game of Life, in April 1985.

17.5   Cellular Automata

The Game of Life is an example of a class of computations that interested Feynman: cellular automata. Like many physicists who had spent their lives going to successively lower levels of subatomic detail, Feynman often wondered what was at the bottom. One possible answer was a cellular automaton. The notion is that the space-time continuum might ultimately be discrete, and that the observed laws of physics might simply be large-scale consequences of the average behavior of tiny cells. Each cell could be a simple automaton that obeys a small set of rules and communicates only with its nearest neighbors — like the points in the lattice calculation for quantum chromodynamics. If the universe in fact works this way, there should be testable consequences, such as an upper limit on the density of information per cubic meter of space.

The notion of cellular automata goes back to John von Neumann and Slanislaw Ulam, whom Feynman had known at Los Alamos. Richard's recent interest in the subject was aroused by his friends Ed Fredkin and Stephen Wolfram, both of whom were fascinated by cellular automata as models of physics. Feynman was always quick to point out to them that he considered their specific models "kooky," but like the Connection Machine, he considered the subject crazy enough to put some energy into. There are many potential problems with cellular automata as a model of physical space and time — for example, finding a set of rules that gives relativistic invariance at the observable scale. One of the first problems is just making the physics rotationally invariant. The most obvious patterns of cellular automata, such as a fixed three-dimensional grid, have preferred directions along the grid axes. Is it possible to implement even Newtonian physics on a fixed lattice of automata?

Feynman had a proposed solution to the anisotropy problem that he attempted (without success) to work out in detail. His notion was that the underlying automata, rather than being connected in a regular lattice like a grid or a pattern of hexagons, might be randomly connected. Waves propagating through this medium would, on average, propagate at the same rate in every direction.

Cellular automata started getting attention at Thinking Machines in 1984 when Wolfram suggested that we should use such automata not as a model of Nature, but as a practical approximation method for simulating physical systems. Specifically, we could use one processor to simulate each cell with neighbor-interaction rules chosen to model something useful, like fluid dynamics. Wolfram was at the Institute for Advanced Study in Princeton, but he was also spending time at Thinking Machines. For two-dimensional problems there was a neat solution to the anisotropy problem. It had recently been shown that a hexagonal lattice with a simple set of rules gives rise to isotropic behavior on the macroscopic scale. Wolfram did a simulation of this kind with hexagonal cells on the Connection Machine. It produced a beautiful movie of turbulent fluid flow in two dimensions. Watching the movie got all of us, especially Feynman, excited about physical simulation. We all started planning additions to the hardware, such as support for floating-point arithmetic, which would make it possible to perform and display a variety of simulations in real time.

17.6   Feynman the Explainer

In the meantime, we were having a lot of trouble explaining to people what we were doing with cellular automata. Eyes tended to glaze over when we started talking about state transition diagrams and finite-state machines. Finally Feynman told us to explain it like this:

> We have noticed in Nature that the behavior of a fluid depends very little on the nature of the individual particles in that fluid. For example, the flow of sand is very similar to the flow of water or the flow of a pile of ball bearings. We have therefore taken advantage of this fact to invent a type of imaginary particle that is especially simple for us to simulate. This particle is a perfect ball bearing that can move at a single speed in one of six directions. The flow of these particles on a large enough scale is very similar to the flow of natural fluids.

This was a typical Feynman explanation. On the one hand, it infuriated the experts who had worked on the problem because it did not even mention all of the clever problems that they had solved. On the other hand, it delighted the listeners because they could walk away with a real understanding of the calculation and how it was connected to physical reality.

We tried to take advantage of Richard's talent for clarity by getting him to criticize the technical presentations we made in our product introductions. Before the commercial announcement of the first Connection Machine, CM-1, and all of our subsequent products, Richard would give a sentence-by-sentence critique of the planned presentation. "Don't say 'reflected acoustic wave.' Say echo." Or, "Forget all that 'local minima' stuff. Just say there's a bubble caught in the crystal and you have to shake it out." Nothing made him angrier than making something simple sound complicated.

Getting Richard to give advice like that was sometimes tricky. He pretended not to like working on any problem that was outside his claimed area of expertise. Often, when one of us asked for him advice, he would gruffly refuse with "That's not my department." I could never figure out just what his department was, but it didn't matter anyway, because he spent most of his time working on these "not my department" problems. Sometimes he really would give up, but more often than not he would come back a few days after his refusal and remark "I've been thinking about what you asked the other day and it seems to me..." This worked best if you were careful not to expect it.

I do not mean to imply that Richard was hesitant to do the "dirty work." In fact he was always volunteering for it. Many a visitor at Thinking Machines was shocked to see that we had a Nobel laureate soldering circuit boards or painting walls. But what Richard hated, or at least pretended to hate, was being asked to give advice. So why were people always asking him for it? Because even when Richard didn't understand, he always seemed to understand better than the rest of us. And whatever he understood, he could make others understand as well. Richard made people feel like children do when a grown-up first treats them as adults. He was never afraid to tell the truth, and however foolish your question was, he never made you feel like a fool.

The charming side of Richard helped people forgive him for his less charming characteristics. For example, in many ways Richard was a sexist. When it came time for his daily bowl of soup, he would look around for the nearest "girl" and ask if she would bring it to him. It did not matter if she was the cook, an engineer or the president of the company. I once asked a female engineer who had just been a victim of this treatment if it bothered her. "Yes, it really annoys me" she said. "On the other hand, he's the only one who ever explained quantum mechanics to me as if I could understand it." That was the essence of Richard's charm.

17.7 A Kind of Game

Richard worked at the company on and off for the next five years. Floating-point hardware was eventually added to the machine, and as the machine and its successors went into commercial production, they were being used more and more for the kind of numerical simulation problems Richard had pioneered with his quantum chromodynamics program. Richard's interest shifted from the construction of the machine to its applications. As it turned out, building a big computer is a good excuse for talking with people who are working on some of the most exciting problems in science. We started working with physicists, astronomers, geologists, biologists, chemists — each of them trying to solve some problem that couldn't have been solved before. Figuring out how to do such calculations on a parallel machine required understanding their details, which was exactly the kind of thing Richard loved to do.

For Richard, figuring out these problems was a kind of game. He always started by asking very basic questions like "What is the simplest example?" or "How can you tell if the answer is right?" (He asked questions until he had reduced the problem to some essential puzzle he thought he could solve.) Then he would set to work, scribbling on a pad of paper and staring at the results. While he was in the middle of this kind of puzzle-solving, he was impossible to interrupt. "Don't bug me. I'm busy" he would say without even looking up. Eventually he would either decide the problem was too hard (in which case he lost interest), or he would find a solution (in which case he spent the next day or two explaining it to anyone who would listen). In this way he helped work on problems in database searching, geophysical modeling, protein folding, image analyzing and the reading of insurance forms.

The last project I worked on with Richard was in simulated evolution. I had written a program that simulated the evolution of populations of sexually reproducing creatures over hundreds of thousands of generations. The results were surprising, in that the fitness of the population made progress in sudden leaps rather than by the expected steady improvement. The fossil record shows some evidence that real biological evolution might also exhibit such "punctuated equilibrium," so Richard and I decided to look more closely at why it was happening. He was feeling ill by that time, so I went out and spent the week with him in Pasadena. We worked out a model of evolution of finite populations based on the Fokker-Planck equations. When I got back to Boston, I went to the library and discovered a book by Motoo Kimura on the subject. Much to my disappointment, all our "discoveries" were covered in the first few pages. When I called Richard and told him what I had found, he was elated. "Hey, we got it right!" he said. "Not bad for amateurs."

In retrospect I realize that in almost everything we worked on together, we were both amateurs. In digital physics, neural networks, even parallel computing, we never really knew what we were doing. But the things that we studied were so new that none of the others working in these fields knew exactly what they were doing either. It was amateurs who made the progress.

17.8   Telling the Good Stuff You Know

Actually, I doubt that it was "progress" that most interested Richard. He was always searching for patterns, for connections, for a new way of looking at something, but I suspect his motivation was not so much to understand the world, as it was to find new ideas to explain. The act of discovery was not complete for him until he had taught it to someone else.

I remember a conversation we had a year or so before his death, walking in the hills above Pasadena. We were exploring an unfamiliar trail, and Richard, recovering from a major operation for his cancer, was walking more slowly than usual. He was telling a long and funny story about how he had been reading up on his disease and surprising his doctors by predicting their diagnosis and his chances of survival. I was hearing for the first time how far his cancer had progressed, so the jokes did not seem so funny. He must have noticed my mood, because he suddenly stopped the story and asked, "Hey, what's the matter?" I hesitated. "I'm sad because you're going to die." "Yeah," he sighed, "that bugs me sometimes too. But not so much as you think." And after a few more steps, "When you get as old as I am, you start to realize that you've told most of the good stuff you know to other people anyway."

We walked along in silence for a few minutes. Then we came to a place where another trail crossed ours and Richard stopped to look around at the surroundings. Suddenly a grin lit up his face. "Hey," he said, all trace of sadness forgotten, "I bet I can show you a better way home." And so he did.

18

CRYSTALLINE COMPUTATION

Norman Margolus

Abstract

Discrete lattice systems have had a long and productive history in physics. Examples range from exact theoretical models studied in statistical mechanics to approximate numerical treatments of continuum models. There has, however, been relatively little attention paid to exact lattice models which obey an *invertible dynamics*: From any state of the dynamical system you can infer the previous state. This kind of microscopic reversibility is an important property of all microscopic physical dynamics. Invertible lattice systems become even more physically realistic if we impose locality of interaction and exact conservation laws. In fact, some invertible and momentum conserving lattice dynamics—in which discrete particles hop between neighboring lattice sites at discrete times—accurately reproduce hydrodynamics in the macroscopic limit.

These kinds of discrete systems not only provide an intriguing information-dynamics approach to modeling macroscopic physics, but they may also be supremely practical. Exactly the same properties that make these models physically realistic also make them efficiently realizable. Algorithms that incorporate constraints such as locality of interaction and invertibility can be run on microscopic physical hardware that shares these constraints. Such hardware can, in principle, achieve a higher density and rate of computation than any other kind of computer.

Thus it is interesting to construct discrete lattice dynamics which are more physics-like both in order to capture more of the richness of physical dynamics in informational models, and in order to improve our ability to harness physics for computation. In this chapter, we discuss techniques for bringing discrete lattice dynamics closer to physics, and some of the interesting consequences of doing so.

18.1 Introduction

In 1981, Richard Feynman gave a talk at a conference hosted by the MIT Information Mechanics Group. This talk was entitled "Simulating Physics with Computers," and is reproduced in this volume.

In this talk Feynman asked whether it is possible that, at some extremely microscopic scale, Nature may operate exactly like discrete computer-logic. In particular, he discussed whether crystalline arrays of logic called *Cellular Automata* (CA) might be able to simulate our known laws of physics in a direct fashion. This question had been the subject of long and heated debates between him and his good friend Edward Fredkin (the head of the MIT Group) who has long maintained that some sort of discrete classical-information model will eventually replace continuous differential equations as the mathematical machinery used for describing fundamental physical dynamics [31, 33].

For classical physics, Feynman could see no fundamental impediment to a very direct CA simulation. For quantum physics, he saw serious difficulties. In addition to discussing well known issues having to do with hidden variables and non-separability, Feynman brought up a new issue: Simulation efficiency. He pointed out that, as far as we know, the only general way to simulate a lattice of quantum spins on an ordinary computer takes an exponentially greater number of bits than the number of spins. This kind of inefficiency, if unavoidable, would make it impossible to have a CA simulation of quantum physics in a very direct manner.

Of course the enormous calculation needed to simulate a spin system on an ordinary computer gives us the result of not just a single experiment on the system, but instead approximates the complete statistical distribution of results for an *infinite* number of repetitions of the experiment. Feynman made the suggestion that it might be more efficient to use one quantum system to simulate another. One could imagine building a new kind of computer, a *quantum spin computer*, that was able to mimic the quantum dynamics of any spin system using about the same number of spins as the original system. Each simulation on the quantum computer would then act statistically like a *single* experiment on the original spin system. This observation that a quantum computer could do some things easily that we don't know how to do efficiently classically, stimulated others to look for and find algorithms for quantum computers that are much faster than any currently known classical equivalents [37, 80]. In fact, if we restrict our classical hardware to perform the "same kind" of computation as the quantum hardware—rather than to just solve the same problem—then we can actually *prove* that some quantum computations are faster. These fast quantum computations present further challenges to hypothetical classical-information models of quantum physics [44].

Despite such difficulties, Feynman did not rule out the possibility that some more subtle approach to the efficient classical computational modeling of physics might yet succeed. He found something very tantalizing about the relationship between classical information and quantum mechanics, and about the fact that in some ways quantum mechanics seems much *more* suited to being economically simulated with bits than classical mechanics: Unlike a continuous classical system, the entropy of a quantum system is finite. The informational economy of quantum systems that Feynman alluded to has of course long been exploited in statistical mechanics, where classical bits are sometimes used to provide finite combinatorial models that reproduce some of the macroscopic *equilibrium* properties of quantum systems [45, 46]. It is natural then to ask how much of the macroscopic *dynamical* behavior of physical systems can also be captured with simple classical information models. This is an interesting question even if your objective is not to revolutionize quantum physics: We can improve our understanding of Nature by making simple discrete models of phenomena.

My own interest in CA modeling of physics stems from exactly this desire to try to understand Nature better by capturing aspects of it in exact informational models. This kind of modeling in some ways resembles numerical computation of differential equation models, where at each site in a spatial lattice we perform computations that involve data coming from neighboring lattice sites. In CA modeling, however, the conceptual model is not a continuous dynamics which can only be approximated on a computer, but is instead a finite logical dynamics that can be simulated *exactly* on a digital computer, without roundoff or truncation errors. Every CA simulation is an exact digital integration of the discrete equations of motion, over whatever length of time is desired. Conservations can be exact, invertibility of the dynamics can be exact, and discrete symmetries can be exact. Continuous behavior, on the other hand, can only emerge in a large-scale average sense—in the *macroscopic limit*. CA models have been developed in which realistic classical physics behavior is recovered in this limit [18, 79].

Physics-like CA models are of more than conceptual and pedagogical interest. Exactly the same general constraints that we impose on our CA systems to make them more like physics also make them more efficiently realizable as physical devices. CA hardware that matches the structure and constraints of microscopic physical dynamics can in principle be more efficient than any other kind of computer: It can perform more logic operations in less space and less time and with less energy dissipation [27, 94]. It is also scalable: A crystalline array of processing elements can be indefinitely extended. Finally, this kind of uniform computer is simpler to design, control, build and test than a more randomly structured machine. The prospect of efficient large-scale CA hardware provides a practical impetus for studying CA models.

18.2   Modeling dynamics with classical spins

From the point of view of a physicist, a CA model is a fully discrete classical field theory. Space is discrete, time is discrete, and the state at each discrete lattice point has only a finite number of possible discrete values. The most essential property of CA's is that they emulate the spatial locality of physical law: The state at a given lattice site depends only upon the previous state at nearby neighboring sites. You can think of a CA computation as a regular spacetime crystal of processing events: A regular pattern of communication and logic events that is repeated in space and in time. Of course it is only the structure of the computer that is regular, not the patterns of data that evolve within it! These patterns can become arbitrarily complicated.

Discrete lattice models have been used in statistical mechanics since the 1920's [45, 46]. In such models, a finite set of distinct quantum states is replaced by a finite set of distinct classical states. Consider, for example, a hypothetical quantum system consisting of n spin-$\frac{1}{2}$ particles arranged on a lattice, interacting locally. The spin behavior of such a system can be fully described in terms of 2^n distinct (mutually orthogonal) quantum states. The *Ising model* accurately reproduces essential aspects of phase-change behavior in such a system using n classical bits—which give us 2^n distinct classical states. In the Ising model, at each site in our lattice we put a *classical spin:* A particle that can be in one of two classical states. We define bond energies between neighboring spins: We might say, for example, that two adjacent spins that are parallel (i.e., are in the same state) have a bond energy of $\epsilon_=$, while two antiparallel (not same) neighbors have energy $\epsilon_{\neq}$. This gives us a classical system which has many possible states, each of which has an energy associated with it. In calculating the equilibrium properties of this system, we simply assume that the dynamics is complicated enough that all states with the same energy as we started with will appear with equal probability. Thus we ignore the actual quantum dynamics of the original spin system, and instead substitute an energy-conserving random pseudo-dynamics.

We could equally well substitute any classical dynamics that has a sufficiently complicated evolution. We will consider one simple CA model that has been successfully used in this manner [19, 42, 74, 98]. We assume that we are dealing with an isolated spin system, not in contact with any heat bath, so that total energy must be exactly conserved. We will also impose the realistic constraint that the microscopic dynamics of an isolated physical system must be exactly invertible: There must always be enough information in the current state to recover any previous state. This constraint helps to ensure that a deterministic dynamics explores its available state-space thoroughly, and so can be analyzed statistically—this issue is discussed in Section 18.4.

We can construct a simple CA that has these properties, in which the next value of the spin at each site on a 2D square lattice only depends upon the current values of its four nearest neighbors. The rule is very simple: A given spin changes state if and only if this doesn't change the total energy associated with its bonds to its four nearest neighbors. Equivalently, a given spin (bit) is flipped (complemented) if exactly two of its four neighbors are zero's, and two are one's. This doesn't change its total bond energy: Both before and after the flip, it will be parallel to half of its neighbors (contributing $2\epsilon_=$ to the total), and antiparallel to the rest (contributing $2\epsilon_{\neq}$).

The rule as stated above would be fine if we updated just one spin on the lattice at a time, but we would like to update the lattice in parallel. To make this work, we will adopt a checkerboard updating scheme: We imagine that our lattice is a giant black and white checkerboard, and we alternately hold the bits at all of the black sites fixed while we update all of the white ones, and then hold the white

 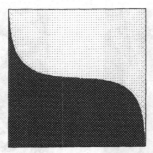 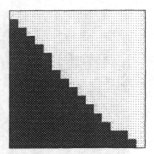

Fig. 18.1. An Ising CA (a) A state that evolved from a random pattern of 90% 0's and 10% 1's. (b) Wave on the boundary between a domain of all 1's, and a domain of all 0's. (c) Closeup of a portion of the boundary.

sublattice fixed while updating the black. In this way, the neighbors of a spin that is changed are not also simultaneously changed, and so our logic about conserving energy remains valid.

Now how can we add invertibility? We already have! If we apply our rule to the same checkerboard sublattice twice in a row, then each spin is either flipped twice or not at all—the net effect is no change. Thus the most recent step in our time evolution can always be undone simply by applying the rule a second time to the appropriate sublattice, and we can recover any earlier state by undoing enough steps.

This example demonstrates that we can simultaneously capture several basic aspects of physics in an exact digital model. First of all, the finite-state character of a quantum spin system is captured by using classical bits. Next, spatial locality is captured by making the rule at each lattice site depend only on nearby neighbors. Finally, both energy conservation and invertibility are captured by splitting the updating process into two phases, and alternately looking at one half of the bits while changing the other half. By making only changes that conserve a bond energy locally, we conserve energy globally. By making the change at each site be a permutation operation that depends only upon unchanged neighbor information, we can always go backwards by taking the same neighbor information and performing the inverse permutation.

Figure 18.1a shows the state of the Ising CA after 100,000 steps of time evolution on a 512×512 lattice, when started from a randomly generated configuration of site values that consisted of 10% 1's and 90% 0's. This is an *equilibrium* configuration: If we compare this to the configuration after 100,000,000 steps, the picture looks qualitatively the same. This equilibrium configuration is divided about equally between large domains that are mostly 0's, and large domains that are mostly 1's.

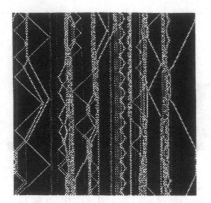 

Fig. 18.2. Ising-like CA's in 1D and 3D. (a) Time history of Bennett's 1D CA. (b) A 3D Ising CA cooled with a heatbath.

Since bond energy is conserved, the total length of boundary between regions of 0's and regions of 1's must be unchanged from that in the initial configuration—the numbers of 0's and 1's are not themselves conserved.

This CA has some surprising behavior when started from a more ordered initial state: It supports the continuum wave equation in an exact fashion. Figure 18.1b illustrates a wave on the boundary between two *pure* domains (all 0's, or all 1's). If we hold the values at the edges of the lattice fixed, then we find that the boundary shown behaves like a standing wave, oscillating in a harmonic fashion that repeats forever without any damping. In fact, it is easy to show that any waveform that we set up along this diagonal boundary—as long as it isn't too steep—exactly obeys the wave equation (*cf.* [43]). To see this, notice (Figure 18.1c) that the boundary *between* the two domains consists of a sequence of vertical and horizontal line-segments each the height or width of one site. If we number these segments sequentially along the boundary, then it is easy to verify that, at each update of the lattice, all of the even-numbered segments move one position along the boundary in one direction, while all of the odd-numbered segments move one position in the opposite direction. Thus the shape of the boundary is exactly the superposition of two discrete waveforms moving in opposite directions.

Similar techniques to those used in the Ising CA give a variety of related CA models [19, 21, 62, 90]. For example, in Figure 18.2a we show the time-history of a 1D rule invented by Charles Bennett that has exactly the same bond-energy conservation that we've just seen [74]. In Bennett's CA, instead of 1-bit at each site we put 2-bits, which we'll call A_i and B_i. The A's and B's will play the roles of the two sublattices in the Ising CA. We first update all of the A's in parallel, holding the B's fixed, and then vice versa. For each A_i, it's neighbors along the 1D chain will be the two B's on either side of it: B_{i-2}, B_{i-1}, B_{i+1} and B_{i+2}. Our rule is the same as before: We complement an A_i if exactly half of its four neighbors are 1's, and half are 0's. Once we have updated all of the A's, then we update the B's in the same manner, using the A's as neighbors. If we consider that there is a bond between each "spin" and its four neighbors, then we are again flipping the spin only if it doesn't change the total bond energy. If we update the same sublattice twice in a row, the net effect is no change: The rule is invertible, exactly like the Ising CA.

In the figure, our 1D lattice is 512 sites wide, with periodic boundaries (joined at the edges). We started the system with all sites empty except for a patch of randomly set bits in sites near the center. Time advances upward in the figure, and we show a segment of the evolution after about 100,000 steps. Rather than show the domains directly, we show all bonds that join antiparallel spins—the number of such "domain boundary" bonds is not changed by the dynamics. Note the variety of "particle" sizes and speeds.

In Figure 18.2b, we show a 3D Ising dynamics with a heat bath. Here the rule is an invertible 3D checkerboard Ising CA similar to our 2D version, except that at every site in our 3D lattice we have added a few extra *heatbath* bits. The heatbath bits at each site record a binary number that is interpreted as an energy. Now our invertible rule is again "flip whenever it is energetically allowed." As long as the heatbath energy at a given site is not too near its maximum, then a spin flip that would lower the bond energy is allowed, because we can put the energy difference into the heatbath. Similarly with transitions that would raise the bond energy. This heatbath-CA technique is due to Michael Creutz [19]. He thought of the bond energy as being potential energy, and the heatbath energy as being the associated kinetic energy. This heatbath CA is perfectly invertible, since applying the dynamics twice to the same sublattice leaves both the spins and the heatbath unchanged.

By adjusting the energy in the heatbath portion of this 3D CA, we can directly control the *temperature* of our system. We simply stop the simulation for a moment while we reach into our system and reset the heatbath values—without changing the spin values. As we *cool* the system in this way, energy will be extracted from bonds, and so if (for example) $\epsilon_{\neq} > \epsilon_{=}$, then there will be fewer domain boundaries—the domains will grow larger. The system shown has been cooled in this manner, and we render the interface between the up and down spins.

Figure 18.3 shows another Ising-like CA defined on a 3D cubic lattice. As in our 2D Ising CA, we have only one bit of state at each lattice site. Each site has a bond with each of its six nearest neighbors, and we perform a 3D checkerboard updating. This time our rule is, "flip a given spin if its six neighbors all have the same value: Six 0's or six 1's." We'll call this the "Same" rule. If we label half of the

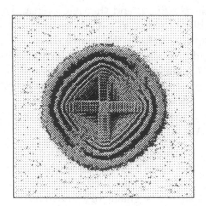

Fig. 18.3. An Ising-like 3D CA. (a) A macroscopic equilibrium configuration. (b) The same configuration, with the front half of the ball removed.

bonds attached to each site as "antiferromagnetic" (i.e., the energy values associated with parallel and antiparallel spins are interchanged for these labeled bonds), then this rule again conserves the total bond energy. Notice, though, that there are many different ways of labeling half of the bonds, and each way corresponds to a different additively conserved energy. We need to use several of these energies simultaneously if we want to express the Same rule as "flip whenever permitted by energy conservation."

The system in Figure 18.3a is $512 \times 512 \times 512$ and was started from an empty space (all 0's) with a small random block of spin values in the center. After about 5000 steps of time evolution, this invertible system settles into the ball shown, which then doesn't change further macroscopically. Microscopically it must keep changing—otherwise if we ran the system backwards it couldn't tell when to start changing again and *unform* the ball. The local density of 1's defines the surface that is being rendered. In Figure 18.3b we remove the front half of the ball to show its interior structure. The analogous rule in 2D does not form stable "balls."

It is easy to define other energy-conserving invertible Ising-like CA's. We could, for example, take any model that has a bond energy defined, find a sublattice of sites that aren't directly connected to each other by bonds, and update those sites in an invertible and energy conserving manner, holding their neighbors fixed. By running through a sequence of such sublattices, we would eventually update all sites. We could also make CA models with the same energy conservations with just two sublattices, by using the technique illustrated in Bennett's CA. Simply duplicate the state at each site in the original model, calling one copy A_i, and the other B_i. If the A's are only bonded to the B's and vice versa, then we can update half of our system in parallel, while holding all neighbors that they depend on fixed. Of course we can construct additional invertible energy-conserving rules by taking any of these examples and forbidding some changes that are energetically allowed.

18.3 Simple CA's with arbitrarily complex behavior

When the Ising model was first conceived in the 1920's, it was not thought of as a computer model: There were no electronic computers yet! It was only decades later that it and other discrete lattice models could begin to be investigated on computers. One of the first to think about such models was John von Neumann [16, 97]. He was particularly interested in using computer ideas to construct a mechanical model that would capture certain aspects of biology that are essential for reproduction and evolution. What he constructed was a discrete world in which one could arrange patterns of signals that act much like the logic circuitry in a computer. Just as computer programs can be arbitrarily complex, so too could the animated patterns in his CA world. Digital "creatures" in his digital universe reproduced themselves by following a digital program. This work anticipated the discovery that biological life also uses a digital program (DNA) in order to reproduce itself.

As we will see below, the level of complexity needed in a CA rule in order to simulate arbitrary patterns of logic and hence *universal computation* is quite low. In physics, this same possibility of building arbitrarily complicated mechanisms out of a fixed set of components seems to be an essential property of the evolution that built us. Is this the *only* essential property of evolution?

In a paradoxical sense, computation universality gives us so much that it really gives us very little. Once we have computation universality, we have a CA that is just as powerful as any conventional computer! By being able to simulate the logic circuitry of any computer, given enough time and space, any universal CA can compute exactly the same things as any ordinary computer. It can play chess. It can simulate quantum mechanics. It can even perform a simulation of a Pentium Processor running Tom Ray's Tierra evolutionary-simulation program [77], and thus we know that it is capable of exhibiting Darwinian evolution. But if we don't put in such an unlikely initial state by hand, is evolution of interesting complexity something that we are ever likely to see? Is it a *robust* property of the dynamics?

Nature has computation universality along with locality, exact conservations and many other constraints. Which of these constraints are important for promoting evolution, and whether it is possible to capture all of the important constraints simultaneously in a CA model, are both interesting questions. Here we will examine a well-known CA rule that is universal, and then discuss some physical constraints that it lacks that might make it a better candidate as a model for Darwinian evolution.

In Figure 18.4 we illustrate Conway's *Game of Life*, probably the most widely

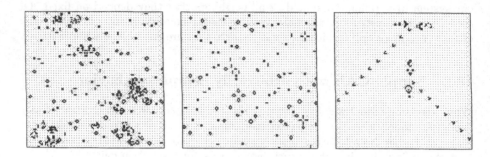

Fig. 18.4. Conway's non-invertible "Game of Life" CA (128×128 closeups taken from a 2K×2K space). (a) One thousand steps into an evolution started from a random configuration of 1's and 0's. (b) The same region after 16 thousand steps—the evolution has settled down into small uncoupled repeating patterns. (c) A configuration that started with two "glider guns."

known CA rule [7]. This is a CA that involves a 2D square lattice with one bit at each lattice site, and a rule for updating each site that depends on the total number of 1's present in its eight nearest neighboring sites. If the total of its neighbors is 3, a given site becomes a 1, if the total is 2, the site remains unchanged; in all other cases the site becomes a 0. This rule is applied to all sites simultaneously.

The Life rule is clearly non-invertible since, for example, an isolated 1 surrounded by 0's turns into a 0: You cannot then tell from the resulting configuration of site values whether that site was a 0 or a 1 in the previous configuration.

If you fill your computer screen with a random pattern of 0 and 1 pixels, and run the Life dynamics on it at video rates, then you see a lively churning and boiling pattern of activity, dying down in places to a scattering of small-period oscillating structures, and then being reignited from adjacent areas (Figure 18.4a). If you speed up your simulation to a few hundred frames per second, then typically after less than a minute for a 2K×2K system all of the interesting activity has died out, and the pattern has settled down into a set of isolated small-period oscillators (Figure 18.4b).

If you watch the initial activity closely, however, and pick out some of the interesting dynamical structures that arise, you can "build" configurations containing constructs such as the ones in Figure 18.4c. These are called *glider guns*. When the Life dynamics is applied to a glider gun, at regular intervals the gun spits out a small pattern that then goes through a short cycle of shapes, with the same shape reappearing every few steps in a shifted position. These *gliders* are the smallest moving objects in the Life universe. By putting together such constructs, one can show how to build arbitrary logic circuits, using sequences of gliders as the signals that travel around and interact [7, 36].

This then is our first example of a universal CA rule. Many other non-invertible universal CA rules are known—the simplest is due to Roger Banks [90]. All of these can support arbitrary complexity if you rig up a special enough initial state. Life is notable because it spontaneously develops interesting complexity starting from *most* initial states. Small structures that do something recognizable occasionally appear briefly, before being sucked back into the digital froth.

18.4 Invertible CA's are more interesting

One problem with Conway's Life as a model of evolution is that it lasts for such a short time when started from generic initial conditions. For a space of 2K×2K bits, there are $2^{4,194,304}$ distinct possible configurations, and this rule typically goes through fewer than 2^{14} of them before repeating a configuration and entering a cycle. This doesn't allow much time for the evolution of complexity! Furthermore, useful computing structures in Life are very fragile: Gliders typically vanish as soon as they touch anything.

The short Life-time problem can be attributed largely to the non-invertible nature of the Life rule—invertible rules do not behave like this. We typically have no idea just how long the cycle times of our invertible CA's actually are, because we have never seen them cycle, except from very special initial states or on very tiny spaces. The reason that invertible CA's have such long cycle-times is actually the same as the reason that essentially all invertible information dynamics have long cycles: *An invertible dynamics cannot repeat any state it has gone through until it first repeats the state it started in.* In other words, if we run an invertible rule for a while, we know what the unique predecessor of every state we have already seen is, except for the first state—its predecessor is still coming up! Thus an invertible system is forced to keep sampling distinct states until it stumbles onto its initial state. Since there is nothing forcing it toward that state as it explores its state space, the cycle of distinct states is typically enormously long: If our invertible CA really did sample states at random without repetition, it would typically have to go through about half of all possible states before chancing upon its initial state [91]. A non-invertible system doesn't have this constraint, and can re-enter its past trajectory at any point [49]. The moral here is that if you want to make a discrete world that lasts long enough to do interesting things, it is a good idea to make it invertible. As a bonus, a more thorough exploration of the available state-space tends to make a system more amenable to statistical mechanical analysis.

To make it easier to capture physical properties in CA's, we will use a technique called *partitioning*, which was developed specifically for this purpose [60, 90]. This technique is closely related to the sublattice technique introduced in Section 18.2 [62, 93]. The idea of partitioning is to divide up all of the bits in our CA system into disjoint local groupings—each bit is part of only one group. Then we update

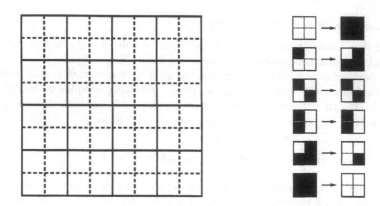

Fig. 18.5. The invertible "Critters" CA. (a) The solid and dotted blockings are used alternately. (b) The Critters rule.

all of the groups independently, before changing the groupings and repeating the process—changing the groupings allows information to propagate between groups. If the updating of each group conserves the number of 1's, for example, then so does the global dynamics. If the updating applied independently to each group is an invertible function, then the global dynamics is also invertible. Since all invertible CA's can be reexpressed isomorphically in a partitioning format—where conservations and invertibility are manifest—this is a particularly convenient format to use for our models [47, 48, 93].

Our first example of a partitioned CA is called "Critters." This is a universal invertible CA that evolves interesting complexity. The Critters rule uses a 2×2 block partition on a 2D square lattice. In Figure 18.5a we show an 8×8 region of the lattice—each square represents a lattice site that can hold a 0 or a 1. The solid lines show the grouping of the bits into 2×2 blocks that is used on the even time-steps, the dotted lines show the odd-time grouping. The Critters rule is shown in Figure 18.5b. This same rule is used for both the even-time grouping and the odd-time grouping. All possible sets of initial values for the four bits in a 2×2 block are shown on the left, the corresponding results are shown on the right. The rule is rotationally symmetric, so not all cases need to be shown explicitly: Each of the four discrete rotations of a block that is shown on the left turns into the same rotation of the corresponding result-block shown on the right.

Notice that each of the 16 possible initial states of a block is turned into a distinct result state. Thus the Critters rule is invertible. Notice also that the number of 1's in the initial state of each block is, in all cases, equal to the number of 0's in the result. Thus this property is true for each update of the entire lattice. If we call 1's *particles* on even steps, and call 0's *particles* on odd steps, then particles

 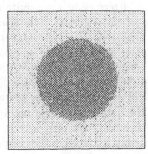 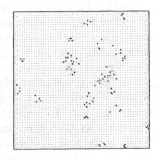

Fig. 18.6. A Critters simulation. (a) The initial state of the full 2K×2K lattice. (b) The state after 1,000,000 steps. (c) A closeup of a region on the right.

are conserved by this dynamics. Notice that the Critters rule also conserves the parity (sum mod 2) along each diagonal of each block, which leads to conservation of parity along every second diagonal line running across the entire space.

It is not interesting to run an invertible rule such as Critters starting from a completely random initial state, as we did in the case of Life. This is because the vast majority of all possible states are random-looking and so, by a simple counting argument, almost all of them have to turn into other random-looking states. To see this, note that any given number of steps of an invertible dynamics must turn each distinct initial state into a distinct final state. Since the set of states with recognizable structure is such a tiny subset of the set of all possible states, almost every random-looking state must turn into another random-looking state.

Thus instead of a random state, a "generic" initial state for an invertible CA will be some easily generated "low-entropy" state—we saw several examples of invertible evolutions from such states in Section 18.2. For the Critters CA, we show a sample simulation started from an empty 2K×2K lattice with a randomly filled 512×512 block of 0's and 1's in the middle (Figure 18.6a). In Figure 18.6b we see the state after 1,000,000 updates of the entire space. In this simulation, opposite edges of the space have been connected together (periodic boundaries). Figure 18.6c shows a closeup of a region near the right edge of the space: All of the structure present has arisen from collisions of small moving objects that emerged from the central region. In analogy to Life, we will call these small moving objects *gliders*. You can see several of these gliders in various phases of their motion in the closeup: They are the compact symmetrical structures composed of four particles, with two particles adjacent to each other, and two slightly separated (see also Figure 18.10). In the Critters dynamics, a glider goes through a cycle of four configurations each time it shifts by two positions.

Unlike the gliders in Life, Critters gliders are quite robust. Consider, for example, what happens when two of these gliders collide in an empty region. At first they form a blob of eight particles that goes through some pattern of contortions. If nothing hits this blob for a while, we always see at least one of the gliders emerge. This property arises from the combination of conservation and invertibility: We can prove, from invertibility, that the blob must break up, but since the only moving objects we can make with eight particles are one or two gliders, then that's what must come out. To see that the blob must break up, we can suppose the opposite. The particles that make up the blob can only get so far apart without the blob breaking up, and so there are only a finite number of possible configurations of the blob. The blob cannot repeat any configuration (and hence enter a cycle of states) because of invertibility: A local cycle of states would be stable going back in time as well as forward, but we know that the blob has to break up if we run backwards past the collision that formed it. Since the blob runs out of distinct configurations and cannot repeat, it must break up. At least one glider must come out. If the collision that formed the blob was rotationally symmetric, then both gliders must come out, since the dynamics is also rotationally symmetric. The robustness of particles that we saw in Figure 18.2a arises in a similar manner.

The Critters rule is fascinating to watch because of the complicated structures that form, with swarms of gliders bouncing around within them and slowly altering them. Sometimes, for example, a glider will hit a little flap left from a previous glider collision, flipping it from one diagonal orientation to another. This will affect what another glider does when it subsequently hits that flap. Gliders will hit obstacles and turn corners, sometimes going one way, sometimes another, depending on the details of the collisions. The pattern must gradually change, because the system as a whole cannot repeat. After a little observation it is clear that there are many ways to build arbitrary logic out of the Critters rule—one simple way is sketched in the next section in order to demonstrate this rule's universality.

Started from an ordered state, the Critters CA accumulates disorder for the same reason that a neat room accumulates disorder: Most changes make it messier. As we have already noted, in an invertible dynamics, a simple counting argument tells us that most messy states don't get neater. Localized patterns of structured activity that may arise within this CA must deal with an increasingly messy environment. No such structure in an invertible world can take in inputs that have statistical properties that are unpredictable by it, and produce outputs that are less messy, again because of our counting argument. Thus its fair to call a measure of the messiness of an invertible CA world the total *entropy*. We can think of this total entropy as being approximated by the size of the file we would get if we took the whole array of bits that fills our lattice and used some standard compression algorithm on it.

It is possible to construct invertible CA's in which a simple initial state turns into a completely random looking mess very quickly. While it is still true that

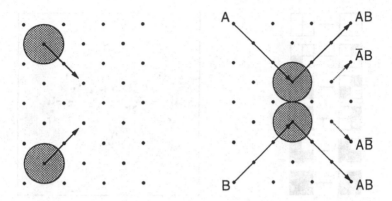

Fig. 18.7. Fredkin's Billiard Ball Model. (a) Balls heading toward a collision. (b) Paths taken in collision are displaced from straight paths.

this invertible CA will probably take forever to cycle, it has found another way to end its interesting activity quickly—what we might call a rapid *heat death*. Of course heat death is the inevitable fate of any CA evolution that has a long enough cycle: Since the vast majority of states are random-looking, very long cycles must consist mostly of such states. We can, however, try to put off the inevitable. In the Critters CA, symmetries and conservation laws act as constraints on the rate of increase of entropy, and so make the interesting low-entropy phase of the dynamics last much longer. It would be interesting to try to capture within CA dynamics other mechanisms that occur in real physics that contribute to metastability and hence delay the heat death.

18.5   A bridge to the continuum

Historically, the partitioning technique used in the previous section was first developed [60] for use in the construction of a very simple universal invertible CA modeled after Fredkin's Billiard Ball Model (BBM) of computation [30].[1] The BBM is a beautiful example of a continuum physical system that is turned into a digital system simply by constraining its initial conditions and the times at which we observe it. This makes it a wonderful bridge between the tools and concepts of continuum mechanics, and the world of exact discrete information dynamics. This model is discussed in [26], but we will review it very briefly here.

In Figure 18.7a we show two finite-diameter billiard balls heading toward a collision, both moving at the same speed. Their centers are initially located at

---

[1]The first universal invertible CA was actually constructed by Toffoli [86], who showed how to take a universal 2D CA that was non-invertible, and add a third dimension that would keep a complete time history of the dynamics, thus rendering it invertible.

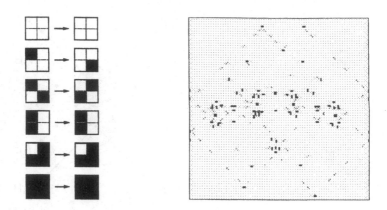

Fig. 18.8. A Billiard Ball Model CA. (a) The BBMCA rule. (b) A BBMCA circuit.

integer coordinates on a Cartesian lattice—we will refer to these as *lattice points*. At regular intervals, the balls will be found at consecutive lattice points, until they collide. In Figure 18.7b we show a collision. The outer paths show the actual course that the balls take after a collision; the inner paths illustrate where each of the two balls would have gone if the other one wasn't there to collide with it. Thus we see that a locus at which a collision might or might not happen performs logic: If the presence of a ball at a lattice point at an integer time is taken to represent a 1, and the absence a 0, then we get 1's coming out on the outer paths only if balls at A AND B came in at the same time. The other output paths correspond to other logical functions of the inputs. It is easy to verify that this collision is a universal and invertible logic element (just reverse all velocities to run BBM circuits backwards). We also allow fixed mirrors in our model to help route ball-signals around the system—these are carefully placed so that the centers of balls are still always found at lattice points at integer times.

In order to make a simple CA model of the BBM, we will represent finite diameter balls in our CA by spatially separated *pairs* of particles, one following the other—the leading edge of the ball followed by the trailing edge. When such a ball collides with something, the front-edge particle collides first, and then the influence is communicated to the trailing edge. This kind of "no rigid bodies" approach to collisions is more consonant with the locality of interaction that we are trying to capture in CA's than a larger-neighborhood model in which distant parts of a fixed-size ball can see what's going on right away.

Figure 18.8a shows the BBMCA rule. Like the Critters rule, this rule is rotationally symmetric and so, again, only one case out of every rotationally equivalent set is shown. Note that the rule conserves 1's (particles), and that only two cases change. This is the complete rule that is applied alternately to the even and odd

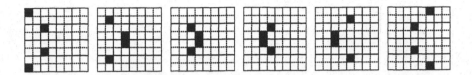

Fig. 18.9. A BBMCA collision. We show succesive snapshots of a small area where a collision is happening. In the first image, the solid-blocks are about to be updated. The blocking alternates in successive images.

2×2 blockings. Note that, much like the Ising CA, this rule is its own inverse: If we simply apply the update rule to the same blocking twice in a row, the net effect is no change.[2]

Figure 18.9 shows a BBMCA collision between two minimum-size balls—the gap between the two particles that make up the front and back of the ball can be any odd number of empty sites. Until the balls get close together the particles that form them all propagate independently: A single 1 in one corner of a block moves to the opposite corner. When we change the blocking, the particle again finds itself alone in a block in the same corner it started in, and again moves in the same direction. When two leading-edge particles find themselves in the same block, the collision begins. These particles are stuck for one step—this case doesn't change. Meanwhile the trailing edge particles catch up, each colliding head-on with a leading-edge particle which was about to head back to meet it (if the gap had been wider). New particles come out at right angles to the original directions, due to the "two-on-a-diagonal" case of the rule, which switches diagonal. Now one of the particles from each head-on collision becomes the new leading edge particle; these are done with the collision and head away from the collision locus, once again propagating independently of the trailing particles. Meanwhile the two new trailing-edge particles are headed toward each other. They collide and are stuck for one step before reflecting back the way they came, each following along the path already taken by a leading edge particle. Each two-particle ball has been displaced from its original path. If the other two-particle ball hadn't been there, it would have gone straight.

Mirrors are built by placing square patterns of four particles straddling two adjacent blocks of the partition. It is easy to verify that such squares don't change under this rule, even if you put them right next to each other. Single particles just bounce back from such mirrors. The collision of a two-particle ball with such a mirror looks just like the collision of two balls that we have already seen; we just replace one of the balls with a mirror whose surface lies along the axis of symmetry

---

[2]The Ising CA is actually very closely related. It can be put into the same 2×2 block-partitioned format if we model bonds instead of sites [90].

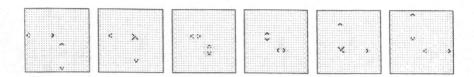

Fig. 18.10. A BBMCA-style collision of pairs of gliders in the Critters CA. The images shown are not consecutive states of the lattice, but are instead spaced in time to correspond (with a 45° rotation) to the images in the previous figure.

of the two-ball collision. The remaining ball can't tell the difference. For more details about the BBMCA, see [62, 90].

Figure 18.8b shows a BBMCA circuit, computing a permutation sequence. Because of their long cycle times, invertible circuits tend to make good pseudo-random number generators. In fact, a perfect random number generator would go through all of its internal states before cycling, and so it would be perfectly invertible. It is also interesting to use the BBMCA to construct circuits that are more directly analogous to thermodynamic systems, since the constraint of invertibility means that *it is impossible to design a BBMCA circuit that, acting on unpredictable statistical inputs that it receives, can reduce the entropy of those data*—for the reasons discussed in the previous section [4, 6, 62]. The BBMCA is simple enough that it provides a good theoretical model to use for other inquiries about connections between physics and computation. For example, one can use its dynamics as the basis of quantum spin models [63].

Using Figure 18.10, we sketch a simple demonstration that the Critters rule of the previous section is universal—we show that pairs of Critters-gliders suitably arranged can act just like the "balls" in the BBMCA, which is universal. Figure 18.10 shows a collision that is equivalent to that shown in Figure 18.9. We don't show every step of the Critters time-evolution; instead we show the pairs of gliders at points corresponding to those in the collision of Figure 18.9. Mirrors can be implemented by two single Critters particles, one representing each end of the mirror.

18.6   Discrete molecular dynamics

Having constructed a CA version of billiard ball dynamics, it seems natural to try to construct CA's that act more like real gases [60, 89]. With enough particle directions and particle speeds, our discrete Molecular Dynamics (MD) should approximate a real gas.

The BBMCA has just four particle directions and a single particle speed. We will make our first MD model by modifying the BBMCA rule. For simplicity, we won't

Fig. 18.11. A simple four-direction lattice gas. (a) A momentum conserving invertible rule. (b) A 512×512 lattice filled randomly with particles, with a square block of ones's in the center. (c) A round pressure wave spreads out from the center.

worry about modeling finite-diameter balls: Single 1's will be our gas molecules. The simplest such rule would be, "During each 2×2 block update, each molecule ignores whatever else is in the block and simply moves to the opposite corner of its block." Then, when we switch partitions, that molecule would again be back in the same kind of corner that it started in and so it would again move in the same direction—moving exactly like an isolated particle in the BBMCA. This simple rule gives us a non-interacting gas, with four directions and one speed.

We would like to add a *momentum conserving* collision to this non-interacting gas rule. We can begin by defining what we mean by momentum. If we imagine that our discrete lattice dynamics is simply a succession of snapshots of a continuum system, as it was in the case of the BBM, then we automatically inherit definitions of energy and momentum from the continuous system. To add a momentum conserving collision to our simple four-direction gas, we should have two molecules that collide head-on come out at right angles. Figure 18.11a shows the non-interacting gas rule with one case modified to add this collision: When exactly two molecules appear on one diagonal of a block, they come out on the other diagonal.

We would not expect such a simple model to behave very much like a real gas. In Figure 18.11b, we show a 512×512 2D space filled with a random pattern of 1's and 0's, with a square block in the center of the pattern that is all 1's. Figure 18.11c shows this system after about 200 updates of the space: We see a round pressure wave. We were amazed when we first ran this simulation in the early 1980's [89]. How could we get such continuous looking behavior from such a simple model? This is the point at which we began to think that perhaps CA MD might be immediately practical for fluid modeling [59, 90]. Discrete lattice models

Fig. 18.12. A slightly modified gas CA. The vertical bar is a density wave traveling to the right, the circle is a region in which waves are slower. We see the wave reflect and refract.

are well adapted to meshes of locally interconnected digital hardware, which can be very large and very fast if the models are simple. It turns out, though, that this particular model is too simple to simulate fluid flow—though it is useful for other purposes. This four-direction *lattice gas automaton* (LGA) is now commonly known as the HPP gas after its originators [41], who analyzed it about a decade before we rediscovered it. Their analysis showed that this four-velocity model doesn't give rise to normal isotropic hydrodynamics.

Notice that the HPP gas is perfectly invertible—like the BBMCA, its rule is its own inverse. Thus we can run our pressure wave backwards, getting back exactly to the square block we started from. This doesn't contradict what we said earlier about entropy in invertible CA's, since a messy state can always be cleaned up if you undo *exactly* the sequence of actions that produced the mess. Normal forward time evolution doesn't do this.

Once we are able to model one macroscopic phenomenon, it is often obvious how to model other phenomena. Starting from a model with sound waves, we can make a model with reflection and refraction of such waves. In Figure 18.12 we show a simulation using a 2-bit variant of the HPP CA. Here we have added a bit to each site, and used it to mark a circular region of the space: One bit at each site is a *gas* bit, and the other bit is a *mark* bit. We now alternate the rule with time so that, for one complete even-time/odd-time update of the lattice we apply the HPP rule to the gas bits at all sites; then we do a complete even/odd update only of unmarked blocks, with all gas particles in blocks containing non-zero mark-bits left unchanged. This gives us two connected HPP systems, one running half as fast as the other. In particular, waves travel half as fast in the marked region, and so we get refraction of the wave that is incident from the left. Notice that the dynamics is still perfectly invertible, and that we can make our "lens" any shape we desire—it is a general feature of CA MD that we can simply "draw" whatever shaped obstacles and potentials we need in a simulation [83]. Related LGA techniques have been used for complex antenna simulations [81].

We can model many other phenomena by coupling 2×2 block rules. We can, for example, use the HPP gas or a finite-impact-parameter variant of it (the TM gas [59, 90], which has better momentum-mixing behavior) as a source of pseudo-randomness in a diffusion model. We start by again putting two bits at each lattice site—one bit will belong to the *diffusing system*, while the other belongs to the *randomizing system*. Let the four bits of the randomizing system in each block simply follow the HPP dynamics described above. Let the four bits of the diffusing system be rotated 90° clockwise or counterclockwise, depending on the parity of the number of 1's in the four "random" bits. This results in a perfectly invertible diffusion in which no more than one diffusing particle can ever land at the same site [90]. Using this approach with enough bits at each site we can, for example, model the diffusion and interaction of different chemical species.

The HPP CA was originally presented in a different format than we have used above [41]. Since this other format is in many cases very natural for discussing MD models, we will describe it here, and then relate it to a partitioned description. We start by putting four bits of state at each site of a square lattice. We will call the bits at the ith site N_i, S_i, E_i and W_i. The dynamics consists of alternating two steps: (1) *move* the data, and then (2) let the groups of bits that land at each site *interact* separately. The first step moves the data in sheets: If we think of the directions on our lattice as being North, South, East and West, then all of the N bits are moved one position North, all the S bits one position South, etc. Our interaction rule at each site combines the bits that came from different directions and sends them out in new directions. A state consisting of two 1's (particles) that came in from opposite directions and two 0's from the other directions, is changed into a state in which the 1's and 0's are interchanged—the particles come out at right angles to their original directions. In all other cases, particles come out in the same directions they came in.

We can think of this as a particular kind of partitioning rule, where the four bits at each site are the groups, and we use the data-movement step to rearrange the bits into new groups. Although in some ways this *site-partitioned* description of the HPP gas is simpler, it also suffers from a slight defect. If we imagine, as we did in our discussion of the Ising model, that our lattice is a giant black and white checkerboard, then we notice that in one data movement step all of the bits that land at black squares came from white squares, and vice versa. No data that is currently on a black square will ever interact with data that is currently on a white square: We have two completely independent subsystems. The 2×2 block version of the HPP rule is isomorphic to just one of these subsystems, and so lets us avoid simulating two non-interacting systems. Of course we can also avoid this problem in the site-partitioned version with more complicated time-dependent shifts: We can always reexpress any partitioned CA as a site-partitioned CA. This fact has been

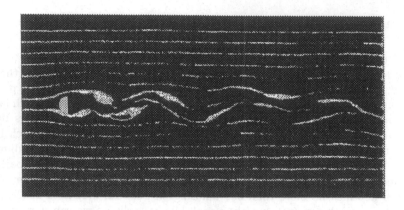

Fig. 18.13. Flow past a half-cylinder using a six-direction lattice gas on a triangular lattice. We simulate "smoke" streamers to visualize the flow.

important in the design of our latest CA machines.

The HPP lattice gas produces a nice round-looking sound wave, but doesn't reproduce 2D hydrodynamics in the large-scale limit. We can clearly make a CA model with more speeds and directions by having molecules travel several lattice positions horizontally and/or vertically at each step—just add more particles at each site, and shift the different momentum fields appropriately during the movement step. With enough speeds and directions, it seems obvious that we can get the right macroscopic limit—this should be very much like a hard-sphere gas, which acts like a fluid. The fact that so many different fluids obey the same hydrodynamic equations also suggests that the details of the dynamics can't matter very much, just the constraints such as momentum and particle conservation.

So how simple a model can work? It was found [28] that we can recover macroscopic 2D hydrodynamics from a model that is only slightly more complicated than the HPP gas. A single-speed model with six particles per site, moving in six directions on a triangular lattice, will do. If all zero-net-momentum collisions cause the molecules at the collision site to scatter into a rotated configuration, and otherwise particles go straight, then in the low speed limit we recover isotropic macroscopic fluid dynamics. Figure 18.13 shows a simulation of a slightly more complicated six-direction lattice gas [12]. The simulation shown is 2K×1K, and we see vortex shedding in flow past a half-cylinder. The white streamers are actually a second gas (more bits per site!), inserted into the hydrodynamic gas as a kind of smoke used to visualize flows in this CA wind tunnel. This is an invertible CA rule, except at the boundaries which are irreversibly being forced (additional bits per site mark the boundaries). Simple single-speed CA's have also been used to simulate 3D hydrodynamics [2, 22].

  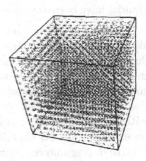

Fig. 18.14. Some CA MD simulations. (a) Flow through a porous medium. (b) A topologically complicated structure within a chemical reaction simulation. (c) Crystallization using irreversible discrete forces.

When it was discovered that lattice gases could simulate hydrodynamic behavior, there was a great deal of excitement in some circles and skepticism in others. The exciting prospect was that by simplifying MD simulations to the point where only the essence of hydrodynamic behavior remained, one could extend the scale of these simulations to the point where interesting hydrodynamics could be done directly with an MD method. This spawned an entire new field of research [10, 13, 22, 23, 52, 71, 76]. This optimistic scenario has not yet been realized. One problem is that simple single speed models aren't well suited for simulating high-speed flows. As in a photon gas, the sound speed in a single-speed LGA is almost the same as the maximum particle speed, making supersonic flows impossible to simulate. You need to add more particle speeds to fix this. The biggest problem, though, is that you need truly enormous CA systems to get the resolution needed for hydrodynamic simulations with high Reynold's numbers [102].

For the near term, for those interested in practical modeling, it makes sense to avoid high-Reynold's numbers and fast fluid flows, and to use MD CA models to simulate other kinds of systems that are hard to simulate by more conventional numerical techniques. Suitable candidates would include systems for which the best current simulation techniques are in fact some form of molecular dynamics, as well as systems for which there are at present no good simulation techniques because traditional MD cannot reach the hydrodynamic regime. An example would be systems with very complicated flows. Figure 18.14a shows a simulated flow through a piece of sandstone. The shape of the sandstone was obtained from MRI imaging of an actual rock, taking advantage of the ability of CA MD simulations to handle arbitrarily shaped obstacles. Shading in the figure indicates flow velocity. Simulations were compared against experiments on the same rock that was imaged, and agreement was excellent [2, 79]. More complicated flows, involving immiscible liquids, have been simulated with this same technique.

CA models of complex systems can be built up by combining simpler models. We simply pile up as many bits as we need at each lattice site, representing as many fluids, random variables, heat baths, and other fields as we desire. Then we update them in some repeated sequence, including extra steps that make different groups of subsystems interact, much as we did in our diffusion and our refraction examples. For practical purposes we will often dispense with invertibility, and be satisfied with irreversible rules coupled to pseudo random subsystems. Figure 18.14b shows an example of a 3D chemical reaction simulation of this sort, which simulates the FitzHugh-Nagumo reaction-diffusion dynamics [56]. The knot and its surroundings are composed of two different chemical phases. The connectivity of the knot in conjunction with domain repulsion keeps the knot from shrinking away. Many kinds of multiphase fluids, microemulsions, and other complex fluids have been simulated using related techniques [9, 14, 79].

We can easily add discrete forces by having particles at a discrete set of vector separations interact. If two such particles are heading away from each other we can point them toward each other and otherwise leave them unchanged—this results in an attraction. This kind of rule isn't invertible, but it is energy and momentum conserving. Figure 18.14c shows a 3D crystallization simulation using a potential built up out of such interactions [103]. This is not currently a very practical way to simulate crystals, but this kind of technique is generally useful [3, 104]. For example, the "smoke" in Figure 18.13 has a weak cohesive force of this kind, which makes the smoke streams thinner.

There are many other ways to build new CA MD models. We often appeal to microdynamical analogy, or to simulating "snapshots" of a hypothetical continuous dynamics. We can take aspects of existing atomistic models and model them statistically at a higher level of aggregation using exact integer counts and conservations, to avoid any possibility of numerical instability [12]. We can combine CA's with more traditional numerical mesh techniques, using discrete particles to handle difficult interface regions [50]. We can adapt various energy-based techniques from statistical mechanics [9, 38]. We can also build useful models in a less systematic manner, justifying their use by simulation and careful measurements [79]. Combining well understood CA MD components to build up simulations of more complex systems is a kind of iterative programming exercise that involves testing components in various combinations, and adjusting interactions.

Although there is already a role for CA MD models even on conventional computers, there is a serious mismatch on such machines between hardware and algorithms. If we are going to design MD simulations to fit into a CA format, we should take advantage of the uniformity and locality of CA systems, which are ideally suited to efficient and large-scale hardware realization.

18.7   Crystalline computers

Computer algorithms and computer hardware evolve together. What we mean by a good algorithm is that we have found some sort of efficient mapping between the computation and the hardware. For example, CA and other lattice algorithms are sometimes "efficiently" coded on conventional machines by mapping the parallelism and uniformity of the lattice model onto the limited parallelism and uniformity of word-wide logical operations—so-called "multi-spin coding." This is a rather grotesque physical realization for models that directly mimic the structure and locality of physics: We first build a computer that hides the underlying spatial structure of Nature, and then we try to find the best way to contort our spatial computation to fit into that mold!

Ultimately all physical computations have to fit into a spatial mold, and so our most efficient computers and algorithms will eventually have to evolve toward the underlying spatial "hardware" of Nature [94]. Because physical information can travel at only a finite velocity, portions of our computation that need to communicate quickly must be physically located close together. Computer architects can only hide this fact-of-life from us for so long. At some point, if we want our computations to run faster, our algorithms must take on the responsibility of dealing with this constraint.

Computer engineers are not unaware of this spatial constraint. Various locally-interconnected parallel computers have been built and studied [53]. Mesh architectures are organized like a kind of CA, but usually with a rather powerful computer with a large memory at each lattice site. Unlike CA's, they normally don't have the same operation occurring everywhere in the lattice at the same time. *SIMD* or *data parallel* mesh machines are more CA-like, since they typically have a simpler processor at each site, and they do normally have the operation of all processors synchronized in perfect lockstep.

Another important spatial computing device is the *gate array*. These regular arrays of logic elements are very much like a universal CA. Initially, we build these chips with arrays of logic elements, but we leave out the wiring. Later, when we need a chip with a specific functionality, we can quickly "program" the gate array by simply adding wires to connect together these elements into an appropriate logic circuit. FPGA's (field programmable gate arrays) make programming the interconnections even easier. What should be connected to what is specified by some bits that we communicate directly to the chip: This rapid transition from bits to circuitry eliminates much of the distinction that is normally made between hardware and software [73].

As general purpose computing devices, none of these CA-like machines are significant mainstream technologies—the evolutionary forces that are pushing us in the CA direction haven't yet pushed hard enough. This was even more true when Tom Toffoli and I first started playing with CA's together, almost two decades ago.

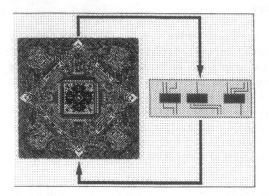 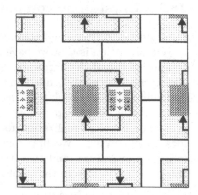

Fig. 18.15. CA machines. (a) Our earliest CA machines scanned their memory like a framebuffer, applying discrete logic to a sequence of neighborhood windows. (b) CAM-8 uses a 3D mesh array of SIMD processors. Each processor handles part of the lattice, alternating data movement with sequential lookup-table updating of the groups of bits that land at each lattice site.

There were no machines available to us then that would let us run and display CA's quickly enough that we could experience them as dynamical worlds. We became increasingly frustrated as we tried to explore the new and exciting realm of invertible CA's on available graphical workstations: Each successive view of our space took minutes to compute.

Tom designed and built our first CA simulation hardware. This CA machine was a glorified frame-buffer that kept the CA space in memory chips. As it scanned a 2D array of pixels out of the memory, instead of just showing them on a screen, it first applied some neighborhood logic to the data, then sent each resulting pixel to the screen while also putting it back into memory. At first the neighborhood logic was on a little prototyping board (this is shown schematically in Figure 18.15a). I learned about digital circuitry by building site-update rules out of TTL chips—each small logic circuit I built became the dynamical law for a different video-rate universe! Eventually, we switched to lookup tables, which largely replaced the prototype circuits. The first few generations of machines, however, all had wires that you could move around to change things—whenever you wanted to demonstrate something, you would invariably find that someone else had rewired your machine [88].

We went through several generations of hardware, playing with models that no one else had ever seen [90]. At a certain point we rediscovered the HPP lattice gas, and our simulations rekindled interest in this kind of model. At this point, our machines became inadequate. They had been designed for interacting and experimenting with small 2D systems that you could watch on a monitor. Real

CA MD was going to require large-scale 3D machines designed for serious scientific simulation, with provisions for extensive data analysis and visualization. A new dedicated CA MD machine could be about 1000 times as cost effective as existing supercomputers for this task, and would provide the interactivity and flexibility of a personal CA workstation.

I designed the new architecture (CAM-8) based on the experience that Tom and I had with our earlier machines [66]. As shown in Figure 18.15b, this machine uses a 3D mesh array of SIMD processors running in perfect lockstep. The lattice being simulated (which can be n dimensional) is divided up evenly among the processors, each processor handling an equal *sector* of the overall lattice. As in our earlier machines, the state of the lattice is kept in ordinary memory chips, and the updating is done by lookup tables—also memory chips. Data cycles around and around between these two sets of memory chips.

Unlike our previous machines, which provided a fixed set of traditional CA neighborhoods, the only neighborhood format supported by CAM-8 is *site-partitioning* (discussed in Section 18.6). Any *bit field* (i.e., set of corresponding bits, one from every site) can be shifted uniformly across the lattice in any direction. Whatever data land at a given lattice site are updated together as a group. This is a particularly convenient neighborhood format from a modeling point of view, since any CA dynamics on any neighborhood can be accomplished by performing an appropriate sequence of data shifts and site updates, each acting on a limited number of site bits at a time. Also, as we've seen, partitioning is a particularly good format for constructing models that incorporate desired physical properties.

Site partitioning is also a very convenient neighborhood format from a hardware standpoint. Since the pattern of data movement is very simple and regular, mesh communication between processors is also very simple. Since the updating is done on each site independently, it doesn't matter what order the sites are updated in, or how many different processors are involved in the updating. All of this can be organized in the manner most efficient for the hardware.

CAM-8 machines were built and performed as expected. All of the simulations depicted in this chapter were done on CAM-8, except for that of Figure 18.14a (which is similar to [2]). CAM-8 has not, unfortunately, had the impact that we hoped it might. First of all, during the time that we were building the machine, it was found that lattice gases weren't as well suited for high Reynold's number hydrodynamic flow simulations as people had hoped. In addition, in the absence of any good CA machines, interest in lattice dynamics calculations had shifted to techniques that make better use of the floating point capabilities of conventional computers. Also, most researchers interested in developing practical applications already had good access to conventional supercomputers, which were 1000 times less cost effective than CAM-8, but had familiar and high-quality software and system support. Finally, the evolutionary forces favoring CA-like machines were temporarily on the wane at the time when CAM-8 was completed, as multiprocessor funding dried up and fine-grained parallel computing companies folded. We didn't build any versions of our indefinitely scalable CAM-8 machine that were large enough to make previously unreachable realms of CA modeling accessible—as our early machines first opened the CA world to us.

In the near term, prospects again look good for CA machines. Although our small personal-computer-scale CAM-8 machines are still about as good as any supercomputer for LGA computations, advances in technology make radical improvements possible. By putting logic directly on DRAM memory chips, which is now routinely done, and by exploiting the enormous memory bandwidth that can be made available on-chip, it is possible today to make a SIMD machine similar to the CAM-8 that is over 10,000 times faster *per memory chip* than the current CAM-8 hardware [69]. Putting together arrays of such chips, qualitatively new physical simulations will become possible. Other SIMD applications such as logic simulation, image processing and 3D bit-map manipulation/rendering will also run at about a trillion bit-operations per second per chip. Whether we manage to make our next dream machine, the time is ripe for commercial SIMD-based CA machines.

What of the more distant future? In the preceding sections, we have emphasized invertible CA's. Aside from their intrinsic interest, they have the virtue that they mimic the microscopic invertibility of physical dynamics. From a macroscopic point of view, this means that these CA's can in principle be implemented using frictionless reversible mechanisms—they don't depend on dissipative thermodynamically irreversible processes in order to operate [27, 100, 106, 107]. Thus 3D machines based on invertible CA's won't have the same problem getting rid of dissipated heat that irreversible machines do [4, 5, 26]. From a more microscopic point of view, we can see the match between invertible computation and invertible quantum physics as making possible direct use of quantum scale elements and processes to do our computations. We can make use of discrete properties of these quantum elements to represent our digital quantities and perform our digital computations.

Thus in the distant future I expect that our most powerful large-scale *general purpose* computers will be built out of macroscopic crystalline arrays of identical invertible computing elements. We would make such large arrays out of identical elements because they will then be easier to control, to design, to build and to test. These will be the distant descendants of todays SIMD and FPGA computing devices: When we need to perform inhomogeneous computations, we will put the irregularities into the program, not the hardware. The problem of arranging the pieces of a computation in space will be part of the programming effort: Architectural ideas that are used today in physical hardware may reappear as data structures within this new digital medium. With molecular scale computing elements, a small chunk of this *computronium* [64] would have more memory and processing power than all of the computers in the world today combined, and high Reynold's number CA MD calculations of fluid flow would be practical on such machines.

Note that I don't expect our highest performance general purpose computers to be quantum spin computers of the sort discussed in Section 18.1. In such a machine, the whole computer operates on a superposition of distinct computations simultaneously. This kind of *quantum parallelism* is very delicate, and the overhead associated with the difficult task of maintaining a superposition of computations over a large spatial scale will be such that it will only be worth doing in very specialized situations—if it is possible at all [75]. This won't be something that we will do in our general purpose computers.

18.8   What makes a CA world interesting?

Future CA machines will make extensive large-scale CA simulations possible—we will be able to study the macroscopic properties of CA worlds that we design. Aside from issues of size and speed, there doesn't seem to be any obvious reason why exact classical information models cannot achieve as high a level of rich macroscopic complexity as we see in our universe. This is a very different modeling challenge than trying to simulate quantum mechanics with CA's. We would like to simulate an interesting macroscopic world which is built out of classical information. It is instructive to try to see what the difficulties might be.

The most important thing in our universe that makes it interesting to us is of course *us*. Or more generally, the existence of complex organisms. Thus let's begin by seeing what it might take to simulate a world in which Darwinian evolution is *likely* to take place. Since no one has yet made a CA that does this, our discussion will be quite speculative.

One of the most successful computer models of evolution is Tom Ray's Tierra [77, 78], which was designed to capture—in an exact digital model—an essential set of physical constraints abstracted from biology. His model did not include spatial locality or invertibility, but we could try to add these features.

Modeling evolution in a robust spatial fashion may, however, entail incorporating some physical properties into our CA systems that are not so obvious [65]. For example, in Nature we have the property that we can take complicated objects and set them in motion. This property seems to be essential for robust evolution: It is hard to imagine the evolution of complex organisms if simpler pieces of them can't move toward each other! No known universal CA has this property (but see [3, 18, 104]). There is nothing in the Life CA, for example, that corresponds to a glider-gun in motion.

The general property of physics that allows us to speak about an object at rest and then identify *the same object* in motion is *relativistic invariance*. The fact that the laws of physics look the same in different relativistic frames means that we can have the same complex macroscopic objects in all states of motion: An organism's chemistry doesn't stop working if it moves, or if the place it lives moves! In a spatial CA dynamics, some sort of spatial macroscopic motion invariance would clearly make evolution more likely. Since our CA's have a maximum speed at which information can travel—a finite speed of light—relativistic invariance is a possible candidate. Full relativistic invariance may be more than we need, but it is interesting to ask, "Can we incorporate relativistic invariance into a universal CA model?"

We have already seen that we can have macroscopic rotational invariance in our lattice gas models, and we know that numerical mesh calculations of relativistic dynamics are possible. Thus achieving a good approximation of relativistic invariance in the macroscopic limit for an exact CA model seems possible [43]. Such a system would, at least in the limit, have the conservations associated with the continuous Lorentz group of symmetries. Although it is not possible to put a continuous symmetry directly into a discrete CA rule, it is certainly possible to put these conservations into the rule, along with a discrete version of the symmetries—just as we did in our lattice gas models.[3]

Thus we might imagine our relativistically invariant CA to be a refinement of lattice gases—we would also like to make it invertible for the reasons discussed in Section 18.4. But we also demand that this CA incorporate computation universality. This may not be easy: Since a relativistically invariant system must have momentum conservation, we will need to worry about how to hold complex interacting structures together. Thus we may need to incorporate some kind of relativistically invariant model of forces into our system.

Simulating forces in an exact and invertible manner is not so easy, particularly if we want the forces to last for a long time [105]. Models in which forces are communicated by having all force-sources continuously broadcast field-particles have the problem that the space soon fills up with these field-particles—which cannot be erased because of local invertibility—and then the force can no longer be communicated. Directly representing field gradients works better, but making this work in a relativistic context may be hard.

At this point, we might also begin to question our basic CA assumptions. We introduced crystalline CA's to try to emulate the spatial locality of physics in our informational models, but we are now discussing modeling phenomena in a realm of physics in which modern theories talk about extra dimensions and variable topology. Perhaps whatever is essential fits nicely into a simple crystalline framework, but perhaps we need to consider alternatives. We could easily be led to informational models in which the space and time of the microscopic computation becomes rather divorced from the space and time of the macroscopic phenomena.

We started this section with the (seemingly) modest goal of using a CA to try to

---

[3]In continuum physics, continuous symmetries are regarded as fundamental and conservations arise as a consequence of symmetry. Fredkin has pointed out that in discrete systems, it must be the conservation that is fundamental.

capture aspects of physics necessary for a robust evolution of interesting complexity, and we have been led to discuss incorporating larger and larger chunks of physics into our model. Perhaps our vision is too limited, and there are radically different ways in which we can have robust evolution in a spatial CA model. Or perhaps we can imitate Nature, but cheat in various ways. We may not need full relativistic invariance. We may not need exact invertibility. On the other hand, it is also perfectly possible that we can't cheat very much and still get a system that's nearly as interesting as our world.

18.9   Conclusion

I was in high school when I first encountered cellular automata—I read an article about Conway's "Game of Life." At that time I was intensely interested in both Physics and Computers, and this game seemed to combine the two of them. I immediately wondered if our universe might be a giant cellular automaton.

The feeling that physics and computation are intimately linked has remained with me over the years. Trying to understand the difficulties of modeling Nature using information has provided an interesting viewpoint from which to learn about physics, and also to learn about the ultimate possibilities of computer hardware. I have learned that many properties of macroscopic physics can be mirrored in simple informational models. I have learned that quantum mechanics makes both the amount and the rate-of-change of information in physical systems finite—all physical systems do finite information processing [68]. I have learned that the non-separability of quantum systems makes it hard to model them efficiently using classical information—it is much easier to construct quantum spin computer models [1]. I have learned that physics-like CA models can be the best possible algorithms when the computer hardware is also adapted to the constraints and structure of physical dynamics. I have learned that developing computer hardware that promotes this viewpoint can consume an enormous amount of time!

Since classical information is much easier to understand than quantum information I have mostly studied classical CA models. In these systems, a macroscopic dynamical world arises from *classical* combinatorics. Continuous classical-physics behavior can emerge in the large-scale limit. We can try to model and understand (and perhaps teach our students about) all sorts of physical phenomena without getting involved in quantum complications. We can also try to clarify our understanding of the fundamental quantities, concepts and principles of classical mechanics *and of classical computation* by studying such systems [32]. The principle of stationary action, for example, must arise in such systems solely from combinatorics—there is no underlying quantum substratum in a CA model. Conversely, we should remember that information (in the guise of entropy) was an important concept in physics long before it was discovered by computer scientists. Just as Ising-like systems have provided intuitive classical models that have helped clarify issues in statistical mechanics, CA's could play a similar role in dynamics.

Since we have focused so much on discrete classical CA models of physics, it might be appropriate to comment briefly on their relationship to discrete quantum models—Feynman's quantum spin computer of Section 18.1. Exactly the same kinds of grouping and sublattice techniques that we have used to construct invertible CA's also allow us to construct quantum CA's—QCA's [61]. We simply replace invertible transformations on groups of bits with unitary transformations on groups of spins. Just as it is an interesting problem to try to recover classical physics from ordinary CA's, it is also interesting to try to find QCA's that recover the dynamics of known quantum field theories in the macroscopic limit [11, 57, 58, 105]. Following our Ising CA example, it might be instructive to investigate classical CA's that are closely related to such QCA's.

Although people have often studied CA's as abstract mathematical systems completely divorced from Nature, ultimately it is their connections to physics that make them so interesting. We can use them to try to understand our world better, to try to do computations better—or we can simply delight in the creation of our own toy universes. As we sit in front of our computer screens, watching to see what happens next, we never really know what new tricks our CA's may come up with. It is really an exploration of new worlds—live television from other universes. Working with CA's, anyone can experience the joy of building simple models and the thrill of discovering something new about the dynamics of information. We can all be theoretical physicists.

18.10 Acknowledgments

Much of what I know of this subject I learned from Edward Fredkin, Tommaso Toffoli and Charles Bennett. They were my close collaborators in the MIT Information Mechanics Group and my dear friends. They provided me with the support and encouragement I needed to begin work in this field. It was a privilege to be a part of their group.

Richard Feynman has been a constant source of inspiration and insight in my physics career. His *Lectures on Physics* [24] sustained me through my undergraduate years, and I spent some wonderful months as a graduate student visiting with him at CalTech and giving lectures on CA's and on reversible computing in his course on computation [26]. I have many fond memories of his warmth, charm, humor and *joie de vivre*.

I learned much from conversations with Mark Smith, Mike Biafore, Gerard Vichniac and Hrvoje Hrgovčić—colleagues in the MIT IM Group. I have also learned much from my colleagues in the lattice gas field who have helped me in my research, particularly Jeff Yepez, who has been a close collaborator on CA research with the CAM-8 machine. The construction of CAM-8 itself was a large effort that involved many people—including Tom Toffoli, Tom Durgavich, Doug Faust, Ken Streeter and Mike Biafore—each of whom contributed in a vital way. My current physics/computation related research has been made possible by the support, collaboration and encouragement of Tom Knight.

This manuscript has benefitted greatly from discussions with and comments by Raissa D'Souza, Lov Grover, David Meyer, and Ilona Lappo. Thanks to Dan Rothman for providing Figure 18.14a, Ray Kapral for providing Figure 18.14b, and Jeff Yepez for providing Figure 18.14c. Figure 18.2b comes from a simulation written by Tom Toffoli and Figure 18.3 comes from an investigation that I'm working on jointly with Raissa D'Souza and Mark Smith.

Support for this work comes from DARPA contract DABT63-95-C-0130, the MIT AI Lab's Reversible Computing Project, and Boston University's Center for Computational Science.

Notes on the references

Much of the material in this chapter is discussed at greater length in [62] and [90]. These documents were both strongly influenced by ideas and suggestions from Edward Fredkin, some of which are also discussed in [30–35]. Some recent books on CA modeling of physics are [79] and [18]. Many of the early lattice-gas and quantum computing papers are reproduced in [22] and [54] respectively. Recent related papers can be found online in the comp-gas and quant-ph archives at http://xxx.lanl.gov, and cross-listed there from other archives at LANL such as chao-dyn. Pointers to papers in these archives are given in some of the references below.

References

[1] D. S. Abrams and S. Lloyd, "Simulation of many-body fermi systems on a universal quantum computer," *Phys. Rev. Lett.* **79**, 2586–2589 (1997) and quant-ph/9703054.

[2] C. Adler, B. Boghosian, E. Flekkoy, N. Margolus and D. Rothman, "Simulating three-dimensional hydrodynamics on a cellular-automata machine," in [76, p. 105–128] and chao-dyn/9508001.

[3] C. Appert, V. Pot and S. Zaleski, "Liquid-gas models on 2D and 3D lattices," in [52, p. 1–12].

[4] C. H. Bennett, "The thermodynamics of computation—a review," in [29, p. 905–940].

[5] C. H. Bennett and R. Landauer, "The fundamental physical limits of computation," *Scientific American* **253**:1, 38–46 (1985).

[6] C. H. Bennett, "Demons, engines, and the second law," *Scientific American* **257**:5, 88-96 (1987).

[7] E. Berlekamp, J. Conway and R. Guy, *Winning Ways For Your Mathematical Plays*, *Volume 2* (Academic Press, New York NY, 1982).

[8] M. Biafore, "Few-body cellular automata," MIT Ph.D. Thesis (1993), available as MIT/LCS/TR-597 (MIT Laboratory for Computer Science, 1993).

[9] B. M. Boghosian, P. Coveney and A. N. Emerton, "A lattice-gas model of microemulsions," *Proc. Roy. Soc. Lond. A* **452**, 1221–1250 (1996) and comp-gas/9507001.

[10] B. M. Boghosian, F. J. Alexander and P. V. Coveney (eds.), *Discrete Models of Complex Fluid Dynamics*, special issue of *Int. J. Mod. Phys. C* **8**:4 (1997).

[11] B. M. Boghosian and W. Taylor IV, "Quantum lattice-gas models for the many-body Schrödinger equation," in [10] and quant-ph/9701016.

[12] B. M. Boghosian, J. Yepez, F. J. Alexander and N. Margolus, "Integer Lattice Gases," *Phys. Rev. E.* **55**:4, 4137–4147 and comp-gas/9602001.

[13] J. P. Boon (ed.), *Advanced Research Workshop on Lattice Gas Automata*, special issue of *J. Stat. Phys.* **68**:3/4 (1992).

[14] J. P. Boon, D. Dab, R. Kapral and A. Lawniczak, "Lattice gas automata for reactive systems," *Phys. Rep.* **273**, 55–147 (1996) and comp-gas/9512001.

[15] R. Brooks and P. Maes (eds.), *Artificial Life IV Proceedings* (MIT Press, Cambridge MA, 1994).

[16] A. Burks, (ed.), *Essays on Cellular Automata* (Univ. Illinois Press, Urbana IL, 1970).

[17] H. Chate (ed.), *International Workshop on Lattice Dynamics*, special issue of *Physica D* **103**:1/4 (1997).

[18] B. Chopard, B. and M. Droz, *Cellular Automata Modeling of Physical Systems*, (Cambridge University Press, Cambridge, 1998).

[19] M. Creutz, "Deterministic ising dynamics," *Annals of Physics* **167**, 62–76 (1986).

[20] D. Farmer, T. Toffoli and S. Wolfram (eds.), *Cellular Automata* (North-Holland, Amsterdam, 1984); reprinted from *Physica D* **10**:1/2 (1984).

[21] R. D'Souza and N. Margolus, "Reversible aggregation in a lattice gas model using coupled diffusion fields," MIT Physics Department, in preparation (will be available in the LANL cond-mat archive).

[22] G. Doolen (ed.), *Lattice-Gas Methods for Partial Differential Equations* (Addison-Wesley, Reading MA, 1990).

[23] G. Doolen (ed.), *Lattice-Gas Methods for PDE's: Theory, Applications and Hardware* (North-Holland, 1991); reprinted from *Physica D* **47**:1/2.

[24] R. P. Feynman, R. B. Leighton and M. Sands, *The Feynman Lectures on Physics* (Addison-Wesley, Reading MA, 1963).

[25] R. P. Feynman, "Simulating physics with computers," in [29, p. 467–488].

[26] R. P. Feynman, *Feynman Lectures on Computation*, edited by A. J. G. Hey and R. W. Allen, (Addison-Wesley, Reading MA, 1996).

[27] M. P. Frank, "Reversibility for efficient computing," MIT Ph.D. Thesis (1999).

[28] U. Frisch, B. Hasslacher and Y. Pomeau, "Lattice-gas automata for the navier-stokes equation," *Phys. Rev. Lett.* **56**, 1505–1508 (1986).

[29] E. Fredkin, R. Landauer and T. Toffoli (eds.), *Proceedings of the Physics of Computation Conference*, in *Int. J. Theor. Phys.*, issues **21**:3/4, **21**:6/7, and **21**:12 (1982).

[30] E. Fredkin and T. Toffoli, "Conservative logic," in [29, p. 219–253].

[31] E. Fredkin, "Digital mechanics: an informational process based on reversible universal CA," in [39, p. 254–270].

[32] E. Fredkin, "A physicist's model of computation," *Proceedings of the XXVIth Recontre de Moriond*, 283–297 (1991).

[33] E. Fredkin, "Finite nature," *Proceedings of the XXVIIth Recontre de Moriond*, (1992).

[34] E. Fredkin, "A new cosmogony," in [70, p. 116–121].

[35] E. Fredkin, "The digital perspective," in [95, p. 120–121].

[36] D. Griffeath and C. Moore, "Life without death is p-complete," Santa Fe Institute working paper 97-05-044, to appear in *Complex Systems* (1998); also available at http://psoup.math.wisc.edu, a general source of information on pattern formation in irreversible CA's.

[37] L. K. Grover, "A fast quantum mechanical algorithm for database search," *Proceedings, 28th Annual ACM Symposium on the Theory of Computing (STOC)*, 212–218 (1996) and quant-ph/9605043; reproduced in [54].

[38] J. R. Gunn, C. M. McCallum and K. A. Dawson, "Dynamic lattice-model simulation," *Phys. Rev. E*, 3069–3080 (1993).

[39] H. Gutowitz, *Cellular Automata: Theory and Experiment* (North Holland, 1990); reprinted from *Physica D* **45**:1/3 (1990).

[40] J. E. Hanson and J. P. Crutchfield, "Computational mechanics of cellular automata: an example," in [17, p. 169–189].

[41] J. Hardy, O. de Pazzis and Y. Pomeau, "Molecular dynamics of a classical lattice gas: transport properties and time correlation functions," *Phys. Rev. A* **13**, 1949–1960 (1976).

[42] H. J. Herrmann, "Fast algorithm for the simulation of Ising models," *J. Stat. Phys.* **45**, 145–151 (1986).

[43] H. J. Hrgovčić, "Discrete representations of the n-dimensional wave-equation," *J. Phys. A* **25**:5, 1329–1350 (1992).

[44] H. J. Hrgovčić, "Quantum mechanics on spacetime lattices using path integrals in a Minkowski metric," MIT Ph.D. Thesis (1992), available online at http://www.im.lcs.mit.edu/poc/hrgovcic/thesis.ps.gz

[45] K. Huang, *Statistical Mechanics*, (John Wiley & Sons, 1987).

[46] E. Ising, "Beitrag zur theorie des ferromagnetismus," *Zeits. für Phys.* **31**, 253–258 (1925).

[47] J. Kari, "Reversibility and surjectivity problems of cellular automata," *Journal of Computer and System Sciences* **48**:1, 149–182 (1994).

[48] J. Kari, "Representation of reversible cellular automata with block permutations," *Mathematical Systems Theory* **29**:1, 47–61 (1996).

[49] S. A. Kauffman, "Requirements for evolvability in complex systems: orderly dynamics and frozen components," in [108, p. 151–192].

[50] T. Kawakatsu and K. Kawasaki, "Hybrid models for the dynamics of an immiscible binary mixture with surfactant molecules," *Physica A* **167**:3, 690–735 (1990).

[51] C. Langton, C. Taylor, J. D. Farmer and S. Rasmussen (eds.), *Artificial Life II, Santa Fe Institute Studies in the Sciences of Complexity, vol. XI* (Addison-Wesley, Reading MA, 1991).

[52] A. Lawniczak and R. Kapral (eds.), *Pattern Formation and Lattice-Gas Automata* (American Mathematical Society, 1996).

[53] F. T. Leighton, *Introduction to Parallel Algorithms and Architectures: Arrays, Trees, Hypercubes* (Morgan Kaufman, 1991).

[54] S. Lloyd (ed.), *Quantum Computation* (John Wiley & Sons, New York, 1999).

[55] H. K. Lo, T. Spiller and S. Popescu (eds.), *Introduction to Quantum Computation and Information* (World Scientific, Singapore, 1998).

[56] A. Malevanets and R. Kapral, "Microscopic model for FitzHugh-Nagumo dynamics," *Phys. Rev. E* **55**:5, 5657–5670 (1997).

[57] D. A. Meyer, "From quantum cellular automata to quantum lattice gases," *J. Stat. Phys.* **85**, 551–574 (1996) and in quant-ph/9604003.

[58] D. A. Meyer, "Quantum lattice gases and their invariants," in [10, p. 717–736] and quant-ph/9703027.

[59] N. Margolus, T. Toffoli and G. Vichniac, "Cellular-automata supercomputers for fluid dynamics modeling," *Phys. Rev. Lett.* **56**, 1694–1696 (1986).

[60] N. Margolus, "Physics-like models of computation," in [20, p. 81–95]; reproduced in [54].

[61] N. Margolus, "Quantum computation," *New Techniques and Ideas in Quantum Measurement Theory* (Daniel GREENBERGER ed.), 487–497 (New York Academy of Sciences, 1986); reproduced in [54].

[62] N. Margolus, "Physics and computation" MIT Ph.D. Thesis (1987). Reprinted as *Tech. Rep. MIT/LCS/TR-415*, MIT Lab. for Computer Science, Cambridge MA 02139 (1988).

[63] N. Margolus, "Parallel quantum computation," in [108, p. 273–287]; reproduced in [54].

[64] N. Margolus, "Fundamental physical constraints on the computational process," *Nanotechnology: Research and Perspectives* (B.C. CRANDALL and J. LEWIS eds.) (MIT Press, 1992).

[65] N. Margolus, "A bridge of bits," in [70, 253–257].

[66] N. Margolus, "CAM-8: a computer architecture based on cellular automata," in [52, p. 167–187] and comp-gas/9509001. See also http://www.im.lcs.mit.edu/cam8.html.

[67] N. Margolus, "An FPGA architecture for DRAM-based systolic computations," in [73, p. 2–11].

[68] N. Margolus and L. Levitin, "The maximum speed of dynamical evolution," in [96, p. 188–195] and quant-ph/9710043.

[69] N. Margolus, "Crystalline computation," to appear in *Proceedings of the Conference on High Speed Computing* (Lawrence Livermore National Laboratory, 1998).

[70] D. Matzke (ed.), *Proceedings of the Workshop on Physics and Computation— PhysComp '92* (IEEE Computer Society Press, 1993).

[71] R. Monaco (ed.), *Discrete Kinetic Theory, Lattice Gas Dynamics and Foundations of Hydrodynamics* (World Scientific, Singapore, 1989).

[72] B. Ostrovsky, M. A. Smith and Y. Dar-Yam, "Simulations of polymer interpenetration in 2D melts," in [10, p. 931–939].

[73] K. Pocek and J. Arnold, *The Fifth IEEE Symposium on FPGA-based Custom Computing Machines* (IEEE Computer Society, 1997).

[74] Y. Pomeau, "Invariant in cellular automata," *J. Phys. A* **17**:8, 415–418 (1984).

[75] J. Preskill, "Fault-tolerant quantum computation," in [55] and quant-ph/9712048.

[76] Y. H. Qian (ed.), *Discrete Models for Fluid Mechanics*, special issue of *J. Stat. Phys.* **81**:1/2 (1995).

[77] T. S. Ray, "An approach to the synthesis of life," in [51, p. 371–408].

[78] T. S. Ray, "Evolution of parallel processes in organic and digital media," in [99, p. 69–91].

[79] D. Rothman and S. Zaleski, *Lattice-Gas Cellular Automata—simple models of complex hydrodynamics*, (Cambridge University Press, Cambridge, 1997).

[80] P. W. Shor, "Algorithms for quantum computation: discrete log and factoring," *Proceedings of the 35th Annual Symposium on the Foundations of Computer Science*, 124–134, (IEEE, 1994) and quant-ph/9508027; reproduced in [54].

[81] N. R. S. Simons, M. Cuhaci, N. Adnani and G. E. Bridges, "On the potential use of cellular-automata machines for electromagnetic-field solution," *Int. J. Numer. Model. Electron. N.* 8:3/4, 301–312 (1995).

[82] K. Sims, "Evolving 3D morphology and behavior by competition," in [15, p. 28–39].

[83] M. A. Smith, "Representations of geometrical and topological quantities in cellular automata," *Physica D* 47, 271–277 (1990).

[84] M. A. Smith, "Cellular automata methods in mathematical physics," MIT Ph.D. Thesis (1994). Reprinted as *Tech. Rep. MIT/LCS/TR-615*, MIT Lab. for Computer Science (1994).

[85] S. Takesue, "Boltzmann-type equations for elementary reversible cellular automata," in [17, 190-200].

[86] T. Toffoli, "Computation and construction universality of reversible cellular automata," *J. Comp. Syst. Sci.* 15, 213–231 (1977).

[87] T. Toffoli, "Physics and computation," in [29, p. 165–175].

[88] T. Toffoli, "CAM: A high-performance cellular-automaton machine," in [20, p. 195–204].

[89] T. Toffoli, "Cellular automata as an alternative to (rather than an approximation of) differential equations in modeling physics," in [20, p. 117–127].

[90] T. Toffoli and N. Margolus, *Cellular Automata Machines—a new environment for modeling* (MIT Press, Cambridge MA, 1987).

[91] T. Toffoli, "Four topics in lattice gases: ergodicity; relativity; information flow; and rule compression for parallel lattice-gas machines," in [71, p. 343–354].

[92] T. Toffoli, "How cheap can mechanics' first principles be?" in [108, p. 301–318].

[93] T. Toffoli and N. Margolus, "Invertible cellular automata: a review," in [39, p. 229–253].

[94] T. Toffoli, "What are nature's 'natural' ways of computing?" in [70, p. 5–9].

[95] T. Toffoli, M. Biafore and J. Leão (eds.), *PhysComp96* (New England Complex Systems Institute, 1996). Also online at http://www.interjournal.org.

[96] T. Toffoli and M. Biafore (eds.), *PhysComp96*, special issue of *Physica D* 120:1/2 (1998).

[97] S. Ulam, "Random processes and transformations," *Proceedings of the International Congress on Mathematics 1950, volume 2*, 264–275 (1952).

[98] G. Vichniac, "Simulating physics with cellular automata," in [20, p. 96–116].

[99] P. D. Waltz (ed.), *Natural and Artificial Parallel Computation* (SIAM Press, Philadelphia, 1996).

[100] C. Vieri, "Reversible computer engineering," MIT Ph.D. Thesis (1998).

[101] S. Wolfram, *Cellular Automata and Complexity* (Addison-Wesley, Cambridge MA, 1994).

[102] V. Yakhot and S. Orszag, "Reynolds number scaling of cellular-automaton hydrodynamics, *Phys. Rev. Lett.* **56**, 1691–1693 (1986).

[103] J. Yepez, "Lattice-gas crystallization," in [76, p. 255–294].

[104] J. Yepez, "A lattice-gas with long-range interactions coupled to a heat bath," in [52, p. 261–274].

[105] J. Yepez, "Lattice-gas dynamics," Brandeis Physics Ph.D. Thesis (1997). Available as *Air Force Research Laboratory Tech. Rep. PL-TR-96-2122(I), PL-TR-96-2122(II) and PL-TR-96-2122(III)*, AFRL/VSBE Hanscom AFB, MA 01731 (1996).

[106] S. G. Younis and T. F. Knight, Jr., "Practical implementation of charge recovering asymptotically zero power CMOS," in *Proceedings of the 1993 Symposium on Integrated Systems*, 234–250 (MIT Press, Cambridge MA, 1993).

[107] S. G. Younis, "Asymptotically zero energy computing using split-level charge recovery logic," MIT Ph.D. Thesis (1994).

[108] W. Zurek (ed.), *Complexity, Entropy, and the Physics of Information* (Addison-Wesley, Reading MA, 1990).

Part V

Fundamentals

19

INFORMATION, PHYSICS, QUANTUM: THE SEARCH FOR LINKS

John Archibald Wheeler * †

Abstract

This report reviews what quantum physics and information theory have to tell us about the age-old question, How come existence? No escape is evident from four conclusions: (1) The world cannot be a giant machine, ruled by any preestablished continuum physical law. (2) There is no such thing at the microscopic level as space or time or spacetime continuum. (3) The familiar probability function or functional, and wave equation or functional wave equation, of standard quantum theory provide mere continuum idealizations and by reason of this circumstance conceal the information-theoretic source from which they derive. (4) No element in the description of physics shows itself as closer to primordial than the elementary quantum phenomenon, that is, the elementary device-intermediated act of posing a yes-no physical question and eliciting an answer or, in brief, the elementary act of observer-participancy. Otherwise stated, every physical quantity, every it, derives its ultimate significance from bits, binary yes-or-no indications, a conclusion which we epitomize in the phrase, *it from bit*.

19.1 Quantum Physics Requires a New View of Reality

Revolution in outlook though Kepler, Newton, and Einstein brought us [1–4], and still more startling the story of life [5–7] that evolution forced upon an unwilling world, the ultimate shock to preconceived ideas lies ahead, be it a decade hence, a century or a millenium. The overarching principle of 20th-century physics, the quantum [8] — and the principle of complementarity [9] that is central idea of the quantum — leaves us no escape, Niels Bohr tells us, [10] from "a radical revision of our attitude as regards physical reality" and a "fundamental modification of all ideas regarding the absolute character of physical phenomena." Transcending Einstein's summons [11] of 1908, "This quantum business is so incredibly important and difficult that everyone should busy himself with it," Bohr's modest words direct us to the supreme goal: *Deduce the quantum* from an understanding of *existence*.

---

*Reproduced from Proc. 3rd Int. Symp. Foundations of Quantum Mechanics, Tokyo, 1989, pp.354-368.

How make headway toward a goal so great against difficulties so large? The search for understanding presents to us three questions, four no's and five clues:

Three **questions**,

- How come existence?

- How come the quantum?

- How come "one world" out of many observer-participants?

Four **no's**,

- No tower of turtles

- No laws

- No continuum

- No space, no time.

Five **clues**,

- The boundary of a boundary is zero.

- No question? No answer!

- The super-Copernican principle.

- "Consciousness"

- More is different.

19.2   "It from Bit" as Guide in Search for Link Connecting Physics, Quantum and Information

In default of a tentative idea or working hypothesis, these questions, no's and clues — yet to be discussed — do not move us ahead. Nor will any abundance of clues assist a detective who is unwilling to theorize how the crime was committed! A wrong theory? The policy of the engine inventor, John Kris, reassures us, "Start her up and see why she don't go!" In this spirit [12-47] I, like other searchers [48–51] attempt formulation after formulation of the central issues, and here present a wider overview, taking for working hypothesis the most effective one that has survived this winnowing: **It from bit**. Otherwise put, every **it** — every particle, every field of force, even the spacetime continuum itself — derives its function, its meaning, its very existence entirely — even if in some contexts indirectly — from the apparatus-elicited answers to yes or no questions, binary choices [52], **bits**.

It from bit symbolizes the idea that every item of the physical world has at bottom — at a very deep bottom, in most instances — an immaterial source and explanation; that what we call reality arises in the last analysis from the posing of yes-no questions and the registering of equipment-evoked responses; in short, that all things physical are information-theoretic in origin and this is a **participatory universe**.

Three examples may illustrate the theme of it from bit. First, the photon. With polarizer over the distant source and analyzer of polarization over the photodetector under watch, we ask the yes or no question, "Did the counter register a click during the specified second?" If yes, we often say, "A photon did it." We know perfectly well that the photon existed neither before the emission nor after the detection. However, we also have to recognize that any talk of the photon "existing" during the intermediate period is only a blown-up version of the raw fact, a count.

The yes or no that is recorded constitutes an unsplitable bit of information. A photon, Wootters and Zurek demonstrate [53, 54], cannot be cloned.

As second example of it from bit, we recall the Aharonov-Bohm scheme [55] to measure a magnetic flux. Electron counters stationed off to the right of a doubly-slit screen give yes-or-no indications of the arrival of an electron from the source located off to the left of the screen, both before the flux is turned on and afterward. That flux of magnetic lines of force finds itself embraced between — but untouched by — the two electron beams that fan out from the two slits. The beams interfere. The shift in interference fringes between field off and field on reveals the magnitude of the flux,

(phase change around perimeter of the included area)
$$= 2\pi \times \text{(shift of interference pattern, measured in number of fringes)} \quad (19.1)$$
$$= \text{(electron charge)} \times \text{(magnetic flux embraced)}/\hbar c$$

Here $\hbar = 1.0546 \times 10^{-27}$ gcm^2/s is the quantum in conventional units, or in geometric units [4, 16] — where both time and mass are measured in the units of length — $\hbar = \hbar c = 2.612 \times 10^{-66}$ cm^2 = the square of the Planck length, 1.616×10^{-33} = what we hereafter term the *Planck area*.

Not only in electrodynamics but also in geometrodynamics and in every other gauge-field theory, as Anandan, Aharonov and others point out [56, 57] the difference around a circuit in the phase of an appropriately chosen quantum-mechanical probability amplitude provides a measure of the field. Here again the concept of it from bit applies [38]. Field strength or spacetime curvature reveals itself through shift of interference fringes, fringes that stand for nothing but a statistical pattern of yes-or-no registrations.

When a magnetometer reads that *it* which we call a magnetic field, no reference at all to a bit seems to show itself. Therefore we look closer. The idea behind the operation of the instrument is simple. A wire of length l carries a current i through

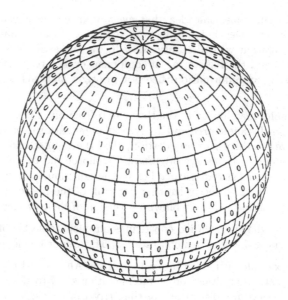

Fig. 19.1. Symbolic representation of the "telephone number" of the particular one of the 2^N conceivable, but by now indistinguishable, configurations out of which this particular blackhole, of Bekenstein number N and horizon area $4N\hbar\log_e 2$, was put together. Symbol, also, in a broader sense, of the theme that *every* physical entity, every it, derives from bits. Reproduced from JGST, p.220.

a magnetic field B that runs perpendicular to it. In consequence the piece of copper receives in the time t a transfer of momentum p in a direction z perpendicular to the directions of the wire and of the field,

$$p = Blit$$
$$= \text{(flux per unit } z) \times \text{(charge, } e, \text{ of the elementary carrier of current)} \qquad (19.2)$$
$$\times \text{(number, } N, \text{of carriers that pass in the time } t)$$

This impulse is the source of the force that displaces the indicator needle of the magnetometer and gives us an instrument reading. We deal with bits wholesale rather than bits retail when we run the fiducial current through the magnetometer coil, but the definition of field founds itself no less decisively on bits.

As third and final example of it from bit we recall the wonderful quantum finding of Bekenstein [58–60] — totally unexpected denouement of earlier classical work of Penrose [61] Christodoulou [62] and Ruffini [63] — refined by Hawking [64, 65] that the *surface area* of the horizon of a blackhole, rotating or not, *measures* the *entropy* of the blackhole. Thus this surface area, partitioned in imagination (Fig. 19.1) into domains each of size $4\hbar\log_e 2$, that is, 2.77... times the Planck area, yields the *Bekenstein number*, N; and the Bekenstein number, so Thorne and Zurek explain [66] tells us the number of binary digits, the number of bits, that would be required to specify in all detail the configuration of the constituents out of which the blackhole was put together. Entropy is a measure of lost information. To no community of newborn outside observers can the blackhole be made to reveal out of which particular one of 2^N configurations it was put together. Its size, an *it*, is fixed by the number, N, of *bits* of information hidden within it.

The quantum, $\hbar$, in whatever correct physics formula it appears, thus serves as lamp. It lets us see horizon area as information lost, understand wave number of light as photon momentum and think of field flux as bit-registered fringe shift.

Giving us its as bits, the quantum presents us with physics as information.

How come a value for the quantum so small as $\hbar = 2.612 \times 10^{-66}$ cm^2? As well as ask why the speed of light is so great as $c = 3 \times 10^{10}$ cm/s! No such constant as the speed of light ever makes an appearance in a truly fundamental account of special relativity or Einstein geometrodynamics, and for a simple reason: Time and space are both tools to measure interval. We only then properly conceive them when we measure them in the same units [4, 16]. The numerical value of the ratio between the second and the centimeter totally lacks teaching power. It is an historical accident. Its occurrence in equations obscured for decades one of Nature's great simplicities. Likewise with $\hbar$! Every equation that contains an $\hbar$ floats a banner, "It from bit". The formula displays a piece of physics that we have learned to translate into information-theoretic terms. Tomorrow we will have learned to understand and express *all* of physics in the language of information. At that point we will revalue $\hbar = 2.612 \times 10^{-66}$ cm^2 — as we downgrade $c = 3 \times 10^{10}$ cm/s today — from constant of Nature to artifact of history, and from foundation of truth to enemy of understanding.

19.3 Four No's

To the question, "How come the quantum?" we thus answer, "Because what we call existence is an information-theoretic entity." But how come existence? Its as bits, yes; and physics as information, yes; but *whose* information? How does the vision of one world arise out of the information-gathering activities of many observer-participants? In the consideration of these issues we adopt for guidelines four no's.

First no: "No tower of turtles," advised William James. Existence is not a globe supported by an elephant, supported by a turtle, supported by yet another turtle, and so on. In other words, no infinite regress. No structure, no plan of organization, no framework of ideas underlaid by another structure or level of ideas, underlaid by yet another level, by yet another, *ad infinitum*, down to a bottomless night. To endlessness no alternative is evident but loop [47, 67], such a loop as this: Physics gives rise to observer-participancy; observer-participancy gives rise to information; and information gives rise to physics.

Existence thus built [68] on "insubstantial nothingness"? Rutherford and Bohr made a table no less solid when they told us it was 99.9... percent emptiness. Thomas Mann may exaggerate when he suggests [69] that "... we are actually bringing about what seems to be happening to us," but Leibniz [70] reassures us, "Although the whole of this life were said to be nothing but a dream and the physical world nothing but a phantasm, I should call this dream or phantasm real enough if, using reason well, we were never deceived by it."

Second no: No laws. "So far as we can see today, the laws of physics cannot have existed from everlasting to everlasting. They must have come into being at the big bang. There were no gears and pinions, no Swiss watch-makers to put things together, not even a pre-existing plan... Only a principle of organization which is no organization at all would seem to offer itself. In all of mathematics, nothing of this kind more obviously offers itself than the principle that 'the boundary of boundary is zero.' Moreover, all three great field theories of physics use this principle twice over... This circumstance would seem to give us some reassurance that we are talking sense when we think of... physics being" [32] as foundation-free as a logic loop, the closed circuit of ideas in a self-referential deductive axiomatic system [71–74].

Universe as machine? This universe one among a great ensemble of machine universes, each differing from the others in the values of the dimensionless constants of physics? Our own selected from this ensemble by an anthropic principle of one or another form [75]? We reject here the concept of universe as machine not least because it "has to postulate explicitly or implicitly, a supermachine, a scheme, a device, a miracle, which will turn out universes in infinite variety and infinite number" [47].

Directly opposite to the concept of universe as machine built on law is the vision of a world self-synthesized. On this view, the notes struck out on a piano by the observer-participants of all places and all times, bits though they are, in and by themselves constitute the great wide world of space and time and things.

Third no: No continuum. No continuum in mathematics and therefore no continuum in physics. A half-century of development in the sphere of mathematical logic [76] has made it clear that there is no evidence supporting the belief in the existential character of the number continuum. "Belief in this transcendental world," Hermann Weyl tells us, "taxes the strength of our faith hardly less than the doctrines of the early Fathers of the Church or of the scholastic philosophers of the Middle Ages" [77]. This lesson out of mathematics applies with equal strength to physics. "Just as the introduction of the irrational numbers... is a convenient myth [which] simplifies the laws of arithmetic... so physical objects," Willard Van Orman Quine tells us [78] "are postulated entities which round out and simplify our account of the flux of existence... The conceptual scheme of physical objects is a convenient myth, simpler than the literal truth and yet containing that literal truth as a scattered part."

Nothing so much distinguishes physics as conceived today from mathematics as the difference between the continuum character of the one and the discrete character of the other. Nothing does so much to extinguish this gap as the elementary quantum phenomenon "brought to a close," as Bohr puts it [10] by "an irreversible act of amplification," such as the click of a photodetector or the blackening of a grain of photographic emulsion. Irreversible? More than one idealized experiment [38] illustrates how hard it is, even today, to give an all-inclusive definition of the term irreversible. Those difficulties supply pressure, however, not to retreat to old ground, but to advance to new insight. In brief, continuum-based physics, no; information-based physics, yes.

Fourth and last no: No space, no time. Heaven did not hand down the word "time". Man invented it, perhaps positing hopefully as he did that "Time is Nature's way to keep everything from happening all at once" [79]. If there are problems with the concept of time, they are of our own creation! As Leibniz tells us, [80] "...time and space are not things, but orders of things... ;" or as Einstein put it, [81] "Time and space are modes by which we think, and not conditions in which we live."

What are we to say about that weld of space and time into spacetime which Einstein gave us in his 1915 and still standard classical geometrodynamics? On this geometry quantum theory, we know, imposes fluctuations [13, 14, 82]. Moreover, the predicted fluctuations grow so great at distances of the order of the Planck length that in that domain they put into question the connectivity of space and deprive the very concepts of "before" and "after" of all meaning [83]. This circumstance reminds us anew that no account of existence can ever hope to rate as fundamental which does not translate all of continuum physics into the language of bits.

We will not feed time into any deep-reaching account of existence. We must derive time — and time only in the continuum idealization — out of it. Likewise with space.

19.4   Five Clues

First clue: The boundary of a boundary is zero. This central principle of algebraic topology [84], identity, triviality, tautology, though it is, is also the unifying theme of Maxwell electrodynamics, Einstein geometrodynamics and almost every version of modern field theory [42], [85–88]. That one can get so much from so little, almost everything from almost nothing, inspires hope that we will someday complete the mathematization of physics and derive everything from nothing, all law from no law.

Second clue: No question, no answer. Better put, no bit-level question, no bit-level answer. So it is in the game of twenty questions in its surprise version [89]. And so it is for the electron circulating within the atom or a field within a space. To neither field nor particle can we attribute a coordinate or momentum until a device operates to measure the one or the other. Moreover any apparatus that *accurately* [90] measures the one quantity inescapably rules out then and there the operation of equipment to measure the other [9, 91, 92]. In brief, the choice of question asked, and choice of when it's asked, play a part — not the whole part, but a part — in deciding what we have the right to say [38, 43].

Bit-registration of a chosen property of the electron, a bit-registration of the arrival of a photon, Aharonov-Bohm bit-based determination of the magnitude of a field flux, bulk-based count of bits bound in a blackhole: All are examples of physics expressed in the language of information. However, into a bit count that one might have thought to be a private matter, the rest of the nearby world irresistibly thrusts itself. Thus the atom-to-atom distance in a ruler — basis for a bit count of distance — evidently has no invariant status, depending as it does on the temperature and pressure of the environment. Likewise the shift of fringes in the Aharonov-Bohm experiment depends not only upon the magnetic flux itself, but also on the charge of the electron. But this electron charge — when we take the quantum itself to be Nature's fundamental measuring unit — is governed by the square root of the quantity $e^2/\hbar c = 1/137.036\ldots$, a "constant" which — for extreme conditions — is as dependent on the local environment [93] as is a dielectric "constant" or the atom-to-atom spacing in the ruler.

The contribution of the environment becomes overwhelmingly evident when we turn from length of bar or flux of field to the motion of alpha particle through cloud chamber, dust particle through 3°K-background radiation or Moon through space. This we know from the analyses of Bohr and Mott [94], Zeh [95, 96], Joos and Zeh [97], Zurek [98–100] and Unruh and Zurek [101]. It from bit, yes; but the rest of the world also makes a contribution, a contribution that suitable experimental design can minimize but not eliminate. Unimportant nuisance? No. Evidence the whole show is wired up together? Yes. Objection to the concept of every *it* from *bits*? No.

Build physics, with its false face of continuity, on bits of information! What this enterprise is we perhaps see more clearly when we examine for a moment a thoughtful, careful, wide-reaching exposition [102] of the directly opposite thesis, that physics at bottom is *continuous*; that the bit of information is *not* the basic entity. Rate as false the claim that the bit of information is the basic entity. Instead, attempt to build everything on the foundation of some "grand unified field theory" such as string theory [103, 104] — or in default of that, on Einstein's 1915 and still standard geometrodynamics. Hope to derive that theory by way of one or another plausible line of reasoning. But don't try to derive quantum theory. Treat it as supplied free of charge from on high. Treat quantum theory as a magic sausage grinder which takes in as raw meat this theory, that theory or the other theory and turns out a "wave equation," one solution of which is "the" wave function for the universe [14, 102, 105–107]. From start to finish accept continuity as right and natural: Continuity in the manifold, continuity in the wave equation, continuity in its solution, continuity in the features that it predicts. Among conceivable solutions of this wave equation select as reasonable one which "maximally decoheres," one which exhibits "maximal classicity" — maximal classicity by reason, not of "something external to the framework of wave function and Schrödinger equation," but something in "the initial conditions of the universe specified within quantum theory itself."

How compare the opposite outlooks of decoherence and it-from-bit? Remove the casing that surrounds the workings of a giant computer. Examine the bundles of wires that run here and there. What is the status of an individual wire? Mathematical limit of bundle? Or building block of bundle? The one outlook regards the wave equation and wave function to be primordial and precise and built on continuity, and the bit to be idealization. The other outlook regards the bit to be the primordial entity, and wave equation and wave function to be secondary and approximate — and derived from bits via information theory.

Derived, yes; but how? No one has done more than William Wootters towards opening up a pathway [108, 109] from information to quantum theory. He puts into connection two findings, long known, but little known. Already before the advent of wave mechanics, he notes, the analyst of population statistics R. A. Fisher proved [110, 111] that the proper tool to distinguish one population from another is not the probability of this gene, that gene and the third gene (for example), but the square roots of these probabilities; that is to say, the two probability amplitudes, each probability amplitude being a vector with three components. More precisely, Wootters proves, the *distinguishability* between the two populations is measured by the angle in Hilbert space between the two state vectors, both real. Fisher, however, was dealing with information that sits "out there". In microphysics, however, the information does not sit out there. Instead, Nature in the small confronts us with a revolutionary pistol, "No question, no answer." Complementarity rules. And complementarity as E.C.G. Stueckelberg proved [112, 113] as long ago as 1952, and as Saxon made more readily understandable [114] in 1964, demands that the probability amplitudes of quantum physics must be complex. Thus Wootters *derives* familiar Hilbert space with its familiar complex probability amplitudes from the twin demands of complementarity and measure of distinguishability.

Try to go on from Wootters's finding to *deduce* the full blown machinery of quantum field theory? Exactly not to try to do so — except as idealization — is the demand laid on us by the concept of it from bit. How come?

Probabilities exist "out there" no more than do space or time or the position of the atomic electron. Probability, like time, is a concept invented by humans, and humans have to bear the responsibility for the obscurities that attend it. Obscurities there are whether we consider probability defined as frequency [115] or defined à la Bayes [116–119]. Probability in the sense of frequency has no meaning as applied to the spontaneous fission of the particular plutonium nucleus that triggered the November 1, 1952 H-bomb blast.

What about probabilities of a Bayesian cast, probabilities "interpreted not as frequencies observable through experiments, but as degrees of plausibility one assigns to each hypothesis based on the data and on one's assessment of the plausibility of the hypotheses prior to seeing the data" [120]. Belief-dependent probabilities, different probabilities assigned to the same proposition by different people [121]? Probabilities associated [122] with the view that "objective reality is simply an interpretation of data agreed to by large numbers of people?"

Heisenberg directs us to the experiences [123] of the early nuclear-reaction-rate theorist Fritz Houtermans, imprisoned in Kharkov during the time of the Stalin terror, "... the whole cell would get together to produce an adequate confession ... [and] helped them [the prisoners] to compose their 'legends' and phrase them properly, implicating as few others as possible."

Existence as confession? Myopic but in some ways illuminating formulation of the demand for intercommunication implicit in the theme of it from bit!

So much for "No question, no answer."

Third clue: The super-Copernican principle [47]. This principle rejects now-centeredness in any account of existence as firmly as Copernicus repudiated here-centeredness. It repudiates most of all any tacit adoption of here-centeredness in assessing observer-participants and their number.

What is an observer-participant? One who operates an observing device and participates in the making of meaning, meaning in the sense of Føllesdal [124], "Meaning is the joint product of all the evidence that is available to those who communicate." Evidence that is available? The investigator slices a rock and photographs the evidence for the heavy nucleus that arrived in the cosmic radiation of a billion years ago [38]. Before he can communicate his findings, however, an asteroid atomizes his laboratory, his records, his rocks and him. No contribution to meaning! Or at least no contribution then. A forensic investigation of sufficient detail and wit to reconstruct the evidence of the arrival of that nucleus is difficult to imagine. What about the famous tree that fell in the forest with no one around [125]? It leaves a fallout of physical evidence so near at hand and so rich that a team of up-to-date investigators can establish what happened beyond all doubt. Their findings contribute to the establishment of meaning.

"Measurements and observations," it has been said, [102] "cannot be fundamental notions in a theory which seeks to discuss the early universe when neither existed." On this view the past has a status beyond all questions of observer-participancy. It from bit offers us a different vision: "Reality is theory" [126]; "the past has no evidence except as it is recorded in the present" [127]. The photon that we are going to register tonight from that four billion-year old quasar cannot be said to have had an existence "out there" three billion years ago, or two (when it passed an intervening gravitational lens) or one, or even a day ago. Not until we have fixed arrangements at our telescope do we register tonight's quantum as having passed to the left (or right) of the lens or by both routes (as in a double slit experiment). This registration like every delayed-choice experiment [21, 40], reminds us that no elementary quantum phenomenon is a phenomenon until, in Bohr's words [10], "It has been brought to a close" by "an irreversible act of amplification." What we call the past is built on bits.

Enough bits to structure a universe so rich in features as we know this world to be. Preposterous! Mice and men and all on Earth who may ever come to rank as intercommunicating meaning-establishing observer-participants will never mount a bit count sufficient to bear so great a burden.

The count of bits needed, huge though it may be, nevertheless, so far as we can judge, does not reach infinity. In default of a better estimate, we follow familiar reasoning [128] and translate into the language of the bits the entropy of the primordial cosmic fireball as deduced from the entropy of the present 2.735°K (uncertainty <0.05°K) microwave relict radiation [129] totaled over a 3-sphere of radius 13.2 × 10⁹ light years (uncertainty <35%) [130] or 1.25 × 10²⁸ cm and of volume $2\pi^2$ radius³,

$$
\begin{aligned}
\text{(number of bits)} &= (log_2 e) \times \text{(number of nats)} \\
&= (log_2 e) \times \text{(entropy /Boltzmann's constant, } k) \\
&= 1.44 \ldots \times [(8\pi^4/45)(radius \cdot kT/\hbar c)^3] \\
&= 8 \times 10^{88}
\end{aligned}
\tag{19.3}
$$

It would be totally out of place to compare this overpowering number with the number of bits of information elicited up to date by observer-participancy. So warns the super-Copernican principle. We today, to be sure, through our registering devices, give a tangible meaning to the history of the photon that started on its way from a distant quasar long before there was any observer-participancy anywhere. However, the far more numerous establishers of meaning of time to come have a like inescapable part — by device-elicited quesstions and registration of answer — in generating the "reality" of today. For this purpose, moreover, there are billions of years yet to come, billions on billions of sites of observer-participancy yet to be occupied. How far foot and ferry have carried meaning-making communication in fifty thousand years gives faint feel for how far interstellar propagation is destined [131, 132] to carry it in fifty billion years.

Do bits needed balance bits achievable? They must, declares the concept of "world as system self-synthesized by quantum networking" [47]. By no prediction does this concept more clearly expose itself to destruction, in the sense of Popper [133].

Fourth clue: "Consciousness". We have traveled what may seem a dizzying path. First, elementary quantum phenomenon brought to a close by an irreversible act of amplification. Second, the resulting information expressed in the form of bits. Third, this information used by observer-participants — via communication — to establish meaning. Fourth, from the past through the billeniums to come, so many observer-participants, so many bits, so much exchange of information, as to build what we call existence.

Doesn't this it-from-bit view of existence seek to elucidate the physical world, about which we know something, in terms of an entity about which we know almost nothing, consciousness [134–137]? And doesn't Marie Sklodowska Curie tell us, "Physics deals with things, not people?" Using such and such equipment, making such and such a measurement, I get such and such a number. Who I am has nothing to do with this finding. Or does it? Am I sleepwalking [138, 139]? Or am I one of those poor souls without the critical power to save himself from pathological science [140–142]?

Under such circumstances any claim to have "measured" something falls flat until it can be checked out with one's fellows. Checked how? Morton White reminds us [143] how the community applies its tests of credibility, and in this connection quotes analyses by Chauncey Wright, Josiah Royce and Charles Saunders Peirce [144]. Parmenides of Elea [145] ($\approx$ 515 B.C.- 450$^+$ B.C.) may tell us that "What is... is identical with the thought that recognizes it." We, however, steer clear of the issues connected with "*consciousness*." The line between the unconscious and the conscious begins to fade [146] in our day as computers evolve and develop — as mathematics has — level upon level upon level of logical structure. We may someday have to enlarge the scope of what we mean by a "who". This granted, we continue to accept — as essential part of the concept of it from bit — Føllesdal's guideline [124], "Meaning is the joint product of all the evidence that is available to those who *communicate*." What shall we say of a view of existence [147] that appears, if not anthropomorphic in its use of the word "who," still overly centered on life and consciousness? It would seem more reasonable to dismiss for the present the semantic overtones of "who" and explore and exploit the insights to be won from the phrases, "communication" and "communication employed to establish meaning."

Føllesdal's statement supplies, not an answer, but the doorway to new questions. For example, man has not yet learned how to communicate with ant. When he does, will the questions put to the world around by the ant and the answers that he elicits contribute their share, too, to the establishment of meaning? As another issue associated with communication, we have yet to learn how to draw the line between a communication network that is closed, or parochial, and one that is open. And how to use that difference to distinguish between reality and poker — or another game [148, 149] — so intense as to appear more real than reality. No term in Føllesdal's statement posses greater challenge to reflection than "communication," descriptor of a domain of investigation [150–152] that enlarges in sophistication with each passing year.

Fifth and final clue: More is different [153]. Not by plan but by inner necessity a sufficiently large number of H_2O molecules collected in a box will manifest solid, liquid and gas phases. Phase changes, superfluidity and superconductivity all bear witness to Anderson's pithy point, more is different.

We do not have to turn to objects so material as electrons, atoms and molecules to see big numbers generating new features. The evolution from small to large has already in a few decades forced on the computer a structure [154, 155] reminiscent of biology by reason of its segregation of different activities into distinct organs. Distinct organs, too, the giant telecommunications system of today finds itself inescapably evolving [151, 152]. Will we someday understand time and space and all the other features that distinguish physics — and existence itself — as the similarly self-generated organs of a self-synthesized information system [156–158]?

19.5   Conclusion

The spacetime continuum? Even continuum existence itself? Except as idealization neither the one entity nor the other can make any claim to be a primordial category in the description of Nature. It is wrong, moreover, to regard this or that physical quantity as sitting "out there" with this or that numerical value in default of question asked and answer obtained by way of appropriate observing device. The information thus solicited makes physics and comes in bits. The count of bits drowned in the dark night of a blackhole displays itself as horizon area, expressed in the language of Bekenstein number. The bit count of the cosmos, however it is figured, is ten raised to a very large power. So also is the number of elementary acts of observer-participancy over any time of the order of fifty billion years. And, except via those time-leaping quantum phenomena that we rate as elementary acts of observer-participancy, no way has ever offered itself to construct what we call "reality." That's why we take seriously the theme of it from bit.

19.6   Agenda

Intimidating though the problem of existence continues to be, the theme of it from bit breaks it down into six issues that invite exploration:

One: Go beyond Wootters and determine what, if anything, has to be added to distinguishability and complementarity to obtain *all* of standard quantum theory.

Two: Translate the quantum versions of string theory and of Einstein's geometrodynamics from the language of continuum to the language of bits.

Three: Sharpen the concept of bit. Determine whether "an elementary quantum phenomenon brought to a close by an irreversible act of amplification" has at bottom

(1) the 0-or-1 sharpness of definition of bit number nineteen in a string of binary digits, or (2) the accordion property of a mathematical theorem, the length of which, that is, the number of supplementary lemmas contained in which, the analyst can stretch or shrink according to his convenience.

Four: Survey one by one with an imaginative eye the powerful tools that mathematics — including mathematical logic — has won and now offers to deal with theorems on a wholesale rather than a retail level, and for each such technique work out the transcription into the world of bits. Give special attention to one and another deductive axiomatic system which is able to refer to itself [159], one and another *self-referential deductive system.*

Five: From the wheels-upon-wheels-upon-wheels evolution of computer programming dig out, systematize and display every feature that illuminates the level-upon-level-upon-level structure of physics.

Six: Capitalize on the findings and outlooks of information theory [160–163], algorithmic entropy [164], evolution of organisms [165–167] and pattern recognition [168–175]. Search out every link each has with physics at the quantum level. Consider, for instance, the string of bits 1111111 ... and its representation as the sum of the two strings 1001110... and 0110001... Explore and exploit the connection between this information-theoretic statement and the findings of theory and experiment on the correlation between the polarizations of the two photons emitted in the annihilation of singlet positronium [176] and in like Einstein-Podolsky-Rosen experiments [177]. Seek out, moreover, every realization in the realm of physics of the information-theoretic triangle inequality recently discovered by Zurek [178].

Finally: Deplore? No, celebrate the absence of a clean clear definition of the term "bit" as elementary unit in the establishment of meaning. We reject "that view of science which used to say, 'Define your terms before you proceed.' The truly creative nature of any forward step in human knowledge," we know, "is such that theory, concept, law and method of measurement — forever inseparable — are born into the world in union [179]." If and when we learn how to combine bits in fantastically large numbers to obtain what we call existence, we will know better what we mean both by bit and by existence.

A single question animates this report: Can we ever expect to understand existence? Clues we have, and work to do, to make headway on that issue. Surely someday, we can believe, we will grasp the central idea of it all as so simple, so beautiful, so compelling that we will all say to each other, "Oh, how could it have been otherwise! How could we all have been so blind so long!"

Acknowledgements

For discussion, advice or judgment on one or another issue taken up in this review, I am indebted to Nandar Balazs, John D. Barrow, Charles H. Bennett, David

Deutsch, Robert H. Dicke, Freeman Dyson and the late Richard P. Feynman as well as David Gross, James B. Hartle, John J. Hopfield, Paul C. Jeffries, Bernulf Kanitscheider, Arkady Kheyfets and Rolf W. Landauer; and to Warner A. Miller, John R. Pierce, Willard Van Orman Quine, Benjamin Schumacher and Frank J. Tipler as well as William G. Unruh, Morton White, Eugene P. Wigner, William K. Wootters, Hans Dieter Zeh and Wojciech H. Zurek. For assistance in preparation of this report I thank E. L. Bennett and NSF grant PHY245-6243 to Princeton University. I give special thanks to the sponsors of the 28-31 August 1989 conference ISQM Tokyo '89 at which the then current version of the present analysis was reported.

This report evolved from presentations at Santa Fe Institute Conferences, 29 May-2 June and 4-8 June 1989 and at the 3rd International Symposium on Foundations of Quantum Mechanics in the Light of New Technology, Tokyo, 28-31 August 1989, under the title "Information, Physics, Quantum: The Search for Links;" and headed "Can We Ever Expect to Understand Existence?", as the Penrose Lecture at the 20-22 April 1989 annual meeting of Benjamin Franklin's "American Philosophical Society, held at Philadelphia for Promoting Useful Knowledge," and the Accademia Nazionale dei Lincei Conference on La Verità nella Scienza, Rome, 13 October 1989; submitted to proceedings of all four in fulfillment of obligation and in deep appreciation for hospitality. Preparation for publication assisted in part by NSF Grant PHY 245-6243 to Princeton University.

Discussion

A discussion followed:

N. G. van Kampen: Did you mean to say that the observer influences the observed object?

J. A. Wheeler: The observer does not influence the past. Instead, by his choice of question, he decides about what feature of the object he shall have the right to make a clear statement.

J.P. Vigier: Two problems.

1. The first is that the QSO raise lots of unsolved problems, i.e. — strange quantized $N_0/log(1 + z)$ relation — correlation with galaxies (Arp) — angular correlation with brightest nearby galaxies (Burbidge et al.)

2. The second is that the idea (Einstein et al.) of the reality of fields has led (assuming that "particles" are field singularities) to the only known justification of the geodesic law. To contest it is to make the meaning of dynamical behaviour purely observer-dependent, i.e., to kill the reality of the physical world.

J. A. Wheeler:

1. The book by Thorne and colleagues, "*Black Holes: The Membrane Paradigm,*" describes how a supermassive black hole, endowed via accretion with great angular momentum inside and an accretion disk outside, produces counter-directed jets and radiation of great power. I know no other mechanism able to produce quasars.

2. No one has discovered a way to get a particle of wave length λ from point A through empty flat space to a point B at a great distance L without its undergoing on the way a transverse spread of the order $\sqrt{L\lambda}$. This spread imposes an inescapable limitation on the classical concept of "worldline."

References

[1] J. Kepler (1571-1630): *Harmonices Mundi,* 5 books (1619). The appendix to Kepler's Book 5 contains one side, the publications of the English physician and thinker Robert Fludd (1574-1637) the other side, of a great debate, analysed by Wolfgang Pauli [W. Pauli: "Der Einfluss archetypischer Vorstellungen auf die Bildung naturwissenschaftlicher Theorien bei Kepler" in *Naturerklärung und Psyche* (Rascher, Zrich, 1952) p.109-194; reprinted in *Wolfgang Pauli: Collected Scientific Papers,* eds. R. Kronig and V. F. Weisskopf (Interscience-Wiley, New York, 1964) Vol.1, p.1023]. Totally in contrast to Fludd's concept of intervention from on high [*Utriusque Cosmo Maioris scilicet et Minoris Metaphysica, Physica atque technica Historia,* 1st ed. (Oppenheim, 1621)] was Kepler's guiding principle, *Ubi materia, ibi geometria* — where there is matter, there is geometry. It was not directly from Kepler's writings, however, that Newton learned of Kepler's three great geometry-driven findings about the motions of the planets in space and in time, but from the distillation of Kepler offered by Thomas Streete (1622-1689), *Astronomia Carolina: A New Theorie of the Celestial Motions* (London, 1661).

[2] I. Newton: *Philosophiae naturalis principia mathematica,* 1st ed. (London, 1687).

[3] A. Einstein: "Zur allgemeinen Relativittstheorie" *Preuss. Akad. Wiss. Berlin,* Sitzber (1915) p.799-801; also (1915) p. 832-839, 844-847; (1916) p.688-696 and (1917) p.142-152.

[4] J. A. Wheeler: *Journey into Gravity and Spacetime* (Scientific American Library, Freeman, New York, 1990), cited hereafter as **JGST,** offers a brief and accessible summary of Einstein's 1915 and still standard geometrodynamics which capitalizes on Élie Cartan's appreciation of the central idea of the theory: the boundary of a boundary is zero.

[5] J. G. Mendel: "Versuche über Pflanzenhybriden" *Verhandlungen des Naturforschenden Vereins* in Brünn 4 (1866).

[6] C. R. Darwin (1809-1882): *On the Origin of Species by Means of Natural Selection, or the Preservation of Favoured Races in the Struggle for Life* (London, 1859).

[7] J. D. Watson and F. H. C. Crick: "Molecular structure of nucleic acids: a structure for deoxyribose nucleic acid" Nature 171(1953) 737-738.

[8] M. Planck: "Zur Theorie des Gesetzes der Energieverteilung im Normalspektrum" Verhand. Deutschen Phys. Gesell. 2 (1900) 237-245.

[9] N. Bohr: "The quantum postulate and the recent development of atomic theory" Nature 121 (1928) 580-590. Reprinted in J. A. Wheeler and W. H. Zurek: Quantum Theory and Measurement (Princeton University Press, 1983) p.87-126, referred to hereafter as WZ. The mathematics of complementarity I have not been able to discover stated anywhere more sharply, more generally and earlier than in H. Weyl: Gruppentheorie und Quantenmechanik (Hirzel, Leipzig, 1928), in the statement that the totality of operators for all the physical quantities of the system in question form an irreducible set.

[10] N.Bohr: "Can quantum-mechanical description of physical reality be considered complete?" Phys. Rev. 48 (1935) 696-702 reprinted in WZ, p. 145-151.

[11] A.Einstein to J.J.Laub, 1908, undated, Einstein Archives; scheduled for publication in The Collected Papers of Albert Einstein, group of volumes on the Swiss years 1902-1914, Volume 5: Correspondence, 1902-1914 (Princeton University Press, Princeton, New Jersey).

[12] J.A.Wheeler: "Assessment of Everett's "relative state" formulation of quantum theory" Rev.Mod.Phys. 29 (1957) 463-465.

[13] J.A.Wheeler: "On the nature of quantum geometrodynamics," Ann. of Phys. 2 (1957) 604-614.

[14] J. A. Wheeler: "Superspace and the nature of quantum geometrodynamics," in Battelle Rencontres: 1967 Lectures in Mathematics and Physics, eds. C. M. DeWitt and J. A. Wheeler, (Benjamin, New York, 1968) p. 242-307; reprinted as "Le superspace et la nature de la géométrodynamique quantique," in Fluides et Champ Gravitationnel en Relativité Genérale, No. 170, Colloques Internationaux (Editions de Centre National de la recherche Scientifique, Paris, 1969) p.257-322.

[15] J. A. Wheeler: "Transcending the law of conservation of leptons," in Atti del Convegno Internazionale sul Tema: The Astrophysical Aspects of the Weak Interactions (Cortona "Il Palazzone," 10-12 Giugno 1970), Accademia Nationale die Lincei, Quaderno N. 157 (1971) p.133-64.

[16] C. W. Misner, K. S. Thorne and J. A. Wheeler: Gravitation (Freeman, San Francisco, now New York, 1973) p. 1217, cited hereafter as MTW; paragraph on participatory concept of the universe.

[17] J. A. Wheeler: "The universe as home for man," in The Nature of Scientific Discovery, ed. O. Gingerich (Smithsonian Institution Press, Washington, 1975) p. 261-296; preprinted in part in American Scientist, 62 (1974) 683-91; reprinted in part as T. P. Snow, The Dynamic Universe (West, St. Paul, Minnesota, 1983) p.108-109.

[18] C. M. Patton and J. A. Wheeler: "Is physics legislated by cosmogony?," in *Quantum Gravity*, eds. C. Isham, R. Penrose and D. Sciama (Clarendon, Oxford, 1975) p. 538-605; reprinted in part in *Encyclopaedia of Ignorance*, eds. R. Duncan and M. Weston-Smith (Pergamon, Oxford, 1977) p.19-35.

[19] J. A. Wheeler: "Include the observer in the wave function?" Fundamenta Scientiae: Seminaire sur les fondements des sciences (Strasbourg) **25** (1976) 9-35; reprinted in *Quantum Mechanics A Half Century Later*, eds. J. Leite Lopes and M. Paty (Reidel, Dordrecht, 1977) p.1-18.

[20] J. A. Wheeler: "Genesis and observership," in *Foundational Problems in the Special Sciences*, eds. R. Butts and J. Hintikka (Reidel, Dordrecht, 1977) p.1-33.

[21] J. A. Wheeler: "The "past" and the "delayed choice" double-slit experiment," in *Mathematical Foundations of Quantum Theory*, ed. A. R. Marlow (Academic, New York, 1978) p.9-48; reprinted in part in WZ, p.182-200.

[22] J. A. Wheeler: "Frontiers of time," in *Problems in the Foundations of Physics, Proceedings of the International School of Physics "Enrico Fermi"* (Course 72), ed. N. Toraldo di Francia (North-Holland, Amsterdam, 1979) p.395-497; reprinted in part in **WZ**, p. 200-208.

[23] J. A. Wheeler: "The quantum and the universe," in *Relativity, Quanta, and Cosmology in the Development of the Scientific Thought of Albert Einstein, Vol. II*, eds. M. Pantalco and F. deFinis (Johnson Reprint Corp., New York, 1979) p.807-825.

[24] J. A. Wheeler: "Beyond the black hole," in *Some Strangeness in the Proportion: A Centennial Symposium to Celebrate the Achievements of Albert Einstein*, ed. H. Woolf (Addison-Wesley, Reading, Massachusetts, 1980) p.341-375; reprinted in part in **WZ**, p.208-210.

[25] J. A. Wheeler: "Pregeometry: motivations and prospects," in *Quantum Theory and Gravitation*, proceedings of a symposium held at Loyola University, New Orleans, May 23-26, 1979, ed. A. R. Marlow (Academic, New York, 1980) p.1-11.

[26] J. A. Wheeler: "Law without law," in *Structure in Science and Art*, eds. P. Medawar and J. Shelley (Elsevier North-Holland, New York and Excerpta Medica, Amsterdam, 1980) p.132-54.

[27] J. A. Wheeler: "Delayed-choice experiments and the Bohr-Einstein dialog," in *American Philosophical Society and the Royal Society: Papers read at a meeting, June 5, 1980* (American Philosophical Society, Philadelphia, 1980) p.9-40; reprinted in slightly abbreviated form and translated into German as "Die Experimente der verzögerten Entscheidung und der Dialog zwischen Bohr and Einstein," in *Moderne Naturphilosophie* ed. B. Kanitscheider (Königshausen and Neumann, Würzburg, 1984) p. 203-222; reprinted in *Niels Bohr: A Profile*, eds. A. N. Mitra, L. S. Kothari, V. Singh and S. K. Trehan (Indian National Science Academy, New Delhi, 1985) p.139-168.

[28] J. A. Wheeler: "Not consciousness but the distinction between the probe and the probed as central to the elemental quantum act of observation," in *The Role of Consciousness in the Physical World*, ed. R. G. Jahn (Westview, Boulder, 1981) p. 87-111.

[29] J. A. Wheeler: "The elementary quantum act as higgledy-piggledy building mechanism," in *Quantum Theory and the Structures of Time and Space, Papers presented at a Conference held in Tutzing, July, 1980*, eds. L. Castell and C. F. von Weizsäcker (Carl Hanser, Munich, 1981) p.27-304.

[30] J. A. Wheeler: "The computer and the universe," Int. J. Theo. Phys. **21** (1982) 557-571.

[31] J. A. Wheeler: "Bohr, Einstein, and the strange lesson of the quantum," in *Mind in Nature*, Nobel Conference XVII, Gustavus Adolphus College, St. Peter, Minnesota, ed. Richard Q. Elvee (Harper and Row, New York, 1982) p.1-30 (also 88, 112, 113, 130-131, 148-40).

[32] J. A. Wheeler: *Physics and Austerity* (in Chinese) (Anhui Science and Technology Publications, Anhui, China, 1982); reprinted in part (Lecture II), in *Krisis*, Vol.1, No.2, ed. I. Masculescu (Klinckscieck, Paris, 1983) p.671-75.

[33] J. A. Wheeler: "Particles and geometry," in *Unified Theories of Elementary Particles*, eds. P. Breitenlohner and H. P. Dürr (Springer, Berlin, 1982) p.189-217.

[34] J. A. Wheeler: "Blackholes and new physics," in Discovery: Research and Scholarship at the University of Texas at Austin, **7**, No.2 (Winter 1982) 4-7.

[35] J. A. Wheeler: "On recognizing law without law," Am. J. Phys. **51** (1983) 398-404.

[36] J. A. Wheeler: "Jenseits aller Zeitlichkeit," in *Die Zeit*, Schrifter der Carl Friedrich von Seiemens-Stiftung, Vol. **6**, eds. A. Peisl and A. Mohler (Oldenbourg, Mnchen, 1983) p.17-34.

[37] J. A. Wheeler: "Elementary quantum phenomenon as building unit," in *Quantum Optics, Experimental Gravitation, and Measurement Theory*, eds. P. Meystre and M. Scully (Plenum, New York and London, 1983) p.141-143.

[38] J. A. Wheeler: "Bits, quanta, meaning," in *Problems in Theoretical Physics*, eds. A. Giovannini, F. Mancini and M. Marinaro (University of Salerno Press, Salerno, 1984) p. 121-141; also in *Theoretical Physics Meeting: Atti del Convegno, Amalfi, 6-7 maggio 1983* (Edizioni Scientifiche Italiane, Naples, 1984) p. 121-134; also in *Festschrift in Honour of Eduardo R. Caianiello*, eds. A. Giovannini, F. Mancini, M. Marinaro, and A. Rimini (World Scientific, Singapore, 1989) p.133-154.

[39] J. A. Wheeler: "Quantum gravity: the question of measurement, ' in *Quantum Theory of Gravity*, ed. S. M. Christensen (Hilger, Bristol 1984) p.224-233.

[40] W. A. Miller and J. A. Wheeler: "Delayed-choice experiments and Bohr's elementary quantum phenomenon," in *Proceedings of International Symposium of Foundations of Quantum Mechanics in the Light of New Technology, Tokyo, 1983*, eds. S. Kamefuchi et al. (The Physical Society of Japan, Tokyo, 1984) p.140-151.

[41] J. A. Wheeler: "Bohr's 'phenomenon' and 'law without law' " in *Chaotic Behavior in Quantum Systems*, ed. G. Casati (Plenum, New York, 1985) p.363-378.

[42] A. Kheyfets and J. A. Wheeler: "Boundary of a boundary principle and geometric structure of field theories," Int. J. Theo. Phys. **25** (1986) 573-580.

[43] J. A. Wheeler: "Physics as meaning circuit: three problems," in *Frontiers of Non-Equililibrium Statistical Physics*, eds. G. T. Moore and M. 0. Scully (Plenum, New York, 1986) p.25-32.

[44] J. A. Wheeler: "Interview on the role of the observer in quantum mechanics," in *The Ghost in the Atom*, eds. P.C.W. Davies and J. R. Brown (Cambridge University Press, Cambridge, 1986) p.58-69.

[45] J. A. Wheeler: "How come the quantum," in *New Techniques and Ideas in Quantum Measurement Theory* ed. D. M. Greenberger (Ann. New York Acad. Sci. **480** (1987) p.304-316.)

[46] J. A. Wheeler: "Hermann Weyl and the unity of knowledge," in *Exact Sciences and their Philosophical Foundations*, eds. W. Deppert et al. (Lang, Frankfurt am Main, 1988) p.469-503; appeared in abbreviated form in American Scientist **74** (1986) 366-375.

[47] J. A. Wheeler: "World as system self-synthesized by quantum networking," IBM J. of Res. and Dev. **32** (1988) 4-15; reprinted, in *Probability in the Sciences*, ed. E. Agazzi (Kluwer, Amsterdam, 1988) p.103-129.

[48] D. M. Greenberger, ed.: *New Techniques and Ideas in Quantum Measurement Theory* (Annals of the New York Academy of Sciences, Vol.480 (1986)).

[49] B. d'Espagnat: *Reality and the Physicist: Knowledge, Duration and the Quantum World* (Cambridge University Press, Cambridge, 1989).

[50] P. Mittelstaedt and E. W. Stachow, eds: *Recent Developments in Quantum Logic* (Bibliographisches Institut, Zrich, 1985).

[51] J. S. Bell: Speakable and Unspeakable in Quantum Mechanics: *Collected Papers in Quantum Mechanics* (Cambridge University Press, Cambridge, 1987).

[52] J. W. Tukey: "Sequential conversion of continuous data to digital data," Bell Laboratories memorandum of 1 September 1947 marks the introduction of the term "bit" reprinted in *Origin of the term bit*, ed. H. S. Tropp (Annals Hist. Computing **6** (1984)152-155.)

[53] W. K. Wootters and W. H. Zurek: "A single quantum cannot be cloned," Nature **279** (1982) 802-803.

[54] W. K. Wootters and W. H. Zurek, "On replicating photons," Nature, **304** (1983)188-189.

[55] Aharonov and D. Bohm: "Significance of electromagnetic potentials in the quantum theory" Phys. Rev. 115 (1959) 485-491; J. D. Bekenstein: *Baryon Number, Entropy, and Black Hole Physics*, Ph.D. thesis, Princeton University (1972); photocopy available from University Microfilms, Ann Arbor, Michigan.

[56] J. Anandan: "Comment on geometric phase for classical field theories," Phys. Rev. Lett. 60 (1988) 2555.

[57] J. Anandan and Y. Aharonov: "Geometric quantum phase and angles," Phys. Rev. D38 (1988) 1863-1870; includes references to the literature of the subject.

[58] J. D. Bekenstein: "Black holes and the second law," Nuovo Cimento Lett. 4 (1972) 737-740.

[59] J. D. Bekenstein: "Generalized second law of thermodynamics in black-hole physics," Phys. Rev. D9 (1973) 3292-3300.

[60] J. D. Bekenstein: "Black-hole thermodynamics" Physics Today 33 (1980) 24-31.

[61] R. Penrose: "Gravitational collapse: the role of general relativity," Riv. Nuovo Cimento 1 (1969) 252-276.

[62] D. Christodoulou: "Reversible and irreversible transformations in black-hole physics," Phys. Rev. Lett. 25 (1970) 1596-1597.

[63] D. Christodoulou and R. Ruffini: "Reversible transformations of a charged black hole," Phys. Rev. D4 (1971) 3552-3555.

[64] S. W. Hawking: "Particle creation by black holes," Commun. Math. Phys. 43 (1975) 199-220.

[65] S. W. Hawking: "Black holes and thermodynamics," Phys. Rev. 13 (1976) 191-197.

[66] W. H. Zurek and K. S. Thorne: "Statistical mechanical origin of the entropy of a rotating, charged black hole," Phys. Rev. Lett. 20 (1985) 2171-2175.

[67] MTW, p.1217.

[68] W. Shakespeare: *The Tempest*, Act IV, Scene 1, lines 148 ff.

[69] T. Mann: *Freud, Goethe, Wagner* (New York, 1937) p.20; trans. by H. T. Lowe-Porter from *Freud und die Zukunft* (Vienna, 1936).

[70] G. W. Leibniz as cited in J. R. Newman: *The World of Mathematics* (Simon and Schuster, New York, 1956).

[71] N. E. Steenrod: *Cohomology Operations* (Princeton University Press, Princeton, New Jersey, 1962).

[72] C. Ehresmann: *Catégories et Structures* (Dunod, Paris, 1965).

[73] D. Lohmer: *Phänomenologie der Mathematik: Elemente einer Phänomenologischen Aufklärung der Mathematischen Erkenntnis nach Husserl* (Kluwer, Norwell, Massachusetts, 1989).

[74] A. Weil: "De la metaphysique aux mathematiques," *Sciences*, p.52-56; reprinted in A. Weil: *Ouevres Scientifiques: Collected Works, Vol. 2*, 1951-64 (Springer, New York, 1979) p.408-412.

[75] J. D. Barrow and F. J. Tipler: *The Anthropic Cosmological Principle* (Oxford University Press, New York, 1986) and literature therein cited.

[76] See for example the survey by F. Feferman: "Turing in the land of O(z)," and related papers on mathematical logic, in R. Herken: *The Universal Turing Machine: A Half Century Survey*, (Kammerer and Unverzagt, Hamburg and Oxford University Press, New York, 1988) p.113-147.

[77] H. Weyl: "Mathematics and logic." A brief survey serving as a preface to a review of *The Philosophy of Bertrand Russell*, Am. Math. Monthly **53** (1946) 2-13.

[78] W. V. O. Quine: p.18 in the essay "On what there is," in *From a Logical Point of View*, 2nd ed. (Harvard University Press, Cambridge, Massachusetts, 1980) p.1-19.

[79] Discovered among graffiti in the men's room of the Pecan Street Cafe, Austin, Texas.

[80] G. W. Leibniz: *Animadversiones ad Joh. George Wachteri librum de recondita Hebraeorum philosophia*, c. 1708, unpublished; English translation in P. P. Wiener, *Leibniz Selections* (Scribners, New York, 1951) p.488.

[81] A. Einstein: as quoted by A. Forsee in *Albert Einstein Theoretical Physicist* (Macmillan, New York, 1963) p.81.

[82] **MTW**, 43.4.

[83] Ref. 22, p.411.

[84] E. H. Spanier: *Algebraic Topology* (McGraw-Hill, New York, 1966).

[85] É. Cartan: *La Geometrie des Espaces de Riemann, Memorial des Sciences Mathematiques* (GauthierVillars, Paris, 1925).

[86] É. Cartan: *Lecons sur la Geometrie des Espaces de Riemann* (Gauthier-Villars, Paris, 1925).

[87] **MTW**, Chap. 15.

[88] M. Atiyah: *Collected Papers. Vol. 5: Gauge Theories* (Clarendon, Oxford, 1988).

[89] Ref. 21, p.41-42; ref. 22, p.397-398.

[90] W. K. Wootters and W. H. Zurek: "Complementarity in the double-slit experiment: quantum nonseparability and a quantitative statement of Bohr's principle," Phys. Rev. **D19** (1979) 473-484.

[91] W. Heisenberg: "Über den anschaulichen Inhalt der quantentheoretischen Kinematik und Mechanik," Zeits. f. Physik **43** (1927) 172-198; English translation in **WZ**, p.62-84.

[92] N. Bohr and L. Rosenfeld: "Zur Frage der Messbarkeit der elektromagnetischen Feldgrössen " Mat.-fys. Medd. Dan. Vid. Selsk. **12**, no.8 (1933); English translation by Aage Petersen, 1979, reprinted in **WZ**, 479-534.

[93] D. J. Gross: "On the calculation of the fine-structure constant," Phys. Today **42**, No.12 (1989).

[94] N. F. Mott: "The wave mechanics of α-ray tracks," Proc. Roy. Soc. London **A126** (1929) 74-84; reprinted in **WZ**, p.129-134.

[95] H. D. Zeh: "On the interpretation of measurement in quantum theory," Found. Phys. **1** (1970) 69-76.

[96] H. D. Zeh: *The Physical Basis of the Direction of Time* (Springer, Berlin, 1989).

[97] E. Joos and H. D. Zeh: "The emergence of classical properties through interaction with the environment," Zeits. f. Physik **B59** (1985) 223-243.

[98] W. H. Zurek: "Pointer basis of quantum apparatus: Into what mixture does the wavepacket collapse?," Phys. Rev. **D24** (1981) 1516-1525.

[99] W. H. Zurek: "Environment-induced superselection rules," Phys. Rev. **D26** (1982) 1862-1880.

[100] W. H. Zurek: "Information transfer in quantum measurements: irreversibility and amplification," in *Quantum Optics, Experimental Gravitation and Measurement Theory*, eds. P. Meystre and M. O. Scully (Plenum, New York, 1983) p.87-116.

[101] W. G. Unruh and W. H. Zurek: "Reduction of a wave packet in quantum Brownian motion," Phys. Rev. **D40** (1989) 1071-1094.

[102] J. B. Hartle: "Progress in quantum cosmology," preprint from Physics Department, University of California at Santa Barbara, 1989.

[103] M. B. Green, J. H. Schwarz and E. Witten: *Superstring Theory* (Cambridge University Press, Cambridge, U.K., 1987).

[104] L. Brink and M. Henneaux: *Principles of String Theory: Studies of the Centro de Estudios Cientificos de Santiago* (Plenum, New York, 1988).

[105] S. W. Hawking: "The Boundary Conditions of the Universe," in *Astrophysical Cosmology*, Pontificia Academic Scientiarum, eds. H. A. Brück, G. V. Coyne and M. S. Longair (Vatican City, 1982) p. 563-594.

[106] A. Vilenkin: "Creation of universes from nothing," Phys. Lett. **B117** (1982) 25-28.

[107] J. B. Hartle and S. W. Hawking: "Wave function of the universe," Phys. Rev. **D28** (1983) 2960-2975.

[108] W. K. Wootters: "The acquisition of information from quantum measurements," Ph.D. dissertation, University of Texas at Austin (1980).

[109] W. K. Wootters: "Statistical distribution and Hilbert space," Phys. Rev. **23** (1981) 357-362.

[110] R. A. Fisher: "On the dominance ratio," Proc. Roy. Soc. Edin. **42** (1922) 321-341.

[111] R. A. Fisher: *Statistical Methods and Statistical Inference* (Hafner, New York, 1956) p.8-17.

[112] E. C. G. Stueckelberg: "Theoreme H et unitarite de S," Helv. Phys. Acta **25** (1952) 577-580.

[113] E. C. G. Stueckelberg: "Quantum theory in real Hilbert space," Helv. Phys. Acta **33** (1960) 727-752.

[114] D. S. Saxon: *Elementary Quantum Mechanics* (Holden, San Francisco 1964).

[115] H. J. Larson: *Introduction to Probability Theory and Statistical Inference*, 2nd ed. (Wiley, New York, 1974).

[116] E. Schrödinger: "The Foundation of the Theory of Probability," Proc. Roy. Irish Acad. **51A** (1947) 51-66 and 141-146.

[117] E. T. Jaynes: "Bayesian methods: General background," in *Maximum Entropy and Bayesian Methods in Applied Statistics*, ed. J. H. Justice (Cambridge University Press, Cambridge, U.K., 1986) p.1-25.

[118] R. Viertl, ed.: *Probability and Bayesian Statistics* (World Scientific, Singapore, 1987).

[119] R. D. Rosenkrantz, ed.: *E. T. Jaynes: Papers on Probability, Statistics and Statistical Physics* (Reidel-Kluwer, Hingham, Massachusetts, 1989).

[120] P. J. Denning: "Bayesian learning," American Sci.**77** (1989) 216-218.

[121] J. O. Berger and D. A. Berry: "Statistical analysis and the illusion of objectivity," American Sci. **76** (1988) 159-165.

[122] J. Burke: *The Day the Universe Changed* (Little, Brown, Boston, Massachusetts, 1985).

[123] F. Beck [pseudonym of the early nuclear-reaction-rate theorist Fritz Houtermans] and W. Godin: translated from the German original by E. Mosbacher and D. Porter, *Russian Purge and The Extraction of Confessions* (Hurst and Blackett, London, 1951).

[124] D. Føllesdal: "Meaning and experience," in *Mind and Language*, ed. S. Guttenplan (Clarendon, Oxford, 1975) p. 25-44.

[125] G. Berkeley: *Treatise Concerning the Principles of Understanding*, Dublin (1710; 2nd ed. 1734); re his reasoning that "No object exists apart from mind," cf. article Berkeley by R. Adamson: *Encyclopedia Brittanica*, Chicago **3** (1959) 438.

[126] T. Segerstedt: as quoted in ref. 22, p.415.

[127] Ref. 21, p.41.

[128] Ya. B. Zel'dovich and I. D. Novikov: *Relativistic Astrophysics, Vol. 1: Stars and Relativity* (University of Chicago Press, Chicago, 1971).

[129] J. Mather et al.: "A preliminary measurement of the cosmic microwave background spectrum by the Cosmic Background Explorer (COBE) Satellite," submitted for publication, Astrophys. J. Lett. (1990).

[130] **MTW**, p.738, Box 27.4; or **JGST**, Chap. 13, p.242.

[131] G. K. O'Neill: *The High Frontier*, 4th ed. (Space Studies Institute, Princeton, New Jersey, 1989).

[132] R. Jastrow: *Journey to the Stars: Space Exploration — Tomorrow and Beyond* (Bantam, New York, 1989).

[133] K. Popper: *Conjectures and Refutations: the Growth of Scientific Knowledge* (Basic Books, New York, 1962).

[134] R. W. Fuller and P. Putnam: "On the origin of order in behavior," General Systems (Ann Arbor, Michigan) **12** (1966)111-121.

[135] R. W. Fuller: "Causal and Moral Law: Their Relationship as Examined in Terms of a Model of the Brain," Monday Evening Papers (Wesleyan University Press, Middletown, Connecticut, 1967).

[136] G.M. Edelman: *Neural Darwinism* (Basic Books, New York, 1987).

[137] W. H. Calvin: *The Cerebral Symphony* (Bantam, New York, 1990).

[138] W. W. Collins: *The Moonstone* (London, 1968).

[139] J. Allan Hobson: *Sleep* (Scientific American Library, Freeman, New York, 1989) p.86, 89, 175, 185, 186.

[140] I. Langmuir: "Pathological Science," 1953 colloquium, transcribed and edited, Phys. Today, **42**, No.12 (1989) 36-48.

[141] N. S. Hetherington: *Science and Objectivity: Episodes in the History of Astronomy* (Iowa State University Press, Ames, Iowa, 1988).

[142] W. Sheehan: *Planets and Perception: Telescopic Views and Interpretations* (University of Arizona Press, Tucson, Arizona, 1988).

[143] M. White: *Science and Sentiment in America: Philosophical Thought from Jonathan Edwards to John Dewey* (Oxford University Press, New York, 1972).

[144] C. S. Peirce: *The Philosophy of Peirce: Selected Writings*, ed. J. Buchler (Routledge and Kegan Paul, London, 1940), passages from p.337, 335, 336, 353 and 358; reprinted in ref. 18, p.593-595. Peirce's position on the forces of Nature, "May they not have naturally grown up," foreshadow though it does the concept of world as self-synthesized system, differs from it in one decisive point, in that it tacitly takes time as primordial category supplied free of charge from outside.

[145] Parmenides of Elea [c. 515 B.C.-450 B.C.], poem Nature, part Truth, as summarized by A. C. Lloyd in article Parmenides, *Encyclopedia Brittanica*, Chicago **17** (1959) 327.

[146] G.E. Pugh: *On the Origin of Human Values* (New York, 1976); chapter Human values, free will, and the conscious mind preprinted in Zygon **11** (1976) 2-24.

[147] F. W. 3. von Schelling [1775-1854]: in *Schellings Werke, nach der Originalausgabe in neuer Anordnung herausgegeben*, 6 vols., ed. M. Schröter, (Beck, München, 1958-1959), esp. Vol.5, p.428-430, as kindly summarized for me by B. Kanitscheider: "dass das Universum von vorn-herein ein ihm immanentes Ziel, eine teleologische Struktur, besitzt und in allen seinen Produkten auf Evolutionäre Stadien ausgerichtet ist, die schliesslich die Hervoybringung von Selbstbewusstsein einschliessen, welches dann aber wiederum den Entstehungsprozess reflektiert und diese Reflexion ist die notwendige Bedingung für die Konstitution der Gegenstände des Bewusstseins."

[148] J. von Neumann and O. Morgenstern: *Theory of Games and Economic Behavior* (Princeton University Press, Princeton, New Jersey, 1944).

[149] J. Wang: *Theory of Games* (Oxford University Press, New York, 1988).

[150] J. R. Pierce: *Symbols, Signals and Noise: The Nature and Process of Communication* (Harper and Brothers, New York, 1961).

[151] M. Schwartz: *Telecommunication Networks: Protocols, Modeling and Analysis* (Addison-Wesley, Reading, Massachusetts, 1987).

[152] M. S. Roden: *Digital Communication Systems Design* (Prentice Hall, Englewood, Cliffs, New Jersey, 1988).

[153] P. W. Anderson: "More is different," Science **177** (1972) 393-396.

[154] C. Mead and L. Conway: *Introduction to VLSI* [very large-scale integrated-circuit design] *Systems* (Addison-Wesley, Reading, Massachusetts, 1980).

[155] P. B. Schneck: *Supercomputer Architecture* (Kluwer, Norwell, Massachusetts, 1987).

[156] F. E. Yates, ed.: *Self-Organizing Systems: The Emergence of Order* (Plenum, New York, 1987).

[157] H. Haken: *Information and Self-Organization: A Macroscopic Approach to Complex Systems* (Springer, Berlin, 1988).

[158] T. Kohonen: *Self-Organization and Associative Memory*, 3rd ed. (Springer, New York, 1989).

[159] C. Smorynski: *Self-reference and Model Logic* (Springer, Berlin, 1985).

[160] G. J. Chaitin: *Algorithmic Information Theory*, rev. 1987 ed., (Cambridge University, Cambridge, 1988).

[161] J. P. Delahaye: "Chaitin's equation; an extension of Gödel's theorem," Notices Amer. Math. Soc. **36** (1989) 948-987.

[162] J. F. Traub, G. W. Wasilkowski and H. Woznaikowski: *Information-Based Complexity* (Academic, San Diego, 1988).

[163] P. Young: *The Nature of Information* (Praeger Greenwood, Westport, Connecticut, 1987).

[164] W. H. Zurek: "Algorithmic randomness and physical entropy," Phys. Rev. **A40** (1989) 4731-4751.

[165] M. Eigen and R. Winkler: *Das Spiel: Naturgesetze steuern den Zufall* (Piper, München, 1975).

[166] W. M. Elsasser: *Reflections on a Theory of Organisms* (Orbis, Frelighsburg, Quebec, 1987).

[167] G. Nicols and I. Prigogine: *Exploring Complexity: An Introduction* (Freeman, New York, 1989).

[168] S. Watanabe, ed.: *Methodolgies of Pattern Recognition* (Academic, New York, 1967).

[169] J. Tou and R. C. Gonzalez: *Pattern Recognition Principles* (Addison-Wesley, Reading, Massachusetts, 1974).

[170] H. Haken, ed.: *Pattern Formation by Dynamic Systems and Pattern Recognition* (Springer, Berlin, 1979).

[171] H. Small and E. Garfield: "The geography of science: disciplinary and national mappings," J. of Info. Sci. **11** (1985) 147-159.

[172] M. Agu: "Field theory of pattern recognition," Phys. Rev. **A37** (1988) 4415-4418.

[173] M. Minsky and S. Papert: *Perceptrons: An Introduction to Computational Geometry*, 2nd ed. (Massachusetts Institute of Technology Press, Cambridge, Massachusetts, 1988).

[174] L. A. Steen: "The science of patterns," Science **240** (1988) 611-616.

[175] B. M. Bennett, D. D. Hoffman and C. Prakash: *Observer Mechanics: A Formal Theory of Perception* (Academic, San Diego, California, 1989).

[176] J. A. Wheeler: "Polyelectrons," Ann. New York Acad. Sci. **46** (1946) 219-238.

[177] D. Bohm: "The paradox of Einstein, Rosen and Podolsky," originally published as section 15-19, Chapter 22 of D. Bohm: *Quantum Theory* (Prentice-Hall, Englewood, Cliffs, N.J., 1950), reprinted in **WZ**, 356-368.

[178] W. H. Zurek: "Thermodynamic cost of computation: Algorithmic complexity and the information metric," Nature **34** (1989)119-124.

[179] E. F. Taylor and J. A. Wheeler: *Spacetime Physics* (Freeman, San Francisco, 1963) p.102.

20

FEYNMAN, BARTON AND THE REVERSIBLE SCHRÖDINGER DIFFERENCE EQUATION

Ed Fredkin

Ed Barton, then an MIT undergraduate, had agreed to work for me. I had had in mind a summer project and was very pleased that Ed was willing to accept the position. My goal was to solve the problem of creating discrete, reversible systems of various equations of physics. What this meant was that a computer programmed to compute the dynamics of such a system would be able to run forward any number of steps (repeatedly computing the future state from the present state) and then to be able to exactly retrace its steps running backwards (repeatedly computing the past state from the present state). The entire problem lay in the one word "exactly". Almost any programmed model of any physical system could be made to reverse its course in time, but such systems never retraced their steps *exactly*! I felt that I had made really good progress recently and the time was at hand to push for a breakthrough; finding powerful, general methods for creating esthetically pleasing exactly reversible equations in most every area of dynamical systems.

To understand what the problem was you need to understand how we then thought about calculations in general. Most workers felt that computers simply could not compute things in an exactly reversible fashion without saving the all the intermediate steps. Rollo Silver had come up with a method (what we would call a "hack") that could compute exactly backwards by saving only the initial conditions and the number of steps. (After going forwards n steps, it could go backwards one step by returning to the beginning and then going forwards $n - 1$ steps. It could continue going back to the beginning and then going forwards $n - 2$ steps, $n - 3$ steps...) Toffoli had figured out how to create a reversible cellular automaton by saving every step in an extra dimension! Both ideas were important but both were esthetically obnoxious. It was 1975 and I had recently returned to MIT after a year spent at Caltech.

I had been invited to Caltech, as a Fairchild Distinguished Scholar to work with Richard Feynman. We had a deal; he was to teach me quantum mechanics (from his perspective) and I was to teach him computer science (from my perspective). At the end of the year he complained to me that I got the better of the deal and I had to agree but the margin was close. It wasn't from a lack of trying on my part; it was very hard to teach Feynman something because he didn't want to let anyone teach him anything. What Feynman always wanted was to be told a few hints as to what the problem was and then to figure it out for himself. When you tried to save him time by just telling him what he needed to know, he got angry because you would be depriving him of the satisfaction of discovering it for himself. I'll never forget what happened when I showed him one of the first HP-35 calculators. This was the first really complicated calculator and it used RPN (reverse polish notation) instead of standard algebraic notation like the TI calculators did. I showed it to Feynman and he was delighted. He grabbed it but he wouldn't allow me to tell him anything about it! It took him many hours of fiddling to figure out how to use most of its features and having cracked its code he was through with it.

As for me, I wanted Feynman to teach me. My trouble was that I was pretty sure of what I wanted to learn and what I didn't want to learn. I was also somewhat hardheaded and too ready to argue. Feynman was wonderful about not bothering me with what I didn't need. He assured me that for my purposes everything I needed was to be found in the Feynman Lectures. I would read something, ask questions and Feynman would answer my questions. It was great fun as I tried to guide him.

"Let's make things simple and first look at the one particle case," I might suggest. Feynman would get fierce. "You'll never get it if you keep trying to look at what happens with just one particle." I would argue for a bit and then realize why Feynman was right. What was so wonderful was that he never gave up on me and always persisted to make sure that I got the basic knowledge without misconceptions.

While at Caltech my assignment to myself other than learning Quantum Mechanics (QM) was to solve the problem of reversible computation. It was a wonderful year and I learned much about QM and I invented Conservative Logic and the so-called "Fredkin Gate". Teaching me QM was almost as hard as teaching Feynman about computers. However, Feynman was able to cope with my foibles, he was persistent and fierce. We had talks, discussions and raging arguments; but it was always Feynman who was raging. Most memorable to me was once when we continued my physics lesson into lunchtime, walking to the greasy spoon cafeteria on campus. It was about an important issue and I was quietly refusing to go along with Feynman's position. He kept his cool until we had just about finished eating and then he couldn't contain his annoyance with me. Suddenly he stood up, so as to tower over me (I was still sitting and eating the last few bites of lunch). He started screaming at the top of his lungs about how stupid I was and why I had to look at it his way etc. I could only smile because I thought the scene was so funny and I was complimented that he cared so much, and anyway, I was sure that I was right.

By this time we had attracted the attention of almost everyone in the restaurant. It was the kind of thing that Feynman loved. People more than a few tables away had started to stand up to get a better view. While it's not true that Feynman was foaming at the mouth, nevertheless bits of saliva were landing on me and my plate.

Even so, it wasn't enough for Feynman; he wasn't towering over me enough. He suddenly looked at his chair, grabbed it and repositioned it so he could stand up on the chair and maybe climb up on the table so as to have a better platform with which to shout down at me. I got alarmed because I always worried that, physically, to some extent, Feynman was fairly fragile. I jumped up grabbed his arm and said, "Let's go." He hesitated for a moment, a bit annoyed at the prospect of losing the opportunity to do something so dramatic. He thought for a second more and then he did a peculiarly Feynman sort of thing. He continued beating on me verbally but the volume returned to normal. However each word was spoken with extreme emphasis, exaggerated stress and stretched out S's that had him hissing like a snake. The various spectators returned to their seats as he continued to wind down; we left the restaurant and walked back to our offices in the Lauritsen building.

The subject of the argument was my insistence on my approach to attacking the problem of creating discrete and reversible models of physical processes as opposed to Feynman's counter proposal. I wanted a general method that started by dealing with the simplest kinds of problems such as point masses following Newton's Laws, and Feynman insisted that I do nothing other than start with the Quantum Mechanical description of physics. My approach was heuristic; though I might have been wrong, I was sure that reversibility was so all encompassing a principle as to apply to every concept of dynamics. I wanted to understand how it dealt with the simplest such systems before looking at QM and Feynman insisted that QM was all there was and I should not think about anything else. The goals were exactly the same. What Feynman couldn't do, however, was to give me any help as to how to start with his approach.

One day towards the end of the year, we were in Feynman's office having another of the usual kinds of arguments when he suddenly got exasperated. He jumped up, went to the blackboard and started posing QM problems and asking me for answers. After a while he suddenly stopped the quiz, broke into a big smile and said, "The trouble with you is not that you don't understand quantum mechanics." From Feynman it was the highest form of compliment.

After I returned to MIT, I and my students made rapid and remarkable progress. With hindsight the solutions that we eventually found were simple and clear (e.g. reversible difference equations, the Billiard Ball Model, the Margolus CA rule, etc.). As is common in such circumstances the problem was not just to find the solutions, rather it was also to figure out what the right questions were. It also turned out that achieving exact reversibility for computational models of physics was equally easy for both Newtonian physics and Quantum Mechanics. Feynman was not wrong; it was just that my intuition functioned better with Newtonian Physics.

So, now that Barton had agreed to work for me and given how bright he was I had to plan out enough work to keep him busy for the summer. I invited Barton to my office.

I liked my office. It was in Tech Square, a modern 9 story building that housed MIT's Laboratory for Computer Science and a few other tenants. It was like $\frac{1}{2}$ block from the main campus. I had done two unusual things to my office. I had put a wall to wall rug on the floor and had obtained an actual blackboard. Other than that it was similarly furnished to other Tech Square offices. When Barton arrived I told him that I wanted to make up a list of the problems we wanted to work on and try to solve over the summer. I had by then already discovered a few ad hoc ways of creating discrete, reversible equations for various systems and I now wanted to do more. I wanted to find powerful general methods and now, at last, attack physics from the QM perspective. I went to the blackboard and started to write down the various things we would work on. And then I got to the Schrödinger Equation; a wonderful differential equation that basically describes how amplitudes change over space and time.

In mathematics we speak of a derivative. It's not too complicated. An example is velocity, which is the first derivative of position with respect to time. "Going north at 60 Miles Per Hour" expresses the notion of velocity, how position will change with respect to time. "Zero to 60 in 6 seconds" expresses the notion of acceleration, how speed will change with time. We say that velocity is the first derivative of position with respect to time and that acceleration is the second derivative of position with respect to time.

Derivatives are not always with respect to time. As we ascend in an airplane, we may notice that the air pressure (which is about 15 PSI at sea level) decreases about $\frac{1}{2}$ PSI per 1,000 feet. The first derivative of pressure with respect to altitude is $-\frac{1}{2}$ PSI per 1,000 feet at sea level. However that seems to imply that the pressure would get to zero by 30,000 feet. It turns out that the rate of decrease changes with altitude! A better rough approximation is that the pressure drops in half for every 18,000-foot increase in altitude. The second derivative of pressure with respect to altitude is about how the rate of change (the first derivative) changes with altitude. In this case, the rate of change also drops in half about every 18,000 feet. This means that at 36,000 feet the pressure is about 3.75PSI and the rate of change is about $-\frac{1}{8}$ PSI per 1,000 feet.

Derivatives are creatures of the Calculus. They assume that the quantities involved can vary continuously. In a computer all the numbers are basically integers. Computers are discrete systems that are the antithesis of continuous systems. Instead of Differential Equations we have Difference Equations. Instead of an equation that can tell us the values at any point, we have equations that can tell us the values at a particular set of discrete points. It's like listening to a clock that ticks every second and is silent the rest of the time. It's how computers work, doing one discrete step at a time.

I had told Feynman that I wanted to work with a discrete version of a particular form of the Schrödinger equation and I wanted some help on how to derive it. He suggested that I take a look at the Feynman Lectures, Volume III on Propagation in a Crystal Lattice. I did and I found a very suggestive version of the Hamiltonian (13.3 and 13.4) but it didn't all jell in my mind.

$$i\hbar\frac{dC_n}{dt}(t) = E_0C_n(t) - AC_{n+1}(t) - AC_{n-1}(t) \qquad (20.1)$$

Feynman Lectures Vol III Eq 13.3

I read on and discovered more of what I wanted in Chapter 16. (Page 16-4, equation 16.10). I was pretty good at seeing a difference equation when looking at a differential equation so I fixated on (16.13) as my starting point.

$$i\hbar\frac{dC}{dt}(x_n) = (E_0 - 2A)C(x_n) + A[2C(x_n) - C(x_n + b) - C(x_n - b)] \qquad (20.2)$$

Feynman Lectures Vol III Eq 16.10

The Schrödinger Equation involves both a first derivative with respect to time and a second derivative with respect to space. To keep it simple I wanted to start with the one-dimensional form of the differential equation.

$$i\hbar\frac{dC}{dt}(x) = -\frac{\hbar^2}{2m}\frac{\partial^2 C(x)}{\partial x^2} \qquad (20.3)$$

Free space one dimensional differential version of the Schrödinger Equation

Writing down the difference equation was easy but annoying. The reason was that the right side (the second derivative with respect to space) came out beautifully symmetrical, but the left side (the first derivative with respect to time) did not.

$$C_{x,t+1} - C_{x,t} = ik(C_{x-1,t} - 2C_{x,t} + C_{x+1,t}) \qquad (20.4)$$

Asymmetric difference version of the Schrödinger Equation

My esthetic sense was not pleased. Well, that was the starting point. I sort of lectured Barton on the equation (though he had taken such courses at MIT and in some ways knew more than me). I had written down about 10 research items on the blackboard. I went on talking and completed my discussion of the task list for the summer project. I told Barton that the list was merely a wish list and that I didn't dream of getting everything done.

Here we must digress for you to understand what was happening. I wanted to present Barton with the best starting point in what was to be his quest for the summer. He was to do a series of projects with the keystone being to find a beautiful, discrete and reversible version of the Schrödinger Equation. I needed to get him started down the right path and also needed to feel that he really understood what I was looking for. When you are searching for the right equation you often are guided by a sense of esthetics; some equations are ugly and some are beautiful. Almost anyone other than me would have no trouble writing down the first difference with respect to time; it's just the difference between the values at 2 points in time. If the Pressure, P, is 15 PSI at sea level ($P_0 = 15$) and 14.5 PSI at 1,000 feet ($P_{1,000} = 14.5$) then the first difference would be $P_{1,000} - P_0 = -\frac{1}{2}$ PSI per 1,000 feet. The trouble was the lack of symmetry; at what altitude is the first difference exactly $-\frac{1}{2}$?

When I finished writing down the difference equation, I was fixated, staring at the blackboard. The chalk marks on my wonderful blackboard glared back. I couldn't send Barton off on his task with that stupid asymmetrical first difference. Suddenly I had an inspiration. I erased the left side of the equation and replaced it with a perfectly symmetrical version of the first difference.

$$\frac{C_{x,t+1} - C_{x,t-1}}{2} = ik(C_{x-1,t} - 2C_{x,t} + C_{x+1,t}) \qquad (20.5)$$

Symmetric difference version of the Schrödinger Equation

I was pleased and proud of myself (despite the basic triviality of what I had done). It was the kind of thing that must have been done by others a thousand times, but it was what erased my sense of unease and I felt good about it. I then finished lecturing Barton on what was to be done with the equation. He was to search for some form that computed approximately the same thing, but that met the criteria necessary for being implemented on a computer in an exactly reversible way. I had given Barton a good start and I felt sure that his work was cut out for him. He would have a busy summer and, I hoped, a productive one. Barton was sitting in a chair near the door to my office while I lectured, and when I was done he remained seated staring at the blackboard. I went to my desk and started to busy myself with other stuff, letting Barton think while I waited to see if he had any questions.

After a long silence with Barton continuously staring at the Schrödinger Equation, I said "What we have to do is to explore all the different forms that the Schrödinger Equation can take until we find one that is reversible." Barton remained silent and continued staring at the equation. I was getting somewhat disconcerted. I thought that Barton was very bright but he seemed to not get the point. "Barton," I said getting more annoyed, "we need a version that's reversible!"

Barton ignored me and kept staring at the board. "It is," he said, quietly. And, after another long pause he again said, "It is."

Now I was really exasperated. "Look Barton, you don't get it. I've written down a simple symmetric version of the Schrödinger Equation and what I want you to do is to spend the summer looking for a reversible version of it. Start with what's on the board and try to make a version that's reversible."

Barton still kept staring at the board and Sphinx-like said again, "It is."

"What do you mean by 'It is'?" This was getting ridiculous. "What are you trying to say?"

Ed Barton finally said "The equation you've written, I think it *is* reversible."

I was actually pissed off. "Ed, try and get this straight. I've done nothing but write down an ad hoc version of the equation just to give you a starting point. It's not the answer; it's the question. Take the question and go find the answer." I had not spent any time thinking about or looking for a reversible version of the Schrödinger Equation. I liked to think of problems that I might be able to solve and refrain from working on them so I could hand them off to students. That way they could have the chance to be the first to solve some real problem.

I could not imagine what Barton was getting at or why, and this was violating my sense of how I wanted this meeting to end. Suffering from an extreme attack of *mental* set, I simply wanted Barton to get started on the right footing and then go off and get to work. Somehow I couldn't hear what he was saying. Barton finally broke through to me "I think that the equation you wrote on the board is the equation you want me to find. I think that it already is exactly reversible."

This time I finally looked at the equation and realized that by simply transposing one term to the other side it was obviously exactly reversible. That this was true was so counter-intuitive to me that I was flabbergasted. You see, in Physics we are normally satisfied with the differential equation. We use the difference form in order to write computer programs which are inexact and almost never reversible. But once the difference equation was written in a symmetrical fashion, it allowed itself to be transformed into an exactly reversible algorithm by merely moving one term to the other side. I had written it down in a moment of inspiration that was driven mostly by esthetics, and had failed to look at and understand what I had written. Ed Barton's immediate observation of the reversibility of that equation remains one of my most striking memories of 18 years on the MIT faculty.

As soon as Barton left my office I did what I had done so often under similar circumstances, I wrote a program in Lisp to run the equation, saw that it was reversible and nevertheless not pathological, then I picked up the phone and called Feynman. "Richard, we've just discovered the most amazing difference form of the Schrödinger Equation." I then proceeded to tell Feynman what we found and how an esthetic principle helped us find it. He was fascinated but had no real comment. I also told him that in a simulation of what I called the "0-Dimensional Schrödinger Difference Equation" I had noticed that the amplitude (calculated in a simple way) did not drift off very far from its original value, which was surprising and serendipitous. For some reason I would always phone Feynman with my latest idea and he would always reply by writing me a letter. A week later I got a letter from Feynman with our new equation and a beautiful proof. Feynman showed that by calculating the amplitude for the 0-Dimensional case in a perfectly time symmetric fashion, keeping to my esthetic principle, then the values computed by the reversible difference equation exactly conserved the value of the amplitude. I was happy and chagrinned. By using one of my favorite tricks on my own equation, Feynman had still managed to one up me!

Appendix

The following equations and *Mathematica* functions illustrate the point of Feynman's contribution with regard to the calculation of probabilities for the reversible difference form of the Schrödinger Equation.

$$i\hbar\frac{\partial C(x)}{dt} = -\frac{\hbar^2}{2m}\frac{\partial^2 C(x)}{\partial x^2}$$

The Schrödinger Equation for motion along a line in free space.

$$\frac{\partial C(x)}{dt} = ik\frac{\partial^2 C(x)}{\partial x^2}$$

Both sides divided by $i\hbar$; constants lumped into k.

$$D_t C_{x,t} = ik D_x^2 C_{x,t}$$

The difference form.

$$C_{x,t+1} - C_{x,t} = ik(C_{x-1,t} - 2C_{x,t} + C_{x+1,t})$$

The explicit form with the asymmetric first difference on the left.

$$C_{x,t+1} = \{ik(C_{x-1,t} - 2C_{x,t} + C_{x+1,t})\} + C_{x,t}$$

The computational form from the above equation. Because of roundoff and truncation errors it is not exactly reversible. (The curly braces signify roundoff to a machine precision number.)

$$\frac{C_{x,t+1} - C_{x,t-1}}{2} = ik(C_{x-1,t} - 2C_{x,t} + C_{x+1,t})$$

$$C_{x,t+1} - C_{x,t-1} = 2ik(C_{x-1,t} - 2C_{x,t} + C_{x+1,t})$$

The explicit form with the symmetric first difference on the left.

$$C_{x,t+1} = \{2ik(C_{x-1,t} - 2C_{x,t} + C_{x+1,t})\} + C_{x,t-1}$$

The forward computational form of the Schrödinger Difference Equation. What is counter-intuitive but absolutely true is that it is exactly reversible when computed on any ordinary computer even though information is seemingly lost at every step due to roundoff and truncation error!

$$C_{x,t-1} = -\{2ik(C_{x-1,t} - 2C_{x,t} + C_{x+1,t})\} + C_{x,t+1}$$

Above, the reverse form of the Schrödinger Difference Equation. It exactly retraces the steps (in the opposite order) of the forward form.

What follows is the derivation of what might be called the simplified, zero dimensional Schrödinger Equation. We start by doubling the time steps from t (as used above) to $2t$, so that we can do half time step calculations. We also assume that the complex amplitudes are divided in time so that all Real values occur at odd time steps $2t+1$, ($C_{x,2t+1}$ is real) and all Imaginary values occur at even time steps of $2t$, ($C_{x,2t}$ is imaginary).

$$C_{x,2t+2} = 2ikC_{x,2t} + C_{x,2t-2} \qquad \text{C complex - double step}$$
$$R_{x,2t+1} = 2ikI_{x,2t} + R_{x,2t-1} \qquad \text{The Real half step}$$
$$I_{x,2t+2} = 2ikR_{x,2t+1} + I_{x,2t} \qquad \text{The Imaginary half step}$$
$$S_{x,t+1} = 2ikS_{x,t} + S_{x,t-1} \qquad \text{Get rid of R and I labels}$$
$$P_{x,t} = S_{x,t}^2 + S_{x,t+1}^2 \qquad \text{Simple minded probability}$$
$$P_{x,t} = S_{x,t}^2 - S_{x,t-1}S_{x,t+1} \qquad \text{Feynman's probability}$$

What follows is a series of *Mathematica* functions that illustrate numerically these properties of reversibility and conservation of probability

```
ClearAll[S]; S[1]=1000; S[0]=0; k = π/256;
S[t_]:= S[t] = Round[2ikS[t-1]] + S[t-2]
```

We define the *Mathematica* function for the zero dimensional Schrödinger Equation. While we compute with integers, the quantities we represent are actually real and complex numbers. The integer 1,000 in the computer represents the number 1.0 in the physical model.

```
Table[S[n], {n,1,10}]
{1000, 25 I, 999, 50 I, 998, 74 I, 996, 98 I, 994, 122 I}
```

The first 10 values computed in the forward direction.

```
ClearAll[S]; Sr[10]=S[10]; Sr[9]=S[9]; k = π/256;
Sr[t_]:= Sr[t] = - Round[2ikSr[t+1]] + Sr[t+2]
```

We define the *Mathematica* function for the zero dimensional Schrödinger Equation running in the reverse direction.

```
Table[Sr[n], {n,10,1,-1}]
{122 I, 994, 98 I, 996, 74 I, 998, 50 I, 999, 25 I, 1000}
```

We get the exact same values going in the reverse direction as in the forward direction.

```
Table[(Abs[S[i] + S[i+1]]²)/1000.0, {i, 1, 24, 2} ]
{1000.62, 1000.5, 1001.48, 1001.62, 1002.92, 1003.4, 1003.7,
   1003.93, 1004.01, 1005.29, 1005.85, 1005.7}
```

The conventional definition of the probabilities hovers near to the value 1,000 (which represents probabilities of 1.0). Since we multiply two scaled numbers in computing the probabilities, we have to also divide by the scale factor of 1,000.

```
ClearAll[Sp]; Sp[1]=SetPrecision[1000.0,40];
Sp[0]=0; k=π/256; Sp[t_] := Sp[t] = 2ikSp[t-1] + Sp[t-2];
Table[(Sp[i]² - Sp[i+1] Sp[i-1])/1000, {i, 1, 24, 2}]
{1000.0000000000000000000000000000000000000,
             1000.0000000000000000000000000000000000000,
   1000.0000000000000000000000000000000000000,
             1000.0000000000000000000000000000000000000,
   1000.0000000000000000000000000000000000000,
             1000.0000000000000000000000000000000000000,
   1000.0000000000000000000000000000000000000,
             1000.0000000000000000000000000000000000000,
   1000.0000000000000000000000000000000000000,
             1000.0000000000000000000000000000000000000,
   1000.0000000000000000000000000000000000000,
             1000.0000000000000000000000000000000000000}
```

Feynman's definition of the amplitudes would always be exactly one if calculated with infinite precision. (Remember, all numbers are scaled up by 1,000)

```
{Sp[10], Sp[11], Sp[12]}
{122.42295112400517290576729397847903102831,
```

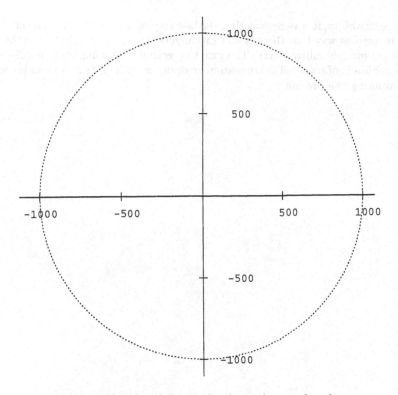

Fig. 20.1. The plot of the amplitudes in the complex plane.

```
990.9768018793071273597310987393469145585,
146.7451811291765735406549221352764396674I}

(Sp[11]² - Sp[10] Sp[12])/1000
1000.0000000000000000000000000000000000000
```

The above sample of data further illustrates the Feynman definition of probability.

```
ComplexToList[phi_] :=
  {Re[phi], Im[phi]};
ListPlot[Table[ComplexToList[S[i] + S[i+1]],
  {i, 1, 512, 2}], AspectRatio → Automatic]
```

N.B. Such reversible computations may be thought of as the computation of any function of the present (with any roundoff and/or truncation error), then we either add or subtract the past (with no error) in order to compute the future. For exact reversibility, it is necessary that the last step of the computation be effectively done in such as way that there is an exact inverse (i.e. integer addition is the exact inverse of integer subtraction). The exact reversibility is completely unaffected by whatever kinds of roundoff or truncation error occur, so long as that error is confined to computing the present.

21

ACTION, OR THE FUNGIBILITY OF COMPUTATION

Tommaso Toffoli

Abstract

We informally explore an emergent interpretation of the action integral in physics, and discuss its connections with the concept of "amount of computation". Much as entropy quantifies the lack of information one has about the *state* of a system, action seems to quantify the lack of information about the system's *law*.

> *From given causes, effects are generated by nature in the most efficient way. ... No natural action can be. abbreviated* [Leonardo]

> *An institution will grow until it uses up all available resources* [Parkinson's law]

21.1   Introduction

Physical *entropy* measures *amount of information*; here I'll argue that physical *action* measures *amount of computation*. My approach will be an impressionistic one: motivation, metaphors, circumstantial evidence, toy examples. This is a still quite unsettled area of research (Jaynes [1] is a particularly illuminating work in this regard); my paper is an invitation, not a review.

We are taught to regard with awe the variational principles of mechanics [2, 3]. There is something miraculous about them, and something timeless too: the storms of relativity and quantum mechanics have come and gone, but Hamilton's principle of least action still shines among our most precious jewels.

But perhaps the reason that these principles have survived such physical upheavals is that after all they are not strictly *physical* principles! To me, they appear to be the expression, in a physical context, of general facts about *computation*, much as the second law of thermodynamics is the expression, in the same context, of general facts about *information* (cf. [4]). More specifically, just as entropy measures, on a log scale, the number of possible microscopic *states* consistent with a given macroscopic description, so I argue that action measures, again on a log scale, the number of possible microscopic *laws* consistent with a given macroscopic behavior.

If entropy measures *in how many different states you could be in detail* and still be substantially the same, then action measures *how many different recipes you could follow in detail* and still behave substantially the same.

Information can be quantified because it is *fungible*.[1] In virtually any circumstances, to the telephone user it does not make a difference whether a conversation travels by copper cable or by microwave link. What the user is charged for is not the message itself or the use of the medium per se, but merely the *variety* of alternative messages that could have been sent. The importance of 'amount of information' in physics is enhanced by the fact that at a microscopic level physics is invertible, and thus 'amount of information' is a conserved quantity; as a consequence, phase space must flow as an incompressible fluid (Liouville's theorem), and this imposes a strict discipline on the possible dynamics.

To what extent is *computation* fungible? Let us imagine two technicians doing numerical integration of a differential equation using Mathematica, one on a MAC and the other on a PC. A pedantic physicist who kept track of every nucleus, electron, and photon would observe totally different histories in the two computers. Even at the coarser level of machine instructions and memory words, what is going on in one computer is very different from the other. Yet one can meaningfully say that the two computers are running "the same program." Of course, the two programs are quite different (try loading and running on the PC the executable file for the MAC!); what we mean is that the differences in the two programs are deliberately engineered to compensate for the differences in the two machines, so that they will yield the same overall *behavior*.

Thus, while at a low level (say, response to a DMA interrupt) the two machines are not interchangeable, at a high enough aggregation level (say, doing numerical analysis) they are substantially *equivalent*. Not only can the two machines simulate one another (that's a consequence of their computation universality, and computation universality is cheap); on almost all macroscopic tasks they can do so with *essentially the same timing*—up to an overall proportionality factor k. At this level, computation is apparently *fungible*. That is, the internal details can be made irrelevant by appropriate programming, and all that remains is a single scalar parameter k—the machine's *computation capacity*. If the MAC has capacity 1.5 times greater than the PC, then on most tasks a laboratory equipped with 30 PCs will be able to do in a month the same amount of computational progress as one equipped with 20 MACs. (Only rare, perversely chosen tasks will be able to show significant performance differences between the two laboratories.) In this scenario, *computation capacity* becomes a tradable commodity that can be sold "by the pound" like potatoes, and thus is a fungible resource (cf. [5]).

---

[1]According to the *American Heritage Dictionary*, "Being of such a nature that one unit or part may be exchanged or substituted for another equivalent unit or part in the discharging of an obligation."

Note that information capacity (e.g., channel capacity or storage capacity), which comes to mind as a typical example of fungible resource, only becomes so in an appropriate macroscopic scenario. To begin with, to take advantage of Shannon's theorems one must handle information in bulk, doing long-term buffering and averaging. Moreover, to arrive at a well-defined value for capacity one must ignore a number of ancillary costs (encoding and decoding resources, latency, etc.).

Our questions, then, are: In what context and to what extent is the concept of 'fungible computation' viable? How can it be used? What is its relevance to physics? What are its connections, if any, to the concept of 'action' and the least-action principle?

21.2 Computation capacity

Computer science already has two quite distinct concepts that relate to "how hard it is to compute a given function f." **Computability** [6] asks whether or not there is an algorithm that for any value x of the argument will produce the corresponding result $y = f(x)$ in a finite number of steps; the issue is really whether the function can be computed at all. **Computation complexity** [7] asks how the number of steps (or other quantities such as the amount of intermediate storage) needed to produce y grows as a function of the *size* of x. To classify an algorithm, one compares the asymptotic rate of growth (as the size of x goes to infinity), or whatever bounds for it can be ascertained, with a number of reference rates (linear, polynomial exponential, etc.).

We are interested in a different concept, namely, the "computational worth" of a computer; intuitively, how much one should charge for its rental in a competitive market in which one and the same computer may be put to different uses (which are not known in advance, at fabrication time or even at rental time) and different computers may well be put to the same use if expedient. We'd like this measure of computational worth to be defined in terms of function, not of structure (i.e., without reference to physical or technological parameters such as area, speed, number of gates, internal wiring); however, if it is to express the (presumed) fungibility of computing resources, it should be approximately *additive* when applied to physical computation in bulk amounts.

For the time being, a **computer** will mean a finite combinational network in which a distinguished subset of input lines constitute the **program** p, the rest of the input lines constitute the **argument** x, and the output lines constitute the **result** y.[2] The **behavior** of the computer is the functional relationship between argument

---

[2] An ordinary computer operated for a fixed number n of clock cycles is a machine in this sense, as the sequential network that makes up the computer can be "unrolled in time" into an n-stage combinational network; the input is represented by the initial state of the computer's memory, and the output by its final state; different values for n yield different "machines". Real-time input

x and result y.

We'll provisionally define the **computation capacity** of a computer, viewed as a *programmable machine*, as the log of the number of distinct *behaviors* the machine can achieve over the range of possible settings or *programs*. While different behaviors must come from different programs, two or more programs may yield the same behavior. We are interested in measuring the variety of actual behaviors, as constrasted to the variety of settings. For example, a "ten-speed" bicycle has two levers, one with two settings (A and B) and one with five, giving ten settings overall. However, the A and B gear-ratio ranges overlap, as indicated in Fig. 21.1, giving a smaller number of effectively distinct "speeds".

		10	8	6	4	5
A	20	2	2.5	3.3	4	5
B	24	2.4	3	4	4.8	6

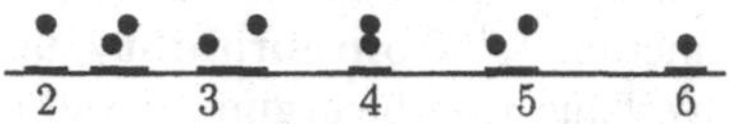

Fig. 21.1. In a "ten-speed" bicycle, the two five-speed ranges substantially overlap, giving fewer than ten effectively distinct gear ratios. Here there are only nine distinct ratios, and, on a coarser grain, only six or seven practically different "speeds".

Much as for the case of channel capacity, this basic definition can be refined and extended in several ways. First, instead of giving equal weights to all behaviors, one may wish to take into account with what *probability* each behavior occurs, which in turn reflects the probabilities of the programs that gave rise to it. Second, one may wish to consider a statistical rather than deterministic dependence of behavior on program ("noisy computer"), and count as computation capacity only that fraction of the variety of behavior that can be attributed to deliberate programming (rather than to chance). This approach may be further extended by including the peculiar statistics of quantum-mechanical behavior as part of the specification of the computer. Third, one may wish to consider summing the number of behaviors over all possible ways of partitioning the input lines into program lines and argument lines, and all possible ways to partition the output lines into result lines and lines to be ignored; this is equivalent to identifying the range of behavior with the set of all possible *subfunctions* that the network can realize. In this way, the distinction between argument and program disappears, and computation capacity becomes associated simply with the overall input/output relationship of the computer. Finally, one may define all sorts of *conditional* capacities by imposing constraints on inputs, outputs, or their relationship; for instance, demand that the argument, seen as a bit string, consist of twice as many zeros as ones.

---

and output, as through a console or a serial port, are represented by lines entering or leaving the network at each of the n stages.

We shall illustrate the basic definition of computation capacity with an example, namely the "canonical" computer consisting of an ordinary PROM (Programmable Read-Only Memory). This is a network consisting of a single node with m input lines collectively called the *address*, n output lines called the *data*, and k more input lines (where $k = n2^m$) called the *program*.

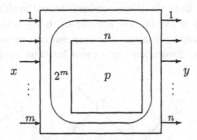

Fig. 21.2. The PROM as a canonical computer, with m lines of address x, n lines of data y, and $n2^m$ bits of program p.

In this canonical machine, each program gives rise to a distinct function $y = p(x)$. The number of functions is thus 2^{n2^m}, and its log (in base 2) is simply $n2^m$; that is, the computation capacity equals the number of PROM program bits.

Note that in a real PROM chip the programmable bits take up most of the chip's real estate (the address decoding tree occupies a much smaller area and consists of a fixed—nonprogrammable—gates). Thus, our measure of computation capacity (log of number of functions) is consistent with the market reality, where PROM chips of different organization (for example, 32- vs 8-bit wide) but with the same number of programmable bits have essentially the same size, cost essentially the same, and have a comparable market share.

If a comprehensive theory of computation capacity along the present lines can be developed at all, it is likely to be much more complex than Shannon's theory of communication (or channel) capacity. In the latter, a channel of given capacity S can be shared between two users who will enjoy capacities S_1 and S_2 respectively. With appropriate encoding and decoding, $S_1 + S_2$ can be made as close as desired to S; however, the computational resources needed for encoding and decoding will in general rise steeply as $S_1 + S_2$ approaches S. Thus, communication capacity is additive only when the cost of computing resources does not enter the picture. If we apply a similar construction to the sharing of a *computer* rather than of a channel, the resources used for encoding and decoding will be of the *same* kind as those that we are trying to share, and cannot be ignored. Thus, we will be confronted with a *nonlinear* theory. We may only hope to get an approximately linear theory when the amount of computation done between the encoding and decoding stages is so large that the pro-rated cost of the latter becomes negligible. In this sense, the fungibility of computation capacity cannot be expected to extend down to the scale of a single gate—much as the price of a pint-bottle of water has little relation to the cost of water in bulk.

For the same reasons, if we try to apply a theory of computation capacity to a distributed computing medium—a network extending over space and time—little insight can be expected from the theory unless the network's interconnection is local and its texture remains uniform over a scale of many node spacings. A local and uniform combinational network extending over one one or more spatial dimensions— and, of course, one time dimension—is a **cellular automaton** (Fig. 21.3).

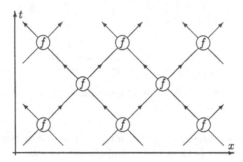

Fig. 21.3. Spacetime representation of a simple one-dimensional cellular automaton. Each logic gate f of the combinational network represents the activity of one cellular automaton "cell" through one clock cycle.

In what follows, the computing networks we'll be interested in will be discrete, fine-grained models of continuous dynamical systems: continuous behavior will emerge from the underlying discrete activity by coarse-grained averaging. Our goal will be to establish a bridge between the variational principles that appear to govern the macroscopic behavior of a system and computation-capacity arguments (and thus, ultimately, merely *counting* arguments) applied to its microscopic description.

Our definition of 'computation capacity' rates the power of a computer in terms of *how many* different tasks it can be programmed to do in a fixed amount of time. What does that have to do with the *difficulty* of the tasks themselves? Why should one pay much for a computer that is able to do zillions of things, if all of them turn out to be trivially simple?

The answer lies in the pigeon-hole principle. The number $n(s)$ of possible tasks having a certain level s of complexity[3] increases very rapidly with the complexity

---

[3]By 'complexity of a computational task' we may take the minimum number of *operations*, on a reference computer architecture, required to complete the task, just as the 'complexity of a piece of information' may be taken as the minimum number of statements, in a reference computer language, required to print out that information.

itself. Thus, if we are told that a computer can do a very large number n of *different* things but we aren't told *which* things, we nonetheless know that there just aren't enough simple things to exhaust its range of behavior, and we conclude that this range must include *some* tasks above the complexity level $s(n)$ (where $s(n)$ is the inverse function of $n(s)$).

More concretely, suppose we want to choose a rental computer for doing a task of complexity s, and all that the rental catalog says about each computer they offer, besides its rental cost, is the number $\bar{n}$ of different things it can do. If for the sake of economy we choose a computer with $\bar{n} \ll n(s)$, we can be almost sure that our task will *not* be among those the computer can complete in the given time. On the other hand, if we pick $\bar{n} \gg n(s)$, then there are good chances that our task *will* be in the computer's repertoire. If we do not want to waste our money, we must be ready to pay enough to get us a computer with a rating at least $n(s)$.

Finally, what about fungibility? If computers are priced proportionally to $\log n$, and by putting together two computers rated n_1 and n_2 I effectively get a rating $n_1 n_2$, then for a price $\log n_1 + \log n_2$ I will be able to do typical tasks of complexity $\approx s(n_1 n_2)$, just as if I rented, for the same price, a single computer of rating $n_1 n_2$. Alternatively, to get a rating $n_1 n_2$ from a low-rating computer I can rent it for a time proportional to $\log n_1 + \log n_2$. In other words, in an ideal market it should be possible to achieve the capacity of a supercomputer—at a comparable cost—by putting together a network of cheap (and weak) computers or by running a small computer for a sufficiently long time.

21.3   Density of computation capacity

While mass is well-defined even for a microscopic sample of material, intensive properties such as density or conductivity are poorly defined and vary with the size and shape of the sample. It is only for macroscopic samples of uniform constitution, in which finite-size effects become negligible, that these quantities converge to a value characteristic of the *material* and independent of sample size and shape. In a similar way, well-characterized intensive properties may emerge when one deals with a large computational network of uniform texture. What we are interested in is fine-grained computing networks for which the concept of *density of computation capacity* is meaningful. Before investing too much in precisely defining this property, it is important to have an idea of how one would *recognize* it and *make use* of it.

Consider an indefinitely extended network such as that of Fig. 21.3, and cut out a swatch of it of width x and depth t, gluing together the two spatial ends of the strip; this will constitute a *computer* according to our definition (§21.2), and will have a certain computation capacity $A(x, t)$. Since information only travels at the speed of one site per step, when $x \gg t$ the swatch is effectively decoupled into a number of almost independent spatial regions. The number of functions computed by the swatch is thus approximately the product of the number of functions computed by the individual regions, and the overall capacity (log of number of functions) will be the sum of the separate capacities. For fixed $t = t_0$ and large enough x, the capacity of the swatch will grow linearly with x.

Let us fix an $x = x_0$ large enough to be in this linear region (the choice of x_0 will depend on t_0, as linearity will set in later when t_0 is larger) and let us start moving t upwards from the reference value t_0. Typically, the computation capacity of the swatch will tend to increase with t, as formerly uncoupled regions of the swatch now have the time to become coupled. However, as t grows to be much larger than x_0, the "computational trajectories" that the data describe as they evolve through time will tend to cycle (*all* trajectories will of course have entered a cycle by time $\approx 2^{x_0}$), and eventually the computation capacity of the swatch will flatten out.

We are interested in the initial relative slope,

$$C = \frac{1}{x} \left. \frac{dA}{dt} \right|_{\substack{t=t_0 \\ x=x_0(t_0)}}$$

If, for a sufficiently large depth t_0 of the swatch, the quantity C is independent of t_0 itself, then this quantity is an intensive property of the network's texture (i.e., of the cellular automaton's local structure), and constitutes the **density of computation capacity** of the cellular automaton; the total computation capacity of a swatch of depth t and width $x \gg t$ will then be simply Cxt, and the cellular automaton can be rented at a flat rate C per unit of spacetime volume.

21.4   Specific ergodicity

The present session is a digression. There still is precious little theoretical understanding of how likely a network texture is to have a well-defined density of computation capacity. Here we give concrete experimental evidence for the existence of an intensive network quantity, namely, *specific ergodicity*, which, though distinct from density of computation capacity, is intimately related to it.

Consider a discrete dynamical system (for instance, a finite-state automaton) whose state set consists of L points. The entropy of the state set is

$$H_{\text{total}} = \log L,$$

in the sense that if you choose a point at random it will take me on average $\log_2 L$ binary questions to correctly guess your point. Consider now an arbitrary invertible dynamics (a permutation) τ on this set. Such a dynamics partitions the set into a collection of orbits, and for any point one can speak of the length ℓ of the orbit to which it belongs. The entropy of (the partition induced by) τ is

$$H_{\text{orbit}} = -\sum_{\text{all orbits}} \frac{\ell}{L} \log \frac{\ell}{L};$$

this is the average number of questions needed to guess just which orbit the points belong to. On ther other hand, if I were told right away on which orbit the chosen point lies, I would have to ask on average only a number of questions equal to

$$H_{\text{phase}} = \sum_{\text{all points}} \frac{\log \ell}{L}$$

in order to locate the point. Of course, $H_{\text{total}} = H_{\text{orbit}} + H_{\text{phase}}$.

The **specific ergodicity** η of the dynamics τ is defined as

$$\eta = \frac{H_{\text{phase}}}{H_{\text{total}}} = \frac{\langle \log \ell \rangle}{\log L}. \tag{21.1}$$

That is, η tells us what fraction of the total uncertainty is uncertainty as to the *phase*—or position on the orbit. Clearly, $0 \le \eta \le 1$. When $\eta = 1$, all points lie on a single orbit, as in a car odometer; the system is *ergodic*. When $\eta = 0$, each orbit consist of a single point, and the system is useless as a counter or a clock. Intuitively, η measures how efficiently the system uses its state variables to "count".

We would like to extend the above considerations to infinite systems such as cellular automata. In this case, though definition (21.1) breaks down (since $H_{\text{total}} = \infty$ and in general also $H_{\text{phase}} = \infty$), it may be replaced by a definition entailing a limit.

Let τ be an invertible cellular automaton. In the one-dimensional case, one can imagine "wrapping around" the system so that site i is identified with site $i + N$ (periodic boundary conditions), yielding a finite system τ_N distributed over N sites and consisting of $L = N^Q$ state points (where Q is the number of states available to each site).

Let η_N be the ergodicity of τ_N. The specific ergodicity of τ itself is the limit, if it exists,

$$\eta = \lim_{N \to \infty} \eta_N. \tag{21.2}$$

Except for a few trivial cases (identity transformation, Bernoulli shifts, etc.) it isn't at all obvious that limit (21.2) should exist. Moreover, if the number of dimensions is greater than one, there are in general different wraparound aspect ratios (e.g., "squarish" vs "long and thin") that yield the same number N of sites, so that a limit, if it exists, may depend on the aspect ratio itself.

We have numerically evaluated the first few terms of the sequence (21.2) for a number of simple cellular automata in one and two dimensions; this was done by randomly sampling the state space and explicitly following the orbit of each sample point until it closed on itself. In all cases, numerical convergence was rapid enough to suggest the existence of a definite limit;[4] moreover, in the two-dimensional cases, the dependence on the aspect ratio was negligible. We'll present here the evidence for a few representative cases, and conclude with a brief discussion.

All the cellular automata considered here are *lattice gases*, where 0 and 1 represent, respectively, the absence or the presence of a particle on a given spacetime track. The one-dimensional ones use the lattice of Fig. 21.4a; the two-dimensional ones, that of Fig. 21.4b.

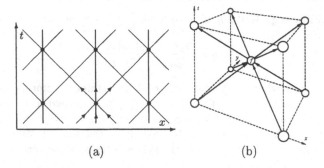

(a)   (b)

Fig. 21.4. Lattices used for the experiments, (a) in one dimension and (b) in two dimensions.

Fig. 21.5 shows a case (FREDS[5]) where the specific ergodicity appears to be close to zero (and indeed can be proved to be zero). The same figure shows two cases where η_N goes up rapidly[6] and appears to have an asymptotic value close to unity.

A more interesting case is illustrated by the rule ROTLR of Fig. 21.6, where η seems likely to settle on a value ($\approx 1/3$) noticeably different from 0 or 1. In this rule, the three inputs are fed straight through or circularly permuted depending on their overall parity.

Fig. 21.7 illustrates the results for two two-dimensional cellular automata discussed in [9]. BBM implements the billiard-ball model of computation [8], and is

---

[4]The experiments where performed using a personal computer running for several days. In view of the exponential complexity of the problem, even moderate improvements in the numerical estimates would require a drastic increase in computing power.

[5]The collision rule embodies the three-input/three-output Fredkin gate [8], with the control line running along the middle track ("rest particle") of each node. This rule is particle-conserving.

[6]FREDR is a variation of FREDS: after performing the Fredkin-gate operation, the three outputs are circularly permuted; RANDC is an invertible rule that was chosen at random, and is not particle-conserving.

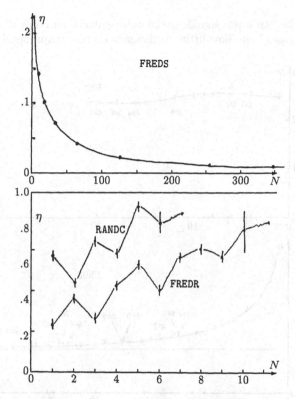

Fig. 21.5. Cases in which η appears to be close to 0 (FREDS) or to 1 (RANDC and FREDR).

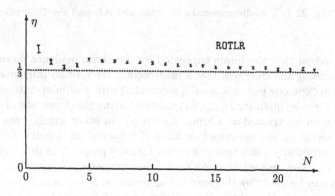

Fig. 21.6. In the ROTLR rule, η_N seems likely to settle on a value close to 1/3.

particle-conserving; TRON implements a synchronization scheme for asynchronous computation: the four input signals are complemented if and only if they are all the same. In both cases, note how little η_N depends on the wraparound aspect ratio.

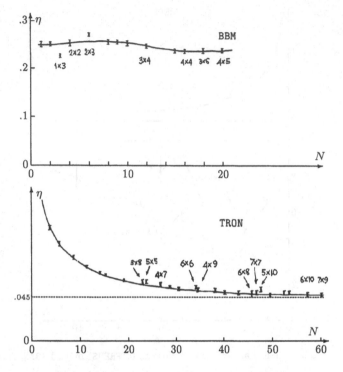

Fig. 21.7. Two-dimensional cases: the BBM rule and the TRON rule.

To the extent that the limits suggested by the above evidence do indeed exist[7] we are witnessing the emergence, at a macroscopic level, of an *intensive* quantity η (measured in "entropy per unit area"), associated with a cellular-automaton *rule* τ, out of the *extensive* quantity H_{phase} (a quantity having the dimension of entropy and associated with τ_N treated as a lumped system). In other words, even though the individual terms η_N are measured on lumped "objects" of definite size and shape, the specific ergodicity η emerges as a computational property of the "material" that makes up a cellular automaton. As far as that property is concerned, the cellular automaton can be sold "by the yard". For instance, if one wants to synthesize an n-bit counter within the cellular automaton, one will have to purchase at least n/η bits of cellular automaton "stuff".

---

[7]See [10] for a little more discussion; the theoretical questions here are similar to those encountered when studying the *topological entropy* of a dynamics [11].

21.5 A Kafkian scenario

We have seen in §21.3 how it may be possible to associate to a computing network a quantity C representing its *density of computation capacity*. Suppose the texture of the network, and thus its macroscopic local properties, vary smoothly over space and time. Then we can view C as a function of x and t. What aspect of network behavior will the knowledge of $C(x, t)$ allow us to predict?

Let us introduce a somewhat Kafkian scenario [12]. Imagine a large bureaucracy (say, an insurance claims department) housed in an office complex. Each room contains a desk, a filing cabinet, a typewriter, an IN/OUT tray, and a staff person. Claims are submitted at an input window. A claim folder will move from room to room through different processing stages, growing and shrinking as evidence is accumulated and digested. Eventually, the claim response will be available at an output window. Through directives, office traditions, and instructions printed on the back of the forms, a folder more or less "knows" what step is due next at any particular stage of processing, and will "seek" appropriate available resources. If a clerk is on vacation his duties may be taken over by an adjacent office, or folders may pile up in his IN tray. Given the large amount of similar work, much parallel processing is present; for example, several rooms may be devoted to time-stamping incoming correspondence. Though at any moment different rooms may be assigned to different tasks, with a little remodeling and retraining both rooms and people are basically interchangeable. As a folder moves through the bureaucracy, its rate of progress $d\tau/dt$ (where τ denotes the processing stage the folder has reached, or the case's "proper time") depends on the local availability of resources. Reassigning a case to a new office if the primary venue is too busy will require work and time and will slow down the case's progress.

The spacetime path of a folder through the bureaucracy will be a goal-oriented random walk. Basically, though the local details are somewhat impredicatable, the folder will tend to progress toward its destination. Similar claim folders may take different times to reach their completion, depending on the congestion of the offices they go through and the accidents they encounter on their path. Suppose, now, that for a productivity study we take a collection of similar cases, submit them at the same time t_0, and check which ones will appear completed at the output window at precisely time t_1. (Some cases will have been completed earlier than t_0 and some will be completed later; these are ignored.) For each case we trace its spacetime trajectory, obtaining a bundle of trajectories as in Fig. 21.8a.

We repeat the experiment the following year (Fig. 21.8b) in similar circumstances, but this time there has been a flu epidemics (indicated by the shaded area) affecting a number of contiguous rooms for a few days. Now, we shall state the basic *Bureaucracy equivalence principle*:

As far as *typical office work* is concerned, a section with a flu epidemics (or pregnant staff members, or preparing for a soccer match, or chronically overloaded, and so forth) is behaviorally *equivalent* to an unaffected section, except for an *appropriate slow-down factor.* (In other words, from a macroscopic viewpoint, the precise details of the disturbance wash out, and all that remains is not a *different kind* of computational resources, but just a *lesser amount.*)

If we look at the distribution of trajectories that *happened* to process a case from beginning to end in the interval (t_0, t_1) we'll find that, in the flu scenario, fewer made it through in that time. Of those which did, most tended to follow a somewhat curved path like in Fig. 21.8b. The reasons for this "deflecting force" are clear. Folders whose zig-zag path passed through the flu epicenter got delayed; only a few rare ones that managed to get stamped between two sneezes made it through by t_1. Folders that strayed far away from the epicenter had to travel a much longer distance, and only few were able to make it through by t_1. There remain those which barely kept clear of the flu area, wasting a little time by this detour but traversing full-efficiency offices all the time.

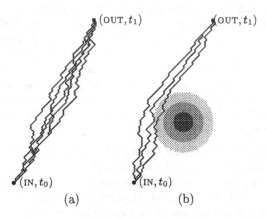

Fig. 21.8. Drift of claim folders through a claims department. Only claims filed at time t_0 and completed at time t_1 are sampled here. In (b), the shaded area had a flu epidemics and had a lower work performance; cases that happened to skirt the affected area had relatively better chances to make it by time t_1.

We can even propose a phenomenological law that "the flu center exerts a gravitational force on the distribution of folder trajectories" and study the dependence of this force on distance. Eventually, we'll be able to explain this phenomenological law through statistical arguments, and show that it is but the expression of a *variational principle.* Let us observe that, for any specific folder path, the amount of effective evolution of the folder from submission to response, which by definition is a constant K (a case will not come out to the output window until its processing is completed) cannot be more than the total amount P of processing it *could* have received along that path, that is,

$$P \equiv \int_{t_0}^{t_1} \left(C(x,t) - W(\dot{x}) \right) dt \geq K, \qquad (21.3)$$

where $C(x,t)$ denotes the density of computation capacity available in office area x on day t, and $W(\dot{x})$ denotes the computation capacity wasted in moving a folder from office to office at a speed $\dot{x}$ (cf. [13]). If we consider all possible paths satisfying inequality (21.3) as equally likely, we'll have a distribution of paths with P bounded above and below,

$$\max P \geq P \geq K,$$

where the max is over all paths from (x_0, t_0) to (x_1, t_1). If the peak of this distribution is in the interior of the interval given by the above bounds, P will be stationary at the peak. If the peak is sharp, then $P_{\text{peak}} \approx P_{\text{mean}}$, and the path for which P is stationary will be a good representative of the entire distribution.

By the same token, if we assume that our universe is run as a "numerical simulation" on a fine-grained parallel processor with fixed texture and a finite density of computation capacity C_0, then we have a qualitative understanding of why a planet known to start at (x_0, t_0) and to arrive (x_1, t_1) will go on a curved orbit around the sun instead of making a straght line. Near the sun, the underlying computational medium is busy simulating the sun's gravitational potential ϕ, and the computational capacity that remains available for simulating a planet is reduced from the base value C_0 by a factor $e^{-\phi}$. For the planet to go too far from a straight path in an attempt to avoid the vicinity of the sun will divert good computational resources into data movement, leading to a point of diminishing returns. Overall, the simulation will take a planet on the path that will provide it with the largest effective amount of "computational services".

Note the continual switching of meaning between 'path' intended as a microscopic history, and 'path' intended as the mean or the mode of a distribution of microscopic histories. To go back to Lamarck and Darwin, we do the same switching when we say that "as the giraffe strove to reach higher leaves, its neck got longer." The giraffes₁ (individuals) were indeed striving to reach higher leaves; some, especially those that had a slightly longer neck, managed more often, fed better, and preferentially transmitted their longer necks to their progeny. The neck length of the giraffe₂ (gene pool distribution) indeed got longer. Though the phrase in quotes, which mixes giraffe₁ and giraffe₂, is strictly speaking a *non sequitur*, it is a convenient abridgement of a valid combinatorial tautology.

Coming back to our story, if "large amount of computation capacity" is defined as "large number of different things (such as individual paths) that can be done," then it is a matter of tautology that a path ensemble will "strive to go" where there is the most "free computation capacity." This is, of course, exactly the same kind of tautology that we use when we say that "heat flows from T_a to T_b when $T_a > T_b$."

21.6   Along the infinite corridor

The "bureaucracy equivalence principle" introduced in the previous section qualitatively corresponds to the equivalence principle of general relativity. We'd like now to investigate in a more quantitative way the amount of computational resources diverted for moving data around (cf. the W term in (21.3)); as we shall see, this quite closely corresponds to the *Lorentz invariance* of special relativity.

We'll introduce another scenario. Along one of MIT's "infinite corridors" [14] there lies a one-dimensional cellular automaton: basically, an indefinitely extended row of identical small computers, or "cells," which communicate only with their immediate neighbors. Each unit has a left and a right input and a left and a right output; at every clock pulse a new value for the outputs is computed as a function ϕ of the current value of the inputs; this function is the same for all cells. Viewed in spacetime, the cellular automaton is but an indefinitely iterated combinational network, as in Fig. 21.3. The activity of a cell through a clock cycle is represented by a logic *gate* ϕ, which can be thought of as spacetime event. Two signals flow into an event, interact through ϕ, and two new signals flow out. To simplify the story, we shall assume that signals travel at lightspeed ($c = 1$) between events, and that they flow through an event in a negligible time.

Our scenario involves Jane—a math professor—and Bob—a computer science student. Jane needs some computations done using the cellular automaton, and asks Bob to write a program for this task. The cellular automaton lies along the corridor on which Jane's office is located. Whenever Jane wants to run the program p on a given argument a she injects the data string pa (the concatenation of p and a) into that portion of cellular automaton that lies just outside her door (Fig. 21.9). From there, computational activity spreads right and left through the cellular automaton as time advances. When the result $b = F(a)$ is ready it will be displayed in the same place (the result will be recognized as such by a distinguished pattern, such as a "result ready" flag). We assume that the sizes of program and argument are negligible with respect to the extent of the computation both in space and time; it is thus immaterial whether data injection is serial or parallel, and we shall treat input injection as a localized event having no spacetime extension. Likewise for output extraction (Fig. 21.9). Clearly, only the portion of the network that is in the absolute past (backward light cone) of the result can affect the latter. Likewise, if the cellular automaton starts in a blank state (except for program and argument), or the program is able to ignore any preexisting data contained in it, then only the portion of the network that is in the absolute future (forward light cone) of the argument can be relevant to the computation. This is suggested by the diamond shape in Fig. 21.9.

Jane waits outside her door for the result. As soon as this appears, she looks at the campus clock on the corridor's wall and records the time t elapsed since the beginning of the computation.

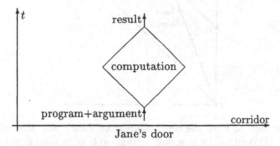

Fig. 21.9. A computation consists of injecting program and argument into the cellular automaton, and extracting the result at a later time. If input and output strings are short with respect to the spacetime extent of the computation (indicated by the diamond), input and output can be treated as localized, pointlike events.

Jane is demanding; she keeps asking Bob to improve the running time of the program. Finally, Bob produces a provenly *optimal* program, guaranteed to give the result in the shortest possible time, t_0.

Now, when she's thinking, Jane likes to pace along the corridor at a speed β; after injecting the argument a she'd like to go on pacing rather than having to wait in place. Thus, she asks Bob to revise the program so that the output will appear wherever she is at that moment, i.e., on the worldline $x = \beta t$ rather than on the line $x = 0$.

What Bob does, as a first try, is add to the program a little "tag" routine: the result b is produced as before on the line $x = 0$ by program p, but as soon as this result is ready the routine takes over and makes the result shift at light speed toward Jane until it reaches her, as illustrated in Fig. 21.10.

Of course, in this way the result does not reach Jane as soon as it is ready, but a a little later, namely, at a time

$$t_{\text{tag}} = \frac{t_0}{1 - \beta} = (1 + \beta + \beta^2 + \cdots)t_0.$$

Jane is not amused: "I want you to write a *new* program, one that aims the result directly along my worldline," she says, "so that I will see it as soon as it is available." Bob obliges, and produces a program p_β that has the same input/output behavior as p, except that the result appears on the $x = \beta t$ line. "Is this the best you can do?" asks Jane. "The time t_{soft} at which the result appears now, though shorter than t_{tag}, is still longer than t_0." Bob, however, claims that also the new program is optimal; the slowdown seems to be a necessary consequence of the constraint $x =$

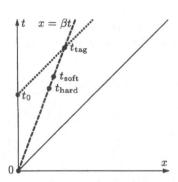

Fig. 21.10. Three ways to have the result appear on Jane's worldline $x = \beta t$ rather than on the line $x = 0$. (1) Add to the program a "tag" so that as soon as the result is obtained (at time t_0) the computation switches to a "shift right" mode which moves the result rightwards at light speed (dotted line) until it encounters Jane's line at a time t_{tag}. (2) Modify the program, possibly in a nontrivial way, so that the result will appear on Jane's worldline at the earliest possible time, t_{soft}. (3) Leave the program unchanged, but slide the whole hardware at a speed βt; according to special relativity, the result will appear at time $t_{\text{hard}} = t_0/\sqrt{1 - \beta^2}$.

βt. "Nonsense," says Jane, "there must be something wrong with your program!" Frustrated, Bob hits upon a brilliant idea. He goes back to the original program p, but, as Jane walks along the corridor at speed β, he pushes the whole cellular automaton along at the same speed: now the result will certainly be at the right place at the right time.

Surprisingly, things do not work quite that way! The result takes a time t_{hard} that is perhaps a bit shorter than t_{soft}, but still longer than t_0 (Fig. 21.10). Bob consults with Karl, a physics students. "Of course," says Karl, "you'll get a longer time; this is an elementary case of relativistic slowdown. In fact, your t_{hard} must equal $t_0/\sqrt{1 - \beta^2}$." Measurements are made, and that is indeed the case.

To sum up, we have a program p for the given cellular automaton, which delivers the result $b = F(a)$ at the same place where the argument was injected ($x = 0$) at some time t_0. If we want the *same* result delivered along Jane's worldline $x = \beta t$ we can either modify the *software*, that is, run a new program p_β on the stationary cellular automaton, or run the old program p on a different *hardware*, that is, on the cellular automaton pushed at speed β. In both cases the result is obtained at a somewhat later time, i.e., at times t_{soft} and t_{hard} respectively. Though for small β the slowdown in the two cases is comparable, the reasons for it appear to be very different. In the hardware solution, we run the same program on a *physically modified hardware*, and as a consequence we have a physical effect, accounted for (though not really explained) by special relativity. In the software solution, in order to produce a moving rather than a stationary result we run a *different program* on the same hardware, and this may well entail a different computational budget.

Are the two reasons *really* different? In other words, couldn't relativity itself be, at bottom, a way to describe how to reprogram the *same* underlying microscopically-grained hardware for different effects? In this case we wouldn't be surprised to arrive at similar resource tradeoffs in the two situations. Lets' take a closer look at the picture.

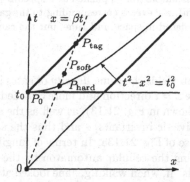

Fig. 21.11. The locus of the point P_{soft}, as β ranges from 0 to 1, must lie in the region bounded by the light cone with vertex at the origin ($t = x$), the constant line $t = t_0$, and the locus of P_{tag} (or the light cone with vertex at P_0). As we shall see, for almost all programs it will lie above the locus of P_{hard} (the relativistic hyperbola with apex at P_0), and will tend to be close to it.

With reference to Fig. 21.11, from relativity we know what must be the curve described by the point P_{hard} as β goes from 0 to 1, namely, the hyperbola $t^2 - x^2 = t_0^2$. Let us derive some bounds for the curve described by P_{soft}. (a) The point P_{soft} must lie inside the light cone (i.e., above the lightlike line $x = t$), because that is where the causal consequences of program and argument (injected at the origin) are felt. (b) It can't lie below the line $t = t_0$, described by P_{tag}, because Bob proved that t_0 is optimal in the absence of constraints. (c) Finally, it can't lie above the line $t = t_0 + \beta t$, because the original program followed by the tag routine supplies a solution with that value of t. The border of the bounded area is indicated by a thick solid line in the figure; note that also the relativistic hyperbola fits well within these bounds.

Let us look first at the P_{hard} case, in which the entire cellular automaton is pushed at a speed β. In this case the metric of the spacetime lattice of Fig. 21.3 is distorted, as shown in Fig. 21.12, but its topological structure (which nodes communicate with which) is not altered. Thus, the causal chaining of events is invariant: for the same program and argument, homologous arcs will carry identical signals, and of course the result will be the same in the two cases.

Moreover, even though position and timing of events will be different, the hard-

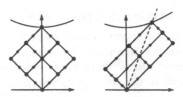

Fig. 21.12. When the cellular automaton is pushed at a speed β (indicated by the dashed line in the spacetime diagram), the events corresponding to the gates of the combinational network of Fig. 21.3 are displaced in space *and* time, but the causal chaining of events is unaltered.

ware solution of Fig. 21.13b yields the same *density* of gates in spacetime (the gates are just less crowded in the $x = t$ direction and more crowded, by the same factor, in the $x = -t$ direction, as shown in Fig. 21.13), as well as the same *spacetime volume* (both quantities are relativistic invariants)—and thus the same overall *number of gates* as the static (P_0) case of Fig. 21.13a. In terms of fungibility of computational resources, if we were leasing the cellular automaton by the hour we couldn't care less if Jane put it on wheels. If, when walking, Jane looked at her own watch instead of the corridor clocks, she would see the result $b = F(a)$ come in a time t_0, as in the static case. It's only when Jane tries to make the machine do double duty—as a telegraph line between different corridor locations as well as a computer—that she loses on one count what she gains on the other.

Let us now look at the P_{soft} case (Fig. 21.13c), in which we run our computation on the same network texture as in the static case, but running a different program and using a different swatch of spacetime, indicated by the dashed outline. We have also indicated, within the dashed area, a solid area of the same size and shape as that used in the P_{hard} case. This spacetime volume contains the same number of gates as in the P_0 and P_{hard} cases, but they are arranged in a less favorable aspect ratio (more depth and less width, corresponding to more emphasis on serial than parallel processing). In this degraded computational context, it is not surprising that a little more spacetime volume may be needed to complete the task, stretching the result as late as P_{soft}. This argument proves[8] that

$$P_{\text{soft}} \geq P_{\text{hard}}. \tag{21.4}$$

Can equality ever be attained in (21.4)? This is another interesting story. Imagine that what travels on the lines of the network (cf. Fig. 21.3) is *tokens*, the 0 state corresponding to the absence of a token and the other $n - 1$ signal states corresponding to tokens of different kinds. A spacetime patch with all 0s in it will then

---

[8]This is, of course, an 'almost always' result. In vanishingly rare computational tasks the new form factor may be an asset rather that a liability.

correspond to empty space. Jane's requirement that Bob's program be optimal entails, of course, that the computation space will be far from empty (using a density of 0s greater than $1/n$ will waste information capacity). On the other hand, if Jane insisted instead on a simple, orthogonal, well-structured program, one that doesn't take advantage of any ad hoc tricks just because they are available (cf. [15]), then the program will most likely leave plenty of empty space. In such a *low-density* limit it can be proved [10] that one can always go from a program p to an equivalent program p_β (cf. Fig. 21.10) by means of a straightforward *transliteration* process, and that the resulting program will make P_{soft} coincide with P_{hard} (equality in (21.4)); moreover, the transliteration process itself consists in just subjecting some of the program variables to a *Lorentz transformation* by a velocity β. In the low-density limit, therefore, the hard and the soft approach are completely equivalent, and computation obeys the same rules of special relativity as physics does. If the density of tokens ("energy density" in the network) is not close to zero, the behavior of this computational model begins to deviate from special relativity—but doesn't physics do that too?

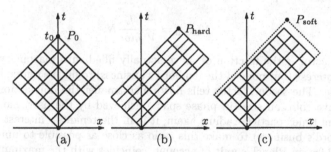

Fig. 21.13. Spacetime volume and computational network texture used in the original computation (a), the "hardware fix" (b), and the "software fix" (c). In the limit of low density of utilization of computational resources, the P_{hard} and P_{soft} solutions can be made to coincide.

21.7 Action and Lagrangian

There are many leads that point to action as the right kind of physical *dimension* to measure "computation capacity".

Suppose somebody is in the business of leasing suitable pieces of spacetime for computational use. The "tenant" shall rent an "apartment" for a time T, and fill the leased volume V with physical machinery in any way he pleases, trying to get the most computation out of his real estate. Consider, for example, a "billiard-ball" model of computation [8] where logic signals (cf. Fig. 21.3) are represented by particles in free motion and logic interactions are realized by collisions between particles. It is reasonable to use the number of collisions as an indicator of the amount of computation performed. If a particle hits a boundary of the leased volume, it is the duty of the landlord to send it back in, since otherwise this particle will intrude into somebody else's real estate and disrupt their computation ("landlord shall assure privacy!"). The lease contract will specify not only V and T, but also what pressure P the walls must withstand: an apartment with harder walls will rent for more. Simple additivity arguments show that an open market rental fee must be of the form PVT, which has the dimensions of *action*. What the tenant actually pays for is a number

$$N = PVT/h$$

(h is Planck's constant) of "unit action cells". How many collisions can he achieve?

Suppose the tenant fills his real estate with n particles of radius r and mass m, traveling at speed v. The pressure will be

$$P = nmv^2/V,$$

and the number of particle collisions will be

$$N_{\text{actual}} = \frac{r^3}{V} \frac{h}{mvr} N.$$

The first factor is the fraction of space actually filled with machinery; it is in the tenant's interest (but none of the landlord's business) to fill the rented space as fully as possible. The second factor tells how much smaller the unit action cell is than the effective collision size (in phase space) achieved by the user, namely, particle momentum times particle radius; again, it is in the tenants's interest (but none of the landlord's business) to make this ratio as close as possible to unity. Thus the fair rental fee, in Planck's unit of account, coincides with the maximum number of collisions the tenant can achieve.

Another argument, due to Margolus and Levitin [16], runs as follows. The average energy E of a quantum system coincides with the maximum rate of dynamical evolution, i.e., with the rate at which the system can successively go through a sequence of *orthogonal* states. Thus the quantity $N = ET$, expressed in Planck's units of account, is the maximum number of distinct states that can be touched during a computation. Since some quantum states may be strung by the dynamics into orbits much shorter than N (say, an orbit of period 3), the average number of distinct states touched by the computation will be less than N. It will be the user's responsibility to design a dynamics with high enough ergodicity (cf. §21.4) to take full advantage of the energy invested in it—a computer that will not enter a cycle too soon.

However, dimensional arguments of this kind are too generic for our purposes. We would like to argue not only that action, as energy×time, is the right yardstick to measure 'amount of computation' with, but that a very specific quantity having the dimension of action, namely, that obtained by integrating a system's *Lagrangian* over time, corresponds to the amount of computational work performed by the system during its time evolution. In turn, we'd like to define this 'computational work' as the log of the *number of different behaviors* that the system could have exhibited—just as the information content of a message is defined as the log of the number of different things that the message could have said. While entropy measures uncertainty *as to state*, we'd like to argue that (Lagrangian) action measures uncertainty *as to law*. Thus, instead of counting states we will be counting laws. The next section is meant for flexing our muscles in this respect.

One might say that the job is already done. In fact, Feynman's theory of path integrals in quantum mechanics [17] provides a satisfying mathematical model of why a classical trajectory should satisfy a principle of stationary action: action affects quantum phase and thus quantum amplitude, quantum amplitudes reinforce or cancel one another according to their phases, the square of the overall amplitude gives a probability, and a sharply peaked probability gives a classical trajectory. The implication of this approach, however, is that the variational principles of classical mechanics can be explained *only* after the introduction of quantum mechanics (which, by the way, is still in search of an "explanation"). Though it may turn out that quantum mechanics is an essential element of the explanation, I rather suspect that the connection between variational principles and counting arguments (which is so solidly established in statistical mechanics) is much more general than that, and may hold at least in a qualitative way in analytical mechanics without bringing in quantum mechanics. Richard Feynman himself, even after having stalked action at closer distance than any other mortal through his theory of path integrals, would still say "I don't know what action is". Apparently, there are more veils to be lifted. By contrast, today we can confidently say that "we know what entropy is."

21.8 $T = dS/dE$ holds for almost any system

Statistical mechanics gets distinguished macroscopic properties (e.g., equilibrium parameters) by applying counting arguments to ensembles of *microscopic states*. Here we give an example of how general properties of analytical mechanical laws may arise by applying counting arguments to "ensembles" of *microscopic laws*.

In re-reading Arnold [18], I started wondering about the "philosophical meaning" of the following. Consider a continuous system with one degree of freedom. Let T be the period of a given orbit of energy E, and dS the volume of phase space swept when the energy of the orbit is varied by an infinitesimal amount dE (Fig. 21.14). As is well known, if the system is Hamiltonian these quantities obey the relation

$$T = dS/dE, \qquad (21.5)$$

Just *how surprising* is this fact?

One way to answer this question is to go to a more accurate physical level

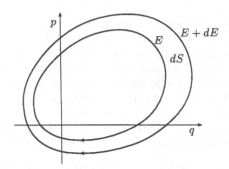

Fig. 21.14. In moving from an orbit of energy E to an infinitesimally near orbit of energy $E + dE$, one sweeps an area dS of the phase space. If the dynamics is Hamiltonian, the orbit period T equals the ratio dS/dE of action-to-energy variation.

(quantum mechanics) and see how energy and rate of evolution are related there (see [16]). Here instead our approach to the question is to consider systems of a more general nature, but for which quantities analogous to T, dE, and dS are still meaningful. Under what conditions will the above relation hold for these systems?

Take, for instance, the class $\mathcal{X}_N$ of all discrete systems having a finite number N of states and an invertible but otherwise arbitrary dynamics. Though continuous quantities such as those appearing in (21.5) may arise from discrete ones in the limit $N \to \infty$, in general relation (21.5) will not hold (or even be meaningful) for every individual system; however, if one considers the entire class, one may ask whether this relation holds approximately for most systems of the class. Alternatively, one may ask whether this relation holds for a suitably-defined "average" system—treated as a representative of the whole class. This kind of approach is routinely used in statistical mechanics;[9] in our context, however, statistical methods are applied to "ensembles" in which the missing information that characterizes the ensemble concerns a system's *law* rather than its initial *state*.

We shall show that relation (21.5) holds for the average element of the "ensemble" $\mathcal{X}_N$. Now, this *is* surprising, as we thought we hadn't told $\mathcal{X}_N$ anything about physics!

The systems of class $\mathcal{X}_N$ have very little structure—basically, just invertibility. Nonetheless, one can still recognize within them the precursors of a few fundamental physical quantities. For instance, the period T of an orbit is naturally identified

---

[9]For example, given a canonical ensemble for a system consisting of an assembly of many identical subsystems, almost all elements of the ensemble display a subsystem energy distribution that is very close to the Boltzmann distribution; the latter can thus be taken as the "representative" subsystem-energy distribution, even though hardly any element of the ensemble displays that distribution *exactly*.

with the number of states that are strung along the orbit. Likewise, a volume S of state space, which in our case is an unstructured "bag" of states, will be measured in terms of how many states it contains. It is a little harder to identify a meaningful generalization of energy; the arguments presented at the end of this section suggest that in this case the correct identification is $E = \log T$, and this is the definition that we shall use below. Armed with the above "correspondence rules," we shall investigate the validity of relation (21.5).

Each system of $\mathcal{X}_N$ will display a certain distribution of orbit lengths; that is, one can draw a histogram showing, for $T = 1, \ldots, N$, the number $n(T)$ of orbits of length T (see Fig. 21.15).

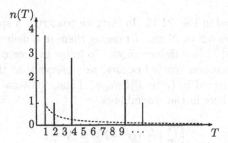

Fig. 21.15. Orbit-length histogram of one particular system. The dashed curve gives the average histogram over the entire set of $N!$ systems.

If in this histogram we move from abscissa T to $T + dT$ we will accumulate a count of $n(T)\, dT$ orbits. Since each orbit contains T points, we will sweep an amount of state space equal to $dS = T\, n(T)\, dT$; thus

$$\frac{dS}{dT} = T\, n(T).$$

On the other hand, since $E = \log T$,

$$\frac{dT}{dE} = T;$$

hence

$$\frac{dS}{dE} = \frac{dS}{dT}\frac{dT}{dE} = T^2 n(T).$$

Therefore, the original relation (21.5) will hold if and only if the orbit-length distribution is of the form

$$n(T) = 1/T.$$

Do the systems of $\mathcal{X}_N$ display this distribution?

Observe that, as N grows, the number of systems in $\mathcal{X}_N$ grows much faster than the number of possible orbit-length distributions: most distributions will occur many times, and certain distributions may appear with a much greater frequency than others. Indeed, as $N \to \infty$, almost all of the ensemble's elements will display a similar distribution. In such circumstances, well-known theoretical and practical considerations recommend defining the "typical" distribution as the *mean distribution* over the ensemble, denoted by $\overline{n(T)}$.

It turns out that for $\mathcal{X}_N$ the mean distribution is *exactly*

$$\overline{n_N(T)} = 1/T \tag{21.6}$$

for any N, as indicated in Fig. 21.15. In fact, we construct a specific orbit of length T by choosing T states out of N and arranging them in a definite circular sequence. This can be done in $\binom{N}{T}\frac{T!}{T}$ different ways. To know in how many elements of the ensemble the orbit thus constructed occurs, we observe that the remaining $N - T$ elements can be connected in $(N - T)!$ ways. Thus, the total number of orbits of length T found anywhere in the ensemble is

$$\binom{N}{T}\frac{T!}{T}(N - T)! = N!\frac{1}{T}.$$

Divide by the size $N!$ of the ensemble to obtain $1/T$.

Thus, the typical discrete system obeys relation (21.5). Intuitively, when N is large enough to make a continuous treatment meaningful, the odds that a system picked at random will closely obey (21.5) are overwhelming.

Why $E = \log T$

Here we motivate the choice $E = \log T$ made above.

Finite systems lack the rich topological structure of the state space found in analytical mechanics. Beside invertibility, in general the only *intrinsic*[10] structure that they are left with is the following: *Given two points a and b, one can tell whether b can be reached from a in t steps; in particular (for $t = 0$), one can tell whether or not $a = b$.* Thus, for instance, one can tell how many orbits of period T are present, but of these one cannot single out an individual one without actually pointing at it, because they all "look the same".

To see whether there is a quantity that can be meaningfully called "energy" in this context, let us observe that physical energy is a function E, defined on the state space, having the following fundamental properties:

---

[10]That is, independent of the labeling of the points, and thus preserved by any isomorphism.

1. *Conservation*: E is constant on each orbit (though it may have the same value on different orbits).

2. *Additivity*: The energy of a collection of weakly-coupled system components equals the sum of the energies of the individual components.

3. *Generator of the dynamics*: Given the constraints that characterize a particular class of dynamical systems, knowledge of the function E allows one to uniquely reconstruct the dynamics of an individual system of that class.

The proposed identification $E = \log T$ obviously satisfies property 1.

As for property 2, consider a finite system consisting of two independent components, and let a_0 and a_1 be the respective states of these two components. Suppose, for definiteness, that a_0 is on an orbit of period 3, and a_1 on one of period 7; then the combined system state (a_0, a_1) is on an orbit of length 21, i.e., $\log T = \log T_0 + \log T_1$. This argument would fail if T_0 and T_1 were not coprime. However, for two randomly chosen integers the expected number of common factors grows extremely slowly with the size of the integers themselves [19] (and, of course, the most likely common factors will be small integers); thus the departure from additivity vanishes in the limit $T \to \infty$.

As for property 3, an individual system of $\mathcal{X}_N$ is completely identified—up to an isomorphism—by its orbit distribution $n(T)$, and thus any "into" function of T (in particular, $E = \log T$) satisfies this property.

21.9 The power of conditioning

Much as in thermodynamics a "state"—i.e., a *macroscopic* state—is in fact a bag of *microscopic* states, we shall entertain the notion that, in analytical mechanics, a "trajectory" of our dynamics is really a bundle of microscopic trajectories of an underlying, fine-grained dynamics. How does the size of this bundle depend of what we know about the (macroscopic) trajectory?

Let us consider a particle that at every tick of the clock can move one notch right or left on a one-dimensional track. The particle is driven by a computer program. To be specific, there is a given computer with a large amount of memory in it; somebody initializes the memory and then lets the computer run. They fix their attention on a specific data bit. Every billionth cycle of the computer they look at the state of this bit: if it is a 1 they will move the particle to the right; if a 0, to the left. At time 0 we put the the particle at notch 0 and leave the room. At time T (say, some 10,000 ticks) we are asked, "Where is the particle now?"

How could we possibly know? We know the computer but we do not know what program it is running or what the initial data were. Actually, we know one hard fact: at time T the particle must be within the interval $[-T, +T]$, since it cannot move more than one notch at a time. Beyond that, what can we say? If hard pressed, we may hazard a guess that the particle is probably closer to the middle of the interval than to the very ends.

They let us go. At time $2T$ they are at it again. This time they tell us where the particle is now: at a certain position $2X$ (with, say, $X = 8000$); but they ask for the same information as before, namely, "Where was the particle at time T?" And, at gunpoint, "A bad answer, and you are dead!"

Let us think rationally. We could simulate the given computer on our laptop PC, trying out all the programs one by one. Perhaps, by the way the computer is designed, there are some positions where the particle can never be at time T no matter what program is running; that might help us narrow down the choice by a tiny bit. But the computer may have a billion bits, and so may be running any of $2^{1,000,000,000}$ programs; this is not a promising approach, since we'll certainly be dead before we've tried out even a small fraction.

Then we realize that, no matter how many *programs* there might be, the number of possible *paths* the particle may have followed from the start to time T is "only" 2^T. Forget the red herring of the $2^{1,000,000,000}$ programs: our chances of survival are no worse than one in 2^T, and T is in the thousands rather than in the billions. Even better, the number of *places* where the particle can be at T (rather than the *paths* from 0 to T) can only grow linearly with T—in fact, it is no more that $T + 1$. That is, our chances cannot possibly be worse that one in T ($\approx 2^{13}$)—even if the particle were equally likely to be in any of the possible slots. With a not too large T in the denominator, perhaps we can put some large constant in the numerator and manage to get an appreciable chance of survival. Let us do some figuring, but without assuming anything more than we know.

The worst scenario—and, by Murphy's law (or Jaynes's law [20])—the one that we owe it to ourselves to entertain unless we want to deceive ourselves, is that all particle paths compatible with our information are equally probable. Forget about the computer program, about which we do not know anything besides that it is large, and concentrate on the kinematic constraints. For a specific position x, how many paths go from event P_0 (particle at the origin at time $t = 0$) through event P_1 (particle at position x at time T) *and* event P_2 (particle at position $2X$ at time $2T$)?

In general, if the net progress during a time interval t is x, and this was done by x_+ steps to the right and x_- to the left, we must have

$$
\begin{array}{ccc}
x_+ + x_- = t, & & x_+ = (t + x)/2 = \frac{1+\beta}{2}t, \\
& \text{or} & \\
x_+ - x_- = x, & & x_- = (t - x)/2 = \frac{1-\beta}{2}t,
\end{array}
$$

where we have called β the average velocity x/t in that interval. Then, the number of paths is[11]

$$N(\beta,t) = \binom{t}{\frac{1+\beta}{2}t \ \frac{1-\beta}{2}t}.$$ (21.7)

Denote by $H(p)$ the **binary entropy** function

$$H(p) = -p\log p - q\log q,$$

that is, the entropy of the binary distribution $\{p,q\}$. This entropy can be rewritten more symmetrically in terms of the the *mean* $\mu = p - q$ of the distribution, as

$$K(\mu) = H\big(p(\mu)\big), \quad \text{where} \quad p = (1+\mu)/2.$$

Using Stirling's approximation, we get from (21.7)

$$\frac{1}{n}\log N(\beta,n) = K(\beta) + O\big(\tfrac{\log n}{n}\big);$$ (21.8)

the "big-oh" term in (21.8) vanishes as $n \to \infty$. Some salient features of $K(\beta)$ are shown in Fig. 21.16.

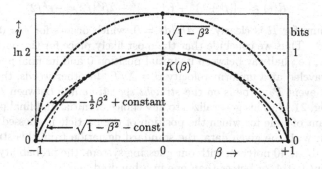

Fig. 21.16. The binary entropy $K(\beta)$ compared with the parabola osculating it at the apex, and with the relativistic factor $\sqrt{1-\beta^2}$ (a half-circle).

If we call β the average velocity of the overall trip (from 0 to $2T$), β_1 that of first lap (from 0 to T), β_2 that the second lap (from T to $2T$), and ϵ the excess of β_1 over β, we have

$$\beta = X/T, \qquad \beta_1 = \beta + \epsilon,$$
$$\text{and}$$
$$\epsilon = (x - X)/T, \qquad \beta_2 = \beta - \epsilon.$$

[11]If $n = h + k$, we may write $\binom{n}{h\ k}$ for $\binom{n}{k}$ to stress the symmetry between h and k.

Coming back to our question, the number of paths from P_0 to P_2 *through* P_1 is

$$N_{012} = N_{01}N_{12} = N(\beta_1, T)N(\beta_2, T)$$
$$= N(\beta + \epsilon, T)N(\beta - \epsilon, T),$$

while that from P_0 to P_2 (with no restrictions on intermediate points) is

$$N_{02} = N(\beta, 2T).$$

Thus, the proability $\mathcal{P}$ that the particle went through P_1, *conditioned* by the knowledge that it started at P_0 and ended up at P_2, is

$$\mathcal{P} = \frac{N_{01}N_{12}}{N_{02}} = \frac{N(\beta + \epsilon, T)N(\beta - \epsilon, T)}{N(\beta, 2T)}.$$

Taking logs and dividing by the number of steps as in (21.8) we obtain a quantity $R(\epsilon)$ proportional to the log of the relative frequency of the paths going through x at T, namely,

$$R(\epsilon) = K(\beta + \epsilon) + K(\beta - \epsilon) - 2K(\beta).$$

Now, it turns out that, for small (and even not *so* small) values of β,

$$K(\beta) \approx -\tfrac{1}{2}\beta^2 + \text{const} \tag{21.9}$$

(cf. Fig. 21.16). Thus, up to an additive constant,

$$R(\epsilon) \approx -[(\beta + \epsilon)^2 + (\beta - \epsilon)^2 - 2\beta^2]/2 = -\epsilon^2/2.$$

The maximum for R is clearly at $dR/d\epsilon = 0$, which occurs for $\epsilon = 0$, that is, for $\beta_1 = \beta_2 \equiv \beta$. Thus we conclude that the most likely place for the particle to be at time T is X, i.e., halfway between its initial position 0 and its final position $2X$, *as if* it had traveled at a uniform velocity $\beta = X/T$. In other words, the most likely position for event P_2 to be is on the *straight spacetime line* between P_0 and P_1, as shown in Fig. 21.17. This generalizes to any choice of initial and final positions, and of any instant of time for which the position of the particle is guessed (Fig. 21.18). Incidentally, with the given data, the standard deviation from this straight line at midtime is about 40 notches; with our guessing system, the probability to come out alive is about 0.014, or better than one in a hundred.

Note that we never suggested that the particle must have had some "momentum" that made it try to preserve its "speed", or that, since the particle moved to the right 9 times out of 10, the computer program, seen as a random-number generator, must have been one characterized by a certain biased "statistics". We did not make hypotheses about where the particle was when we were not seeing it; all we used was the little we were told, namely, where the particle was at $t = 0$ and $t = 2T$, and that it was hopping right or left one notch at the time (the latter is a crucial piece of information). Our prediction is the "flattest" that one could make out of the available data, and thus is just a *tautological rewording*—though an illuminating one—of these data.

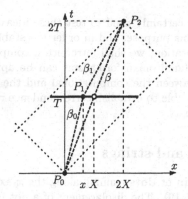

Fig. 21.17. For a particle that performs a symmetric random walk, starts at $P_0 = (0,0)$, and ends up at $P_2 = (2T, 2X)$, what is the spatial probability distribution at $t = T$? All paths are comprised within the forward light-cone from P_0 and the backward light-cone from P_2; thus, the range of positions at T is restricted to the solid line in the middle. The particle's mean velocity from P_0 to P_2 is $\beta = X/T$; the mean for the subpath P_0P_1 is β_1; that for P_1P_2, β_2. The requested distribution has both mean and maximum at $x = X$ (hollow dot).

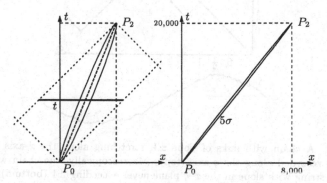

Fig. 21.18. For any given time t, the probability distribution for the particle to be at x has a mean indicated by the line of slope β and a standard deviation σ indicated by the width of the surrounding ellipse. The figure on the right uses the data from the example ($T = 10,000$, $X = 8,000$); even when showing a 5σ deviation, the width of the ellipse is barely noticeable.

21.10 On two levels[12]

You may think it impertinent that I took the random walk—no inertia, total dependence on external whims—as a model of the motion of a free particle governed

---

[12]This phrase is borrowed from Kadanoff [21].

purely by inertia. We certainly need to test these ideas with more realistic models of dynamics. To this purpose, and in order to establish a link between action and amount of computation, we shall introduce a computer-like system to which the Lagrangian and Hamiltonian formalisms can be applied verbatim. By moving back and forth between the computational and the physical interpretation of this system we will be able to establish correspondence rules between physical and computational constructs.

21.10.1   Chains and strings

Consider a linear **chain** of **dots** running along the x-axis with two dots for every unit of length (Fig. 21.19). The displacement of a dot from a horizontal reference line will be recorded along the q-axis. A dot is connected to its two neighbors by **links** of slope ± 1. Thus, on a macroscopic scale the chain will appear as a continuous string with slope in the x–q plane never exceeding ± 1.

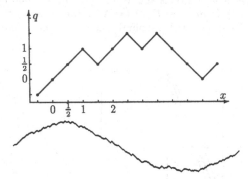

Fig. 21.19. A chain with links of slope ± 1, stretching along the x-axis and with displacements recorded along the q-axis (top). Macroscopically, the chain will look like a continuous string with slope in the x–q plane never exceeding ± 1 (bottom).

For brevity, coordinates having integer values (1,2,3, ...) will be called **even**; those having half-integer values $(\frac{1}{2}, \frac{3}{2}, \frac{5}{2} \dots)$, **odd**. We are interested in the class of chain dynamics obeying the following constraints (Fig. 21.20):

- At even times, the candidates for a move are the dots at even places (solid dots); at odd times, those at odd places (hollow dots).

- If a candidate can hop up or down one unit in the q direction while retaining links of slope ± 1 with its neighbors, then it is up to the specific dynamics to decide whether it will do so (**flip**). Otherwise, the candidate will remain in place (**rest**).

Thus a dynamics is just a table that, for any q, x, t that are all three odd or all three even, specifies whether the candidate dot is *permitted* to flip. Whether a dot will actually flip depends not only on whether the adjacent links allow it to flip, but also on whether the permission bit allows it to take advantage of this possibility.

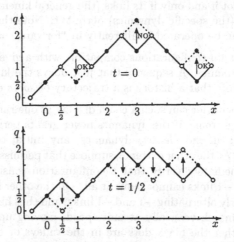

Fig. 21.20. Two successive configurations of a discrete chain with links of slope ± 1. At even steps, the even dots (solid) try to move; at odd steps, the odd dots (hollow). The arrows indicate the moves that are possible without breaking the links. A possible move will actually be made (the dot will **flip**) only if the dynamics gives permission; otherwise the dot will **rest**. In this figure, permission was not granted at $t = 0$ for the dot at $x = 3$, so that at time $t = \frac{1}{2}$ we find it at the same position.

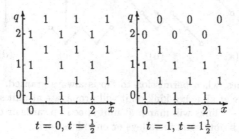

Fig. 21.21. At its finest level, the program of the chain computer is just a binary table containing a permission bit for each point (q, x, t) such that all three coordinates are even or all odd ("body-centered cubic lattice"). Each frame display, in an interleaved fashion, data for two consecutive values of t (even t at even q, x; odd t, at odd q, x). For $q < 2$, this dynamics always grants permission; for $q \geq 2$, it grants permission on two half-steps and then witholds it for two more half-steps.

One can think of the present setting as a special-purpose *computer*; the *data* are the states (±1) of the chain's links; the *program* is the permission table itself. That is, we view spacetime (here including for convenience q as another spatial coordinate) as a read-only memory containing a bit at each point. The *processor* is just the mechanism that interprets the data as a linked chain and the program as a dynamics, flipping a dot if and only if its links (the general kinematic constraints) and the permission bit (the specific dynamics) permit it. Note that this is a *reversible* computer, which can be operated indifferently in "forward" or "reverse".[13]

A sequence of chain configurations compatible with a given dynamics is a **trajectory** of that dynamics. A sequence that just obeys the kinematic links will be called a **history** (note that a history is a trajectory of *some* dynamics).

What kinds of behavior can such a chain display as one ranges over the possible dynamics? At one extreme, if the dynamics never grants permission, then no flips can possibly occur: in the *identity* dynamics, any initial configuration remains forever unchanged. At the other extreme, suppose that permission is always granted. Then one can immediately verify that a configuration consisting of all $+1$ links (Fig. 21.22a) or all -1 links cannot move at all. On the other hand, a configuration consisting of regularly alternating $+1$ and -1 links, and thus having an average slope of 0, will steadily march vertically at unit speed, moving upwards or downwards depending on whether the black dots are in the valleys or on the peaks at even times(Fig. 21.22c,d). In turns out that, for an arbitrary chain configuration, this permissive, "flip-whenever-you-can" dynamics gives exact *wave equation* behavior; this will be proved in §21.10.2, where we also discuss other dynamics that can be "programmed" in our chain computer.

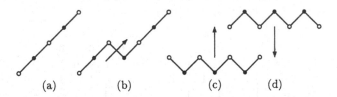

(a)   (b)   (c)   (d)

Fig. 21.22. Examples, when permission to flip is always granted: A chain with slope $+1$ (or -1) cannot move at all (a); a kink on it will propagate at unit speed (b). A chain with alternating $+1$ and -1 links will march at unit speed upwards or downwards—depending on whether the black dots start in valleys or on ridges (c,d).

In sum, we have defined a whole class of dynamics; the general format is given by the kinematic constraints (try to flip at even or odd places at, respectively, even or odd times, and only if the links allow it), while the specifics is given by the permission table (flip only if the permission bit is set).

---

[13]More than that, given appropriate interlocks it can go forward, backward, or idle independently at each point of the chain.

21.10.2   Means and flows

To get from a discrete chain to a continuous string, so as to get differentiable quantities which we can use in variational arguments, we must go to a *macroscopic level*; i.e., consider space averages over lengths that are large with respect to the lattice spacing, and time averages over a large number of steps. Ideally, coarse-grained averaging should commute with the dynamics; that is, if we average the microscopic state variables and then let this average evolve for a certain time using an appropriate macroscopic dynamics we should get the same result as if we had let them evolve over the same time using the microscopic dynamics on the microscopic data and then done the averaging.

At the microscopic level, knowledge of the *positions* $q(x)$ completely specifies the system's state; that is, the assignment of $q(x)$ at a given instant completely determines the subsequent microscopic evolution of the system under a given dynamics. More specifically, initial rate-of-change information—which dots are going to flip at the initial step—is not part of the initial specifications but is dictated by the dynamics itself.

The situation is different at the macroscopic level. As a smeared-out, macroscopic state variable, $q(x)$ does not capture enough information about the state of the system to uniquely determine its macroscopic evolution. In fact, as one can see in Fig. 21.22c,d, two chains with the same $q(x)$ may move one up and the other down! More macroscopic data than just $q(x)$, namely, the rates of change $\partial q(x,t)/\partial t$ (or some equivalent information) are needed to give a self-contained macroscopic dynamics, and, as we shall see, they are indeed sufficient.

Coming back to the microscopic level, let us fix our attention on one candidate dot (Fig. 21.20) and the unit-space cell surrounding it; this cell contains two links, one on the right and one on the left of the dot itself, whose values we shall call respectively $\rho_\rightarrow$ and $\rho_\leftarrow$. From these we construct, by symmetrization, the new quantities

$$\rho = (\rho_\rightarrow + \rho_\leftarrow)/2,$$
$$j = (\rho_\rightarrow - \rho_\leftarrow)/2. \tag{21.10}$$

On a unit-cell scale, the possible values for ρ and j are

ρ	$\rho_\leftarrow$ -1	$+1$		j	$\rho_\leftarrow$ -1	$+1$
$\rho_\rightarrow$ -1	-1	0	,	$\rho_\rightarrow$ -1	0	-1
$+1$	0	$+1$		$+1$	$+1$	0

.

Thus, along the chain, we have the two microscopic sequences of links, $\rho_\rightarrow(x)$ and $\rho_\leftarrow(x)$, and the two derived sequences $\rho(x)$ and $j(x)$; for example, for Fig. 21.20 we

have

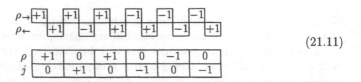

$$\tag{21.11}$$

Under coarse graining, $\rho(x,t)$ represents the average slope of the of chain (in the x–q plane) while $-j(x,t)$ represents its average velocity (the "slope" in the t–q plane):

$$
\begin{aligned}
\frac{\partial q}{\partial x} &= \rho = \frac{\rho_{\rightarrow} + \rho_{\leftarrow}}{2}, \\
\frac{\partial q}{\partial t} &= -j = -\frac{\rho_{\rightarrow} - \rho_{\leftarrow}}{2}.
\end{aligned}
\tag{21.12}
$$

To integrate the dynamics we have to determine the evolution of ρ and j, or, equivalently, or $\rho_{\rightarrow}$ and $\rho_{\leftarrow}$. It turns out that if permission to flip is always granted, then the two sequences $\rho_{\rightarrow}$ and $\rho_{\leftarrow}$ of (21.11) shift respectively rightwards and leftwards one position at every step *without interacting*. In other words, the dependency on x and t is such that

$$
\begin{aligned}
\rho_{\rightarrow}(x,t) &= \rho_{\rightarrow}(x - t), \\
\rho_{\leftarrow}(x,t) &= \rho_{\leftarrow}(x + t).
\end{aligned}
\tag{21.13}
$$

In this case, then, the sequence ρ can be thought of as the superposition of two traveling strings that move at unit speed in opposite directions. As we know, this is the general of the one-dimensional wave equation. Thus, we conclude that under this "permissive" dynamics *the string $q(x,t)$ strictly obeys the one-dimensional wave equation*

$$
\frac{\partial^2 q}{\partial x^2} - \frac{\partial^2 q}{\partial t^2} = 0.
\tag{21.14}
$$

on any scale, with no damping whatsoever.[14]

If permissions are assigned randomly and independently with a density $1 - s$, then, with reference to Fig. 21.23, where the two trains $\rho_{\rightarrow}$, $\rho_{\leftarrow}$ are shown running on opposite tracks, when two cars pass by one another they will swap contents with probability η. The result is that, for small η, these two quantities are related by

$$
\begin{aligned}
\partial_t \rho_{\rightarrow} &= -\partial_x \rho_{\rightarrow} - s(\rho_{\rightarrow} - \rho_{\leftarrow}), \\
\partial_t \rho_{\leftarrow} &= -\partial_x \rho_{\leftarrow} + s(\rho_{\rightarrow} - \rho_{\leftarrow}),
\end{aligned}
\tag{21.15}
$$

---

[14]Norman Margolus and I came upon this model around 1983. On a computer simulation, it is impressive to see the full interplay of elasticity and inertia emerge from such simple discrete primitives.

and thus ρ, j, and q all satisfy the *telegraph* equation

$$\partial_{tt}u = \partial_{xx}u - 2s\partial_t u,$$

which is essentially the wave equation with damping proportional to s. For fixed s, if we scale t by a and x by $\sqrt{a}$, then in the limit as $k \to \infty$ the telegraph equation turns into the *diffusion* equation,

$$\partial_t u = \frac{1}{2s}\partial_{xx}u.$$

Nonrandom tables, and especially tables with a fractal structure, need not lead to an emerging macroscopic dynamics. But in all cases we get a *linear* dynamics, since the $\rho_\rightarrow$ and $\rho_\leftarrow$ tokens of Fig. 21.23 are switched without even being looked at—they do not interact! Note that, in a measure-theoretical sense, *almost all* permission tables yield a uniform and independent distribution of permissions with density $1/2$. In this sense, the typical dynamics of our class is that of a heavily damped string.

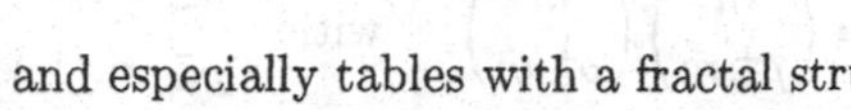

$$\rho_\rightarrow \boxed{+1} \rightarrow \boxed{+1} \rightarrow \boxed{+1} \rightarrow \boxed{-1} \rightarrow \boxed{-1} \rightarrow \boxed{-1} \rightarrow$$
$$\rho_\leftarrow \leftarrow \boxed{+1} \leftarrow \boxed{-1} \leftarrow \boxed{+1} \leftarrow \boxed{+1} \leftarrow \boxed{-1} \leftarrow \boxed{+1}$$

Fig. 21.23. The sequences $\rho_\rightarrow$ and $\rho_\leftarrow$ may be thought of as trains traveling in opposite directions at unit speed. A request to flip (in the original chain) corresponds to two train cars passing by one another; if the permission is denied, then the two cars swap contents, so that a bit that was traveling rightwards now starts moving leftwards and vice versa. This leads to scattering in the $\rho_\rightarrow$-$\rho_\leftarrow$ picture, corresponding to *damping* in the q picture.

21.10.3 Inertia of the lumped string

We shall now focus our attention on the harmonic string behavior, as supported by the permissive dynamics in our toy computer.

Let us close the string into a loop of length m ("periodic boundary conditions" $q(x+m) = q(x)$), and consider the position of its center of mass (on the q-axis) as a function of time. Whatever the configuration of the string, this point will move at a constant velocity β (since $\int j\,dx$ is strictly conserved) in the range $-1 \le \beta \le +1$, with $2n + 1$ discrete possible values

$$\beta = i/m, \quad \text{for } i = -m, \ldots -1, 0, +1, \ldots +m.$$

We will have

$$\begin{aligned} \rho_\rightarrow + \rho_\leftarrow &= 0, \\ \rho_\rightarrow - \rho_\leftarrow &= -2\beta, \end{aligned} \quad \text{or} \quad \begin{aligned} \rho_\rightarrow &= -\beta, \\ \rho_\leftarrow &= \beta. \end{aligned} \tag{21.16}$$

Of the m elements of the train $\rho_\rightarrow$, a fraction $\rho_\rightarrow^+$ will consist of $+1$s and the rest, $\rho_\rightarrow^-$, of -1s, that is,

$$\rho_\rightarrow^+ + \rho_\rightarrow^- = 1,$$
$$\rho_\rightarrow^+ - \rho_\rightarrow^- = \rho_\rightarrow, \qquad (21.17)$$

and similarly for $\rho_\leftarrow$. Thus, the number of configurations compatible with velocity β is (using the same notation as in §21.9)

$$\mathcal{N} = \binom{m}{\rho_\rightarrow^\pm m}\binom{m}{\rho_\leftarrow^\pm m}, \quad \text{with} \quad \begin{aligned} \rho_\rightarrow^+ &= (1-\beta)/2, \\ \rho_\rightarrow^- &= (1+\beta)/2, \end{aligned} \qquad (21.18)$$

or

$$\ln \mathcal{N} \approx 2mK(\beta) \approx -m\beta^2 + \text{const}, \qquad (21.19)$$

which is essentially the same dependency on β as (21.9). This will of course give the correct results as the Lagrangian of the "lumped-string" free particle; moreover, the appearance of a mass factor m makes it possible to study the interactions between particles of different masses[15] and see how the number of trajectories varies as a function of different masses and velocities.

21.10.4   Refraction

Here we'll briefly consider the case of a nonhomogeneuos medium. If the permission table witholds permission for the two consecutive steps that make up a time unit (cf. top part of Fig. 21.21), then the evolution of the chain is frozen for that time unit. These "suspensions" may be randomly scattered through time, reducing the effective speed by a factor of n (see [9] and [22] for examples of this approach in lattice-gas dynamics). Suppose (Fig. 21.24) that the particle starts at the origin, traverses free space ("index of refraction" $k = 1$) until time t, and then traverses a stretch of time with "index of refraction" $k = n$, finally landing at Q. What is the most likely value for q when the particle enters the denser medium? Equivalently, what is the "refraction angle"?

We will derive the Lagrangian directly from the microscopic dynamics, always assuming that the Lagrangian must be an indicator of the number of trajectories compatible with the given data. In the denser medium, whatever the macroscopic velocity v, the particle must be following a sequence of configuration appropriate for a velocity $v' = nv$, with the only difference that on time slots when permissions are witheld the current configuration will be held frozen. Thus, in the denser medium, the Lagrangian must reflect the path statistics of this higher "internal" velocity rather than that of the apparent velocity v.

---

[15] Interactions between chains require a slightly more complex CPU for our computer, able to look further than just first neighbors when considering a flip.

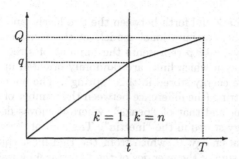

Fig. 21.24. A lumped-string particle travels until t through a 100% permissive medium ("vacuum"); thereafter, the permission rate is granted only every nth time slot ("index of refraction" n). If the average velocity in the denser medium is v, the string must be moving at an actual speed $v' = nv$ when moving at all, and the path statistics that enters in the Lagrangian must be that corresponding to this higher speed v'.

Multiplying the velocity by a factor of n in (21.16) shifts the statistics in (21.18) so that a factor of n^2 appears in (21.19); rescaling time by a factor $1/n$, and then moving this factor from the dt term to the L term in $L\,dT$ adds to the Lagrangian a factor $1/n$, so that the correct Lagrangian for index of refraction n is

$$L_n(v) = \frac{1}{n}n^2 L_{\text{vacuum}}(v) = nmv^2.$$

This expression for the Lagrangian indeed gives, upon extremizing the action integral, the correct actual trajectory with $q/t = n(Q - q)/(T - t)$.

21.10.5 The string as an extended system

We shall now look at a chain of indefinite length undergoing harmonic motion as a spatially extended system.

If we denote by $\mathcal{K}$, $\mathcal{U}$, $\mathcal{H}$, and $\mathcal{L}$ the densities of respectively kinetic energy, potential energy, total mechanical energy (Hamiltonian), and Lagrangian, the correspondence with analytical mechanics for the harmonic chain, given by (21.12) and (21.14), is completed by setting

$$\begin{aligned} \mathcal{K} &= j^2, & &\text{and} & \mathcal{H} &= \mathcal{K} + \mathcal{U}, \\ \mathcal{U} &= \rho^2, & & & \mathcal{L} &= \mathcal{K} - \mathcal{U}, \end{aligned} \tag{21.20}$$

whence

$$\mathcal{H} = j^2 + \rho^2 = \frac{1}{2}\left(\rho_\rightarrow^2 + \rho_\leftarrow^2\right), \tag{21.21}$$

$$\mathcal{L} = j^2 - \rho^2 = -\rho_\rightarrow \rho_\leftarrow. \tag{21.22}$$

We can now go back and forth between the two levels, and find the microscopic interpretation, in this computational model, of the usual macroscopic concepts. For example, the quantity ρ represents the amount of stretch of the chain from the flat configuration in which links are randomly oriented up and down—and its square represent the energy stored in this "spring". The quantity j represents the momentum of the string (the difference between the number of the "valleys", where the chain moves up, and that of "peaks", where it moves down), and its square represents the energy stored in the "inertia". The sum of these squares is of course the total mechanical energy $\mathcal{H}$, which from the rightmost equality in (21.21) can also be seen as the sum of the energies of the two traveling waves $\rho_\rightarrow$ and $\rho_\leftarrow$.

For a combinatorial interpretation of $\mathcal{H}$, consider a piece of chain of length m and proceed as in (21.18), but with $\rho_\rightarrow^+$ and $\rho_\leftarrow^+$ more generally given by

$$\rho_\rightarrow^+ = (1 + \rho_\rightarrow)/2,$$
$$\rho_\leftarrow^+ = (1 + \rho_\leftarrow)/2. \qquad (21.23)$$

We obtain

$$\mathcal{N} = \binom{m}{\rho_\rightarrow^+ m}\binom{m}{\rho_\leftarrow^+ m}$$

or

$$\frac{1}{n}\log\mathcal{N} = K(\rho_\rightarrow)K(\rho_\leftarrow)$$
$$\approx -\frac{1}{2}(\rho_\rightarrow^2 + \rho_\leftarrow^2) = -\mathcal{H} + \text{constant}.$$

Thus, in this model of the elastic string, the energy $\int \mathcal{H}(x)dx$ measures, on a log scale, the number of microscopic chain configurations compatible with a given macroscopic assignment of positions and velocities. A low-energy state is "cheap" because it is "common"—there are so many ways to achieve it. Conversely, high-energy states are "rare". In fact, the four states of maximal energy ($j = 0$, $\rho = \pm 1$, as in Fig. 21.22a; or $j = \pm 1$, $\rho = 0$, as in Fig. 21.22c,d) are each represented by a single microscopic configuration.

Since the underlying cellular automaton is microscopically reversible, its fine-grained entropy is strictly constant—rare states map into rare states, common ones into common ones; in this model, thence, energy conservation is just a macroscopic expression of microscopic reversibility.

Note that $\mathcal{H}$ depends on the coarseness of graining; vibrations of a wavelength shorter than this grain do not contribute to $\mathcal{H}$, and may be viewed as thermalized degrees of freedom.

From (21.22), the wave equation (21.14) follows immediately by the Euler-

Lagrange equation

$$\frac{d}{dt}\frac{\partial \mathcal{L}}{\partial \frac{\partial q}{\partial t}} + \frac{d}{dx}\frac{\partial \mathcal{L}}{\partial \frac{\partial q}{\partial x}} = 0. \tag{21.24}$$

By (21.22), $\mathcal{L} = -\rho_{\rightarrow}\rho_{\leftarrow}$. Expanding $\rho_{\rightarrow}\rho_{\leftarrow}$ by (21.17) yields

$$\rho_{\rightarrow}\rho_{\leftarrow} = \left(\rho_{\rightarrow}^{+}\rho_{\leftarrow}^{+} + \rho_{\rightarrow}^{-}\rho_{\leftarrow}^{-}\right) - \left(\rho_{\rightarrow}^{+}\rho_{\leftarrow}^{-} + \rho_{\rightarrow}^{-}\rho_{\leftarrow}^{+}\right). \tag{21.25}$$

In the above equation, the four terms in parentheses represent the probabilities that two consecutive chain links, the first from $\rho_{\rightarrow}$ and the other from $\rho_{\leftarrow}$, form

$$\begin{array}{ll} \rho_{\rightarrow}^{+}\rho_{\leftarrow}^{+} & \text{a } +1 \text{ slope} \\ \rho_{\rightarrow}^{-}\rho_{\leftarrow}^{-} & \text{a } -1 \text{ slope} \\ \rho_{\rightarrow}^{+}\rho_{\leftarrow}^{-} & \text{a ridge} \\ \rho_{\rightarrow}^{-}\rho_{\leftarrow}^{+} & \text{a valley} \end{array}$$

(cf. Fig. 21.22).

If a chain's evolution obeys the permissive rule, a ridge yields a downward flip (cf. Fig. 21.20); a valley, an upward flip; and ±1 slopes yield a rest. Thus, on any small patch of a proposed macroscopic spacetime history,

$$\begin{aligned} \mathcal{L}_{\text{actual}} &= \text{density of flips} - \text{density of rests} \\ &= 2(\text{density of flips}) + \text{constant}. \end{aligned} \tag{21.26}$$

Consequently, given spacetime boundary conditions for the macroscopic string configuration (e.g., the entire string at t_0 and t_1), the histories that are actual solutions of the permissive dynamics are those that fill in the intervening spacetime area with the *least number of flips*.

One might be tempted to say, "Aha! Here is an example of nature's parsimony!" and stop here. But, of course, with a different rule the actual trajectory may not be that which minimizes the number of flips. Something deeper and more general is going on, which is better seen by turning from the figure to the background. A trajectory that minimizes the number of flips is by the same token one that maximize the *number of rests*. Now, in this dynamics, the rests points are all and only those spacetime points where no motion was possible to begin with, because of kinematic constraints. Thus a rest is a point where witholding permission would not have made any difference! For any particular rest point in any particular history, a table with a 0 at that point instead of a 1 would have yielded the *same history*; consequently, that history is also a trajectory of this variant table. The actual data flow is that which *maximizes* the number of possible tables that could have given the same data flow—in other words, it is the behavior that is *maximally indifferent to the rule*. If the microscopic rule is not known precisely or, equivalently, is subject to noise, those macroscopic behaviors that are less dependent on the details of the rule will be more stable and will recur more often.

We have seen that, in the Hamiltonian, an increase in either potential or kinetic energy leads to *fewer available states*. In the Lagrangian, the potential energy enters with the opposite sign as in the Hamiltonian. If we were looking for maximum number of *states*, that would lead us in the wrong direction. But states of high U are rare because they are special, and they are special in that they have high symmetry in relation to the dynamics; for this reason, a greater number of microscopic implementation of the dynamics leaves them unaffected.[16]

21.11   Conclusions

If we extrapolate from the above examples we are led to the following ideas:

- The Lagrangian, which is a function not only of the independent parameters x and t but most essentially of the dependent variables q and their first derivatives, measures the multiplicity of microscopic laws that are consistent (a) with an agreed-upon set of kinematic constraints, (b) with the specified dynamics, which must be viewed merely as a specification of macroscopic trajectories, and thus encompasses a whole ensemble of possible microscopic dynamics, and (c) with the proposed segment of macroscopic trajectory. The two terms U and K express different kinds of contribution to this multiplicity count.

- A segment of trajectory with high potential energy U is one whose shape is heavily constrained by *the general constraints specifying the class of laws*, and thus is less affected by the choice of a particular law of this class.

- A segment of trajectory with high kinetic energy K is one in which much computation capacity is spent in transporting data, and thus less capacity remains for bringing about internal evolution of the data themselves.

- The action integral measures the computation capacity (multiplicity of microscopically accessibly paths under the above constraints) offered by the substrate to a proposed macroscopic trajectory.

- The more this amount of computation capacity exceeds what is strictly needed for the proposed macroscopic evolution, the greater the number of microscopic ways that the latter can be fulfilled (typically, by wasting extra capacity in aimlessly thrashing)—and thus the greater the chances that the proposal will be carried out.

---

[16]For an analogy, a circle is more symmetric under rotations than a triangle, and for this reason there are fewer circle *shapes* than triangle shapes. But for the same reason there are more *rotations* that will leave a circle unaffected.

In sum, by postulating a fine-grained dynamical substrate underlying the given dynamics, one turns analytical mechanics into a form of statistical mechanics, though applied to *microlaws* rather than to *microstates*. One views the actual trajectory as the peak of a distribution of microscopic trajectories, and interprets the "principle of least action" as a concise expression of combinatorial tautologies, much as the "survival of the fittest" and the "law of large numbers".

If all of this stands review, the principle of least action is an expression not of nature's *parsimony* [2] but of nature's *prodigality*: a system's natural trajectory is the one that will hog the *most* computational resources. Would you have expected otherwise?

Do not read me wrong. I am duly impressed by the enormous power of variational principles; but when I use them I would like to know where it is that they get their power from (*caveat emptor!*)—I just don't believe in *magic*.

21.12 Acknowledgments

This research was funded in part by NSF (DMS-9596217) and by the I.S.I. Foundation (Turin, Italy). Part of this work was carried out during the 1997 Elsag-Bailey–I.S.I. Foundation research meeting on quantum computation.

I would like to thank Lev Levitin, Mark Karpovsky, Sandu Popescu, and Zac Walton for stimulating discussions.

References

[1] E.T. Jaynes, "The minimum entropy production principle", *Annual Rev. Physical Chem.* (S. Rabinovitch ed.), Annual Reviews Inc., Palo Alto, CA (1980), reprinted and commented in E. T. Jaynes, *Papers on Probability, Statistics and Statistical Physics*, 2nd edition, 401–424 (Kluwer, New York, 1989).

[2] S. Hildebrandt and A. Tromba, *The Parsimonious Universe*, xiv+330 (Springer-Verlag, Berlin, 1996).

[3] C. Lanczos, "The Variational Principles of Mechanics", 4th edition, xxix+418 pp (Dover, New York, 1970).

[4] R. Landauer, "Information is physical", *Physics Today* 44, 23–29 (1991).

[5] B. Hayes, "Collective Wisdom", *American Scientist* 86, 118–122 (1998).

[6] M. Davis, *Computability and Unsolvability* (McGraw-Hill, New York, 1958); reprinted (Dover, New York, 1983).

[7] M. Garey and D. Johnson, *Computers and Intractability: A Guide to the Theory of NP-Completeness*, (Freeman, San Francisco CA, 1979).

[8] E. Fredkin and T. Toffoli, "Conservative logic," *Int. J. Theor. Phys.* **21**, 219–253 (1982).

[9] T. Toffoli and N. Margolus, *Cellular Automata Machines—a new environment for modeling* (MIT Press, Cambridge, 1987).

[10] T. Toffoli, "Four topics in lattice gases: Ergodicity; Relativity; Information flow; and Rule compression for parallel lattice-gas machines," *Discrete Kinetic Theory, Lattice Gas Dynamics and Foundations of Hydrodynamics* (R. Monaco ed.), 343–354 (World Scientific, Singapore, 1989).

[11] R. L. Adler, A. G. Konheim and M. H. McAndrew, "Topological entropy," *Trans. Amer. Math. Soc.* **114**, 309–319 (1965).

[12] F. Kafka, *The Trial* (Knopf, New York, 1992).

[13] A. Tyagi, "A principle of least computational action", *Workshop on Physics and Computation*, 262–266 (IEEE Computer Society Press, Los Alamitos CA, 1993).

[14] F. Hapgood, *Up the Infinite Corridor: MIT and the Technical Imagination* (Addison-Wesley, Reading MA, 1994).

[15] G. Chaitin, *The Limits of Mathematics: A Course on Information Theory and Limits of Formal Reasoning* (Springer, New York, 1998) in press.

[16] N. Margolus and L. Levitin, "The maximum speed of dynamical evolution", to appear in *Physica D*.

[17] R. Feynman and A. Hibbs, *Quantum Mechanics and Path Integrals* (McGraw-Hill, New York, 1965).

[18] V. Arnold, *Mathematical Methods of Classical Mechanics* (Springer-Verlag, New York, 1978).

[19] M. Schroeder, *Number Theory in Science and Communication*, second enlarged edition (Springer-Verlag, Berlin, 1986).

[20] E. T. Jaynes, "Information theory and statistical mechanics", *Phys. Rev.* **106**, 620–630 (1957), and **108**, 171–189 (1957).

[21] L. Kadanoff, "On Two Levels," *Physics Today*, 7–9 (September 1986).

[22] M. Smith, "Representation of geometrical and topological quantities in cellular automata," *Physica D* **45**, 271–277 (1990).

22

ALGORITHMIC RANDOMNESS, PHYSICAL ENTROPY, MEASUREMENTS, AND THE DEMON OF CHOICE

W. H. Zurek *

Abstract

Measurements — interactions which establish correlations between a system and a recording device — can be made thermodynamically reversible. One might be concerned that such reversibility will make the second law of thermodynamics vulnerable to the designs of the *demon of choice*, a selective version of Maxwell's demon. The strategy of the demon of choice is to take advantage of rare fluctuations to extract useful work, and to reversibly undo measurements which do not lead to such a favorable but unlikely outcomes. I show that this threat does not arise as the demon of choice cannot operate without recording (explicitly or implicitly) whether its measurement was a success (or a failure). Thermodynamic cost associated with such a record cannot be, on the average, made smaller than the gain of useful work derived from the fluctuations.

22.1 Feynman

When I was asked to write for a volume dedicated to Richard Feynman, I decided that I should select the subject in which I was influenced by him the most, and which would still be consistent with the overall theme of computation and physics. And these influences started well before I met him in person: I got Feynman's "Lectures on Physics" more than a quarter century ago, in Polish translation, from my father. As a finishing high school student I was accompanying him on a hunting expedition in the lake district of Poland — a remote corner of the country. Every few days we drove for supplies to the provincial capital, and there I noticed the volumes in the local bookstore. My father asked why (the expense was considerable), but surprisingly easily gave way to my arguments. I spent much of the rest of the hunting vacation (a couple of weeks altogether) getting through volume I.

Over the years I have developed a habit of treating the "Lectures" sort of like a collection of poems. I like some "poems" more than others, and I return to the favorites now and again. And when I am stuck with a physics problem, reading a few of the relevant "poems" is often the best way to get "unstuck". But there

*©by Wojciech Hubert Zurek, March 12, 1998

are a few chapters which have been read over and over again without any such an ulterior motive, for sheer pleasure. Amongst them, I would certainly include the discussion of the fluctuations and the second law (the famous "ratchet and pawl" argument [1]).

Thermodynamic concerns and arguments have often pre-saged the deepest developments in physics. I suspect this is because thermodynamics "knows" about the physical relevance of information, and hence, it knew about the Planck constant, stimulated emission, black hole entropy, and so on. When I met Feynman in person for the first time (at a small workshop organised near Austin, Texas, by John Archibald Wheeler in the Spring of 1981), I remember — amongst other things — a thermodynamic argument he used to great effect to prove that one cannot accelerate elementary particles by shaking them together with a bunch of heavier objects, so that they could acquire equipartition kinetic energies (and therefore, because of their small mass, enormous momenta). This idea (credible at first sight, as it is akin to the Fermi acceleration of cosmic rays) was brought up by one of the participants. It would not work — Feynman argued — because all sorts of other modes of the vacuum would have to get their fair share of energy, creating an equilibrium heat bath, with approximate equipartition between all the modes (rather then with the energy in the elementary particles one really wanted to accelerate in the first place).

But that was not the most vivid memory of that first encounter with the man whose "Lectures" I had acquired a decade or so earlier. Rather, I remember best that he showed up at the first lecture unshaved and uncombed, with dry grass in his hair. It turned out that he spent the night outside — apparently, he decided the accomodations for the speakers (which were in the posh tennis club) were too opulent, returned the key to his apartment at the reception, and decided to "camp out". During the morning coffee he has also reported in detail (and with great gusto) how he had trouble breaking the code to get into his briefcase (where he had the sweater — it got cold). He knew the code, of course, by heart, but it was middle of the night, so he somehow had to dial it in complete darkness. He clearly relished the challenge. I do not remember how did he solve the problem, but the flavor of the adventure and of his report was very much in the spirit of the "adventures of a curious character". And all of this was a few months after his (first) cancer operation.

I came to talk to Feynman regularly, more or less once a month, during my Tolman Fellowship at Caltech (which started in the Fall of 1981), and a bit less often for a few years afterwards. I have also sat occasionally in the class on physics and computation he taught with John Hopfield. And I remember discussing with him (among other subjects) the connection between physics, information and computation. In fact, this was a recurring theme. For me, it became somewhat of an obsession early on — I really liked the universality of Turing machines, the halting problem, and the algorithmic view of information. While I was in Austin the fascination with these ideas and their possible relevance for physics was reinforced under the influence of John Wheeler. Which brings me, at long last, to the *algorithmic information content*, *measurements*, and various *thermodynamic demons* which probe the utility of acquired information.

22.2 Algorithmic Information Content, Measurement and the Second Law

Maxwell's demon — a hypothetical intelligent entity capable of performing measurements on a thermodynamic system and using their outcomes to extract useful work — was considered a threat to the validity of the second law of thermodynamics for over a century [2, 3]. Feynman was fascinated with the subject, and his discussion of ratchet and pawl [1] banished forever the "unintelligent" trapdoor version of the demon by clarifying and updating the influential argument put forward by Smoluchowski [4] much earlier, and in a rather different setting [1]. However, Smoluchowski's trapdoor carries out no (explicit) measurements. Therefore, trapdoors and ratchets and pawls can be analysed without reference to information [1, 4].

The complete Maxwell's demon should be able to measure, and it (...?; he? she?!) should be of course intelligent. Smoluchowski's trapdoor does not fit this bill. Measurements were incorporated into the discussion by Szilard [6], Landauer [7], and Bennett [8] who have argued, in a setting involving ensembles of demons, that the acquisition of information is only possible when the demon's memory is repeatedly erased, to prepare it for the new data. The *cost of erasure* eventually offsets whatever thermodynamic advantages of the demon's information gain might offer. This point (which has come to be known as "Landauer's principle") is now widely recognised as a key ingredient of thermodynamic demonology. This originally classical reasoning has been since extended to quantum physics [9, 10], and may even be experimentally testable [11].

However, the widespread fascination with Maxwell's demon is ultimately due to its intelligence. A demon will record a specific outcome of the measurement, and — using its intelligence — will try to make an optimal decision about the best possible action, which would maximize the work extracted from a given recorded phase space configuration. This is very much the course of action we take (although, fortunately for us, in a far-from-equilibrium setting). How can one convince an intelligent demon that, all cleverness notwithstanding, its attempts at defeating the second law are doomed? This is hard to accomplish at the level of ensembles: Each demon knows

---

[1]Smoluchowski's original trapdoor was a hole surrounded by hairs combed so that they all come out on the same side of the partition between the two chambers (rather than a real trap*door*). Naively, this arrangement of hairs should favor molecules passing in the direction in which the hair is combed, and impede the reverse motion. Smoluchowski pointed out that thermal fluctuations will "ruffle the hair" and make this arrangement ineffective as a rectifier of fluctuations when the whole system is at the same fixed temperature. Numerical simulations of trapdoors confirm these conclusions [5]. They also show why our intuition based on far-from-equilibrium behavior of trapdoors can be easily misled.

nothing but its own record, and need not care about the other members of "its
ensemble" that have found out something else in their measurements — it will find
out its solution to its own problem.

The ultimate analysis of Maxwell's demon must involve a definition of intelligence, a characteristic which has been all too consistently banished from discussions
of demons carried out by physicists. On the other hand, intelligence has been —
since Turing and his famous test — often invoked in the discussions of computer
scientists. To convince ourselves (and the intelligent demon) of the limits imposed
by the second law we shall, following Ref. [12], adopt an operational definition of
intelligence which arose in the context of the theory of computation. It is based
on the so-called Church–Turing thesis [13] — which in effect formalizes Turing's
expectations about the "mental" capabilities of computers and states that intelligence is equivalent to the same kind of information processing that is in principle
implementable on a universal computer.

Using the Church–Turing thesis as a point of departure, the present author has
demonstrated that even this intelligent threat to the second law can be eliminated —
the original "smart" Maxwell's demon can be exorcized. This is easiest to establish
when one recognizes that the net ability of demons to extract useful work from
systems depends on the sum of measures of two distinct aspects of disorder [12]:

(i) The usual *statistical entropy* given by:

$$H(\rho) = -Tr\rho \lg \rho \qquad (22.1)$$

where ρ is the density matrix of the system, determines the ignorance of the observer.

(ii) The *algorithmic information content:* [14–20]

$$K(\rho) = |p_\rho^*| \qquad (22.2)$$

is given by the size ("|...|"), in bits, of the shortest algorithm (p^*) which, for an "op-
erating system" of a given Maxwell's demon, can reproduce the detailed description
(ρ) of the state of the system. $K(\rho)$ quantifies the cost of storing of the acquired
information, which is related to the randomness inherent in the state of the system
revealed by the measurement.

The Church–Turing thesis enters in this second algorithmic ingredient, as it
involves an assumption that the intellectual abilities of Maxwell's demons can be
regarded as equivalent to those of a universal Turing machine: It is assumed that
demons can execute programs (such as p_ρ^*) to reconstruct records of past measurements out of their optimally compressed versions, or to carry out other logical
operations in optimizing performance. Algorithmic information content provides a
well-defined measure of the storage space required to register the known characteristics of the system.

Physical entropy [12] is the sum of the statistical entropy and of the algorithmic information content:

$$\mathcal{Z}(\rho) = H(\rho) + K(\rho) \qquad (22.3)$$

Above, it is assumed that the base for the logarithm in Eq. 22.1 is the same as the size of the alphabet used by the computer which constitutes the operating system of the Maxwell's demon. In practice, it is customary and convenient to employ a binary alphabet, so that both $H(\rho)$ and $K(\rho)$ are measured in bits.

In order to appreciate the physical significance of the algorithmic randomness contribution, it is useful to discuss the behavior of H, K and $\mathcal{Z}$ in the course of measurements and to follow the operations of the engines controlled by demons. In short, the two measures turn out to be complementary — not in the quantum sense, but a bit like kinetic and potential energy — and their sum is, on the average, conserved under optimal measurements carried out on an equilibrium ensemble. Analysis which leads to this conclusion was carried out by this author [10, 12] and extended by Caves [21]. Below we offer only a brief summary of the salient points.

In the course of ideal measurement on an equilibrium ensemble the decrease of ignorance is, on the average, compensated for by the increase of the size of the minimal record [12]:

$$\Delta H \simeq - < \Delta K > . \qquad (22.4)$$

Consequently, physical entropy $\mathcal{Z}$ plays a role analogous to a constant of motion. The transformation of the state of the system is now, however, brought about by a *demonical* (rather than *dynamical*) evolution, by the act of acquisition of information. This "conservation law" can be demonstrated within the context of the algorithmic theory of information [10, 12, 21, 22]. However, its validity can be traced to coding theory [12, 21–23]. According to the noiseless coding theorem of Shannon [23], the minimal size $\mathcal{L}$ of the message required to encode information which corresponds to a decrease of entropy by ΔH is, on the average over all of the messages, bounded by:

$$\Delta H \leq \mathcal{L} < \Delta H + 1$$

This inequality is used in the proof of Eq. 22.4 and is ultimately responsible for the constancy of the physical entropy $\mathcal{Z}$ in the course of the measurement [12, 21].

The role of $\mathcal{Z}$ in determining the efficiency of demon-operated engines is the ultimate reason for regarding $\mathcal{Z}$ as physical entropy. For the total amount of work which can be extracted from a physical system in contact with a heat reservoir of temperature T in the course of a cycle which involves a measurement ($\rho \to \rho_i$) and isothermal expansion ($\rho_i \to \rho$) can be made as large as, but no larger than:

$$\Delta W = k_B T(\mathcal{Z}(\rho) - \mathcal{Z}(\rho_i)) \qquad (22.5)$$

To justify this last assertion, I shall appeal to Landauer's principle [7] which formalizes earlier remarks of Szilard [6] and states that erasure of one bit of information from the memory carries a thermodynamic price of $k_B T$. Although Landauer's principle assigns a definite price to the storage of information, this price need not be paid right away: A demon with a large unused memory can continue to carry out measurements as long as it has room to store information. However, such a demon poses no threat to the second law: Its operation is not truly cyclic. In effect, it operates by employing its initially empty memory as a low temperature (zero entropy) heat sink.

Erasure of the results of used up measurements carries a price tag of

$$\Delta W^- = T < (K(\rho_i) - K(\rho)) > , \qquad (22.6a)$$

which must be subtracted from the gain of useful work

$$\Delta W^+ = T(H(\rho) - H(\rho_i)) , \qquad (22.6b)$$

to obtain the net work extracted by the demon. This immediately justifies Eq. 22.5. The hybrid $\mathcal{Z}$ is the physical entropy which provides the demon with an individual, personal measure of the potential for thermodynamic gains due to the information in its possession. It also demonstrates that a demon operating on a system in thermodynamic equilibrium will never be able to threaten the second law, for the ensemble average of $\mathcal{Z}$ is at best conserved, so that $< \Delta \mathcal{Z} > \leq 0$ in course of the process of acquisition of information.

22.3   Physical Entropy and the Demon of Choice

This last assertion is, however, justified only if the demon is forced to complete each measurement-initiated cycle. One can, by contrast, imagine a *demon of choice*, an intelligent and selective version of Maxwell's demon, who carries out to completion only those cycles for which the initial state of the system is sufficiently nonrandom (concisely describable, or *algorithmically simple*) to allow for a brief compressed record (small $K(\rho)$). This strategy appears to allow the demon to extract a sizeable work (ΔW^+) at a small expense (ΔW^-). Moreover, if the measurements can be reversibly undone, then the ones with disappointing outcomes could be reversed at no cost. Such demons would still threaten the second law, even if the threat is somewhat more subtle than in the case of Smoluchowski's trapdoor.

Caves [22] has considered and partially exorcised such a demon of choice by demonstrating that in any case the net gain of work cannot exceed $k_B T$ per measurement. Thus, the demons would be, at best, limited to exploiting thermal fluctuations. Moreover, in a comment [24] on Ref. [22] it was noted that taking advantage of such fluctuations is not really possible. Here I shall demonstrate that the only decision-making process free of inconsistencies necessarily leaves in the observer's (demon's) memory a "residue" which requires eventual erasure. The least cost of erasure of this residue is just enough to restore the validity of the second law. The aim of this paper is to make this argument (first put forward by this author at the meeting of the *Complexity, Entropy, and the Physics of Computation* network of the Santa Fe Institute in April of 1990) more carefully and more precisely.

To focus on a specific example consider *Gabor's engine* [25] illustrated in Fig. 22.1. There, the unlikely but profitable fluctuation occurs whenever the gas molecule is found in the small compartment of the engine. The amount of extractable work is:

$$\Delta W_p^+ = k_B T \ \lg(L/\ell) \tag{22.7}$$

The expense (measured by the used up memory) is only:

$$\Delta W^- = k_B T, \tag{22.8}$$

so that the net gain of work per each successful cycle is:

$$\Delta W_p = k_B T \ (\lg(L/\ell) - 1) \tag{22.9}$$

The more likely "uneconomical" cycles would allow a gain of work:

$$\Delta W_u^+ = k_B T \ \lg L/(L - \ell) \ , \tag{22.10}$$

so that the cost of memory erasure (still given by Eq. 22.8) outweighs the profit, leaving the net gain of work:

$$\Delta W_u = -k_B T(1 - \lg L/(L - \ell)). \tag{22.11}$$

When each measurement is followed by the extraction and erasure routine, the averaged net work gain per cycle is negative (i.e., it becomes a loss):

$$< \Delta W > = \frac{\ell}{L}\Delta W_p + \frac{L-\ell}{L}\Delta W_u = -k_B T[1 + (\frac{\ell}{L}\lg\frac{\ell}{L} + \frac{L-\ell}{L}\lg\frac{L-\ell}{L})] \tag{22.12}$$

The break even point occurs for the case of Szilard's engine [6], where the partition divides the container in half. In the opposite limit, $\ell/L \ll 1$, almost every measurement leads to an unsuccessful case which results in a negligible amount of extracted work but undiminished cost of erasure per cycle.

The design of the demon of choice attempts to capitalize on precisely this otherwise unprofitable limit by *undoing* all of the likely (and unprofitable) measurements at no thermodynamic cost, thus avoiding the necessity for erasure of the unused outcomes. It is important to emphasize that a measurement of the thermodynamic quantities can be indeed undone at no cost: A prejudice that measurement must be thermodynamically expensive goes back at least to the ambiguities in the original paper of Szilard [6] (who has hinted at, but failed to clearly identify erasure as the only thermodynamically expensive part of the measuring process), and was further reinforced by the popular (but incorrect) discussion of Brillouin [26]. Fig. 22.2 demonstrates how to carry out a measurement on a particle in the Gabor's engine (such measurement becomes reversible when the operations indicated are carried out infinitesimally slowly).

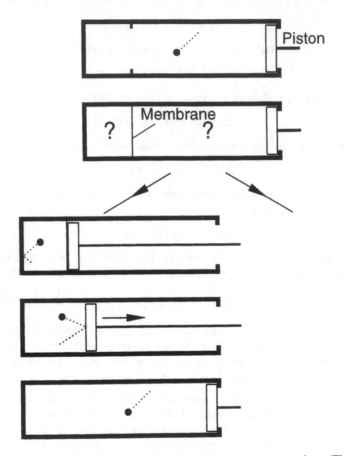

Fig. 22.1. Gabor's engine[25]. See text for the standard operating procedure. The decision between the two branches (of which only one — the profitable one — is shown) can be made reversibly with the help of the device shown in Fig. 22.2.

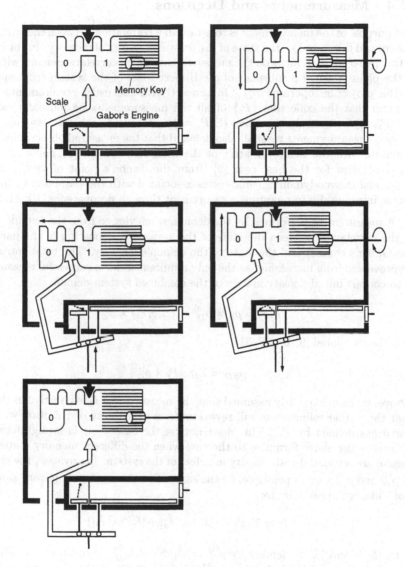

Fig. 22.2. Blueprint of a reversible measuring device for Gabor's engine. The measurements can be done (or undone) by turning the crank on the right in the appropriate direction and pushing in or pulling out the "scale". Thermodynamic reversibility is achieved in the limit of an infinitesimally slow operation. Faster controlled-not like measurements can be carried out on a dynamical timescale by implementing the unitary evolution given by Eq. 22.14. The design shown above is similar to the Szilard's engine contraption devised in Ref. [28].

22.4 Measurements and Decisions

The purpose of the measurement is to establish a correlation between the state of the system and the record — the state of the few relevant bits of memory. In the context of this paper we shall focus on the measurements which correlate memory with a cell in the phase space or a subspace of the Hilbert space of the system (corresponding to the projection operator P_i). In concert with the usual requirements I shall demand that the collection $\{P_i\}$ of all the measurements be mutually exclusive ($Tr(P_i, P_j) = 0$), and exhaustive ($\Sigma_i P_i = 1$). To avoid problems associated with quantum measurements we shall also demand that the measured observables should commute with the density matrix of the measured system $[P_i, \rho_S] = 0$. Thus, we shall allow for the best case [9] (from the demon's point of view), with no additional thermodynamic inefficiencies associated with the reduction of the state vector introduced into quantum measurement through decoherence [10, 11, 28–31].

A measurement performed by the demon, when viewed from the outside, results in the correlation between the state of the system (i.e. location of the particle in the Gabor's engine) and the state of the demon's memory. The total entropy can be prevented from increasing, as the only requirement for a successful measurement is to convert initial density matrix of the combined system-demon:

$$\rho_{SD}^{(o)} = \rho_S \times \rho_D^{(o)} = (\Sigma_i p_i P_i) \times \rho_D^{(o)} \tag{22.13a}$$

into the correlated [9, 10, 28–31]:

$$\rho_{SD} = \Sigma_i p_i (P_i \times \rho_D^{(i)}) \tag{22.13b}$$

Above, we have implicitly assumed that the measurement is exhaustive in the sense that the further refinements will reveal uniform probability distribution within the partitions defined by P_i. This need not be the case — it is straightforward to generalize the above formulae to the case when the different memory states of the demon are correlated with density matrices of the system. In any case, the entropies of $\rho_D^{(i)}$ and $\rho_D^{(o)}$ can, in principle, be the same: For, there exists a unitary *controlled-not* - like evolution operator:

$$U = \Sigma_i P_i \times (|\delta_i ><\delta_o| + |\delta_o ><\delta_i|) \tag{22.14}$$

with $|\delta_i >$ and $|\delta_o >$ defined by $\rho_D^{(i)} = |\delta_i >< \delta_o|\rho_D^{(o)}$, providing that $\rho_D^{(i)}$ correspond to distinguishable (orthogonal) memory states of the demon — a natural requirement for a successful measurement.

The statistical entropy of the system-demon combination is obviously the same before and after measurement, as, by construction of U, $H(\rho_D^{(i)}) = H(\rho_D^{(o)})$. Moreover, the measurement is obviously reversible: Applying the unitary evolution operator, Eq. 22.14, twice, will restore the pre-measurement situation.

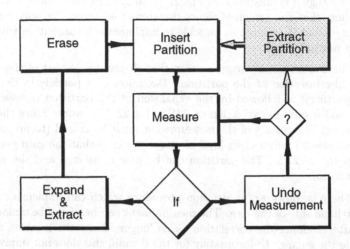

Fig. 22.3. Decision flowchart for the demon of choice. The branch on the left is profitable (and it is followed when the particle is "caught" in the small left chamber, see Fig. 22.1). The branch on the right is unprofitable, and as it is explained in the text in more detail, the demon of choice cannot be "saved" by reversing only the unprofitable measurements.

From the viewpoint of the outside observer, the measurement leads to a correlation between the system and the memory of the demon: The ensemble averaged increase of the ignorance about the content of demon's memory;

$$\Delta H_D = H(\rho_D) - H(\rho_D^{(o)}) = -\Sigma_i p_i \lg p_i , \qquad (22.15)$$

(where $\rho_D = Tr_S \rho_{SD}$ and $H(\rho) = -Tr\rho \lg \rho$) is compensated for by the increase of the mutual information defined as;

$$I_{SD} = H(\rho_D) + H(\rho_S) - H(\rho_{SD}), \qquad (22.16)$$

so that $\Delta H_D = \Delta I_{SD}$ (see Refs. [29] and [33] for the Shannon and algorithmic versions of this discussion in somewhat different settings).

From the viewpoint of the demon the acquired data are definite: The outcome is some definite demon state $\rho_D^{(n)}$ corresponding to the memory state n, and associated with the most concise record — increase of the algorithmic information content — given by some $\Delta K(n) = K(\rho_S^{(n)}) - K(\rho_S^{(o)})$.

The demon of choice would now either; (i) proceed with the expansion, extraction and erasure, providing that his estimate of the future gain:

$$\Delta W = k_B T(\Delta H - \Delta K) = k_B T \Delta \mathcal{Z} \qquad (22.17)$$

was positive, or, alternatively; (ii) undo the measurement at no cost, providing that $\Delta W < 0$. An algorithm that attempts to implement this strategy for the case of Gabor's engine is illustrated in Fig. 22.3. To see why this strategy will not work, we first note that the demon of choice threatens the second law only if its operation is cyclic — that is, it must be possible to implement the algorithm without it coming to an inevitable halt.

There is no need to comment on the left-hand side part of the cycle: it starts with the insertion of the partition. Detection of a particle in the left-hand side compartment is followed by the expansion of the partition (converted into a piston) and results in extraction of ΔW_p^+, Eq. 22.7, of work. Since the partition was extracted, the results of the measurement must be erased (to prepare for the next measurement) which costs $k_B T$ of useful work, so that the gain per useful cycle is given by Eq. 22.9. The partition can be now reinserted and the whole cycle can start again.

There is, however, no decision procedure which can implement the goal of the right-hand side of the tree. The measurement can be of course undone. The demon — after undoing the correlation — no longer knows the location of the molecule inside the engine. Unfortunately for the demon, this does not imply that the state of the engine has also been undone. Moreover, the demon with empty memory will immediately proceed to do what demons with empty memory always do: It will measure. This action is an "unconditional reflex" of a demon with an empty memory. It is inevitable, as the actions of the demon must be completely determined by its internal state, including the state of its memory. (This is the same rule as for Turing machines.) But the particle in the Gabor's engine is still stuck on the unprofitable side of the partition. Therefore, when the measurement is repeated, it will yield the same disappointing result as before, and the demon will be locked forever into the measure - unmeasure "two-step" within the same unprofitable branch of the cycle by its algorithm, which compels it to repeat two controlled-not like actions, Eq. 22.14, which jointly amount to an identity.

This vicious cycle could be interrupted only if the decision process called for extraction and reinsertion of the partition *before* undoing the measurement (and thus causing the inevitable immediate re-measurement) in the unprofitable right branch of the decision tree. Extraction of the partition before the measurement is undone increases the entropy of the gas by $k_B[\lg(L-\ell)/L]$ and destroys the correlation with the demon's memory, thus decreasing the mutual information: The molecule now occupies the whole volume of the engine. Moreover it occurs with no gain of useful work. Consequently, reversibly undoing the measurement *after* the partition is extracted is no longer possible: The location on the decision tree (extracted partition, "full" memory) implicitly demonstrates that the measurement has been carried out and that it has revealed that the molecule was in the unprofitable compartment — it can occurr only in the right hand branch of the tree.

The opening of the partition has resulted in a free expansion of the gas, which squandered away the correlation between the state of the gas and the state of the memory of the demon. Absence of the correlation eliminates the possibility of undoing the measurement. Thus, now erasure is the only remaining option. It would have to be carried out before the next measurement, and the price of $k_B T$ per bit would have to be paid [6, 7].

One additional strategy should be explored before we conclude this discussion: The demon of choice can be assumed to have a large memory tape, so that it can put off erasures and temporarily store the results of its $\mathcal{N}$ measurements. The tape would then contain $\sim \mathcal{N} \cdot (\ell - L)/L$ 0's (which we shall take to signify an unprofitable outcome) and $\sim \mathcal{N}\ell/L$ 1's. In the limit of large $\mathcal{N}$ ($\mathcal{N}\ell/L \gg 1$) the algorithmic information content of such a "sparse" binary sequence s is given by [14–20]:

$$K(s) \simeq -\mathcal{N}[\frac{\ell}{L} \lg \frac{\ell}{L} + \frac{L-\ell}{L} \lg \frac{L-\ell}{L}] \qquad (22.18)$$

Moreover, a binary string can be, at least in principle, compressed to its minimal record (s^* such that $K(s) = |s^*|$) by a reversible computation [12]. Hence, it is possible to erase the record of the measurements carried out by the demon at a cost of no less than

$$< \Delta W^- > = k_B T[K(s)/\mathcal{N}] . \qquad (22.19)$$

Thus, if the erasure is delayed so that the demon can attempt to minimize its cost before carrying it out, it can at best break even: The $-k_B T$ in Eq. 22.12 is substituted by the $< \Delta W^- >$, Eq. 22.19, which yields:

$$< \Delta W > = < \Delta W^+ > + < \Delta W^- > = 0. \qquad (22.20)$$

It is straightforward to generalize this lesson derived on the example of Gabor's engine to other situations. The essential ingredient is the "noncommutativity" of the two operations: "undo the measurement" can be reversibly carried out only before "extract the partition." The actions of the demon are, by the assumption of the Church–Turing thesis, completely determined by its internal state, especially its memory content. Demons are forced to make useless re-measurements. Santayana's famous saying that "those who forget their history are doomed to relive it" applies to demons with a vengance! For, when the demon forgets the measurement outcome, it will repeat the measurement and remain stuck forever in the unprofitable cycle. One could consider more complicated algorithms, with additional bits and instructions on when to measure, and so on. The point is, however, that all such strategies must ultimately contain explicit or implicit information about the branch on which the demon has found itself as a result of the measurement. Erasure of this information carries a price which is on the average no less than the "illicit" gains which would violate the second law.

22.5   Conclusions

The aim of this paper was to exorcise the demon of choice — a selective version of Maxwell's demon which attempted to capitalize on large thermal fluctuations by reversibly undoing all of the measurements which did not reveal the system to be sufficiently far from equilibrium. I have demonstrated that a deterministic version of such a demon fails, as no decision procedure is capable of both (i) reversibly undoing the measurement, and, also, of (ii) opening the partitions inserted prior to the measurement to allow for energy extraction following readoff of the outcome.

Our discussion was phrased — save for an occassional reference to density matrices, Hilbert spaces, etc. — in a noncommital language, and it is indeed equally applicable in the classical and quantum contexts. As was pointed out already some time ago [9, 10], the only difference arises in the course of measurements. Quantum measurements are typically accompanied by a "reduction of the state vector". It occurs whenever an observer measures observables that are not co-diagonal with the density matrix of the system. It is a (near) instantaneous process [34], which is nowadays understood as a consequence of decoherence and einselection [19, 28, 30–34]. The implications of this difference are minor from the viewpoint of the threat to the second law posed by the demons (although decoherence is paramount for the discussion of the interpretation of quantum theory). It was noted already some time ago that decoherence (or, more generally, the increase of entropy associated with the reduction of the state vector) is not necessary to save the second law [9]. Soon after the algorithmic information content entered the discussion of demons [12, 21] it was also realised that the additional cost decoherence represents can be conveniently quantified using the "deficit" in what this author knew then as the 'Groenewold–Lindblad inequality' [35, 36], and what is now more often (and equally justifiably) called the 'Holevo quantity [37];

$$\chi = H(\rho) - \sum_i p_i H(\rho_S^{(i)}) \, , \tag{22.21}$$

which is a measure of the entropy increase due to the "reduction". The two proofs [36, 37] involving essentially the same quantity have appeared almost simultaneously, independently, and were motivated by — at least superficially — quite different considerations.

We shall not repeat these discussions here in detail. There are however several independently sufficient reasons not to worry about decoherence in the demonic context which deserve a brief review. To begin with, decoherence cannot help the demon as it only adds to the "cost of doing business". And the second law is apparently safe even without decoherence [9]. Moreover, especially in the context of Szilard's or Gabor's engines, decoherence is unlikely to hurt the demon either, since the obvious projection operators to use in Eq. 22.14 correspond to the particle being on the left (right) of the partition, and are likely to diagonalise the density matrix of the system in contact with a typical environment [9] (heat bath). (Superpositions of states corresponding to such obvious measurement outcomes are very Schrödinger cat-like, and, therefore, unstable on the decoherence timescale [34].) Last not least, even if demon for some odd reason started by measuring some observable which does not commute with the density matrix of the system decohering in contact with the heat bath environment, it should be able to figure out what's wrong and learn after a while what to measure to minimise the cost of erasure (demons are supposed to be intelligent, after all!).

So decoherence is of secondary importance in assuring validity of the second law in the setting involving engines and demons: Entropy cannot decrease already without it! But decoherence can (and often will) add to the *measurement* costs, and the cost of decoherence is paid "up front", during the measurement (and not really during the erasure, although there may be an ambiguity there — see a quantum calculation of erasure-like process of the consequences of decoherence in Ref. [38]). However, in the context of dynamics decoherence is the ultimate cause of entropy production, and, thus, the cause of the algorithmic arrow of time [33]. Moreover, there are intriguing quantum implication of the interplay of decoherence and (algorithmic) information that follow: Discussions of the interpretational issues of quantum theory are often conducted in a way which implicitly separates the information observers have about the state of the systems in the "rest of the Universe" from their own physical state — their identity. Yet, as the above analysis of the observer-like demons demonstrates, there can be *no information without representation*. The observer's state (or, for that matter, the state of its memory) determines its actions and should be regarded as an ultimate description of its identity. So, to end with one more "deep truth" *existence* (of the observers state, and, especially, of the state of its memory) *precedes the essence* (observer's information, and, hence their future actions).

Acknowledgements

I have benefited from discussion on this subject with many, including Andreas Albrecht, Charles Bennett, Carlton Caves, Murray Gell-Mann, Chris Jarzynski, Rolf Landauer, Seth Lloyd, Michael Nielsen, Bill Unruh, and John Wheeler, who, in addition to stimulating the initial interest in matters concerning physics and information, insisted on my monthly dialogues with Feynman. This has led to one more "adventure with a curious character": In the Spring of 1984 I participated in the "Quantum Noise" program at the Institute for Theoretical Physics, UC Santa Barbara. It was to end with a one-week conference on various relevant quantum topics. One of the organisers (I think it was Tony Leggett), aware of my monthly escapades at Caltech, and of Feynman's (and mine) interests in quantum computation asked me whether I could ask him to speak. I did, and Feynman immediately agreed.

The lectures were held in a large conference room at the campus of the University of California at Santa Barbara. For the "regular speakers" and for most of the talks (such as my discussion of the decoherence timescale which was eventually published as Ref. [34]) the room was filled to perhaps a third of the capacity. However, when I walked in in the middle of the afternoon coffee break, well in advance of Feynman's talk, the room was already nearly full, and the air was thick with anticipation. A moment after I sat down in one of the few empty seats, I saw Feynman come in, and quietly take a seat somewhere in the midst of the audience. More people came in, including the organisers and the session chairman. The scheduled time of his talk came... and went. It was five minutes after. Ten minutes. Quarter of an hour. The chairman was nervous. I did not understand what was going on — I clearly saw Feynman's long grey hair and an occasional flash of an impish smile a few rows ahead.

Then it struck me: He was just being "a curious character", curious about what will happen... He did what he had promised — showed up for his talk on (or even before) time, and now he was going to see how the events unfold.

In the end I did the responsible thing: After a few more minutes I pointed out the speaker to the session chairman (who was greatly relieved, and who immediately and reverently led him to the speaker's podium). The talk (with the content, more or less, of Ref. [39]) started only moderately behind the schedule. And I was immediately sorry that I did not play along a while longer — I felt as if I had given away a high-school prank before it was fully consummated!

References

[1] R. P. Feynman, R. B. Leighton, and M. Sands, *The Feynman Lectures on Physics*, vol. 1, pp 46.1 – 46.9 (Addison-Wesley, Reading, Massachussets, 1963).

[2] J. C. Maxwell, *Theory of Heat*, 4th ed., pp. 328-329 (Longman's, Green, & Co., London 1985).

[3] H. S. Leff and A. F. Rex, *Maxwell's Demon: Entropy, Information, Computing* (Princeton University Press, Princeton, 1990).

[4] M. Smoluchowski in *Vortgäge über die Kinetische Theorie der Materie und der Elektizität* (Teubner, Leipzig 1914).

[5] P. Skordos and W. H. Zurek, "Maxwell's Demons, Rectifiers, and the Second Law" *Am. J. Phys.* **60**, 876 (1992).

[6] L. Szilard, *Z. Phys.* **53** 840 (1929). English translation in Behav. Sci. **9**, 301 (1964), reprinted in *Quantum Theory and Measurement*, edited by J. A. Wheeler and W. H. Zurek (Princeton University Press, Princeton, 1983); Reprinted in Ref. 3.

[7] R. Landauer, *IBM J. Res. Dev.* **3**, 183 (1961); Reprinted in Ref. 3.

[8] C. H. Bennett, *IBM J. Res. Dev.* **17** 525 (1973); C. H. Bennett, *Int. J. Theor. Phys.* **21**, 905 (1982); C. H. Bennett, *IBM J. Res. Dev.*, **32**, 16-23 (1988); Reprinted in Ref. 3.

[9] W. H. Zurek, "Maxwell's Demons, Szilard's Engine's, and Quantum Measurements", Los Alamos Preprint LAUR 84-2751 (1984); pp. 151-161 in *Frontiers of Nonequilibrium Statistical Physics*, G. T. Moore and M. O. Scully, eds., (Plenum Press, New York, 1986); reprinted in Ref. 3.

[10] For a quantum treatement which employs the Groenewold-Lindblad/Holevo inequality and uses the "deficit" χ in that inequality to estimate of the price of decoherence, see W. H. Zurek, pp 115-123 in the *Proceedings of the 3rd International Symposium on Foundations of Quantum Mechanics*, S. Kobayashi *et al.*, eds. (The Physical Society of Japan, Tokyo, 1990).

[11] S. Lloyd, *Phys. Rev.* **A56**, 3374-3382 (1997).

[12] W. H. Zurek, *Phys. Rev.* **A40**, 4731-4751 (1989); W. H. Zurek, *Nature* **347**, 119-124 (1989).

[13] For an accessible discussion of Church–Turing thesis, see D. R. Hofstadter, Gödel, Escher, Bach, chapter XVII (Vintage Books, New York, 1980).

[14] R. J. Solomonoff, *Inf. Control* **7**, 1 (1964).

[15] A. N. Kolmogorov, *Inf. Transmission* **1**, 3 (1965).

[16] G. J. Chaitin, *J. Assoc. Comput. Mach.* **13**, 547 (1966).

[17] A. N. Kolmogorov, *IEEE Trans. Inf. Theory* **14**, 662 (1968).

[18] G. J. Chaitin, *J. Assoc. Comput. Mach.* **22**, 329 (1975); G. J. Chaitin, *Sci. Am.* **23**(5), 47 (1975).

[19] A. K. Zvonkin and L. A.Levin, *Usp. Mat. Nauk.* **25**, 602 (1970).

[20] A. K. Zvonkin and L. A. Levin, *Usp. Mat. Nauk.* **25**, 602 (1970).

[21] C. M. Caves, "Entropy and Information", pp. 91-116 in *Complexity, Entropy, and Physics of Information*, W. H. Zurek, ed. (Addison-Wesley, Redwood City, CA, 1990).

[22] C. M. Caves, *Phys. Rev. Lett.* **64**, 2111-2114 (1990).

[23] W. Shannon and W. Weaver, *The Mathematical Theory of Communication* (University of Illinois Press, Urbana, 1949).

[24] C. M. Caves, W. G. Unruh, and W. H. Zurek, *Phys. Rev. Lett.*, **65**, 1387 (1990).

[25] D. Gabor, *Optics* **1**, 111-153 (1964).

[26] L. Brillouin, *Science and Information Theory*, 2nd ed. (Academic, London, 1962).

[27] C. H. Bennett, *Sci. Am.* **255** (11), 108 (1987).

[28] W. H. Zurek, *Phys. Rev.* **D24**, 1516 (1981); *ibid.* **D26**, 1862 (1982); *Physics Today* **44**, 36 (1991).

[29] W. H. Zurek, "Information Transfer in Quantum Measurements: Irreversibility and Amplification"; pp. 87-116 in *Quantum Optics, Experimental Gravitation, and Measurement Theory*, P. Meystre and M. O. Scully, eds. (Plenum, New York, 1983).

[30] E. Joos and H. D. Zeh, *Zeits. Phys.* **B59**, 223 (1985).

[31] D. Giulini, E. Joos, C. Kiefer, J. Kupsch and H. D. Zeh, *Decoherence and the Appearance of a Classical World in Quantum Theory* (Springer, Berlin, 1996).

[32] W. H. Zurek, *Progr. Theor. Phys.* **89**, 281-312 (1993).

[33] W. H. Zurek, in the *Proceedings of the Nobel Symposium 101 'Modern Studies in Basis Quantum Concepts and Phenomena'*, to appear in *Physica Scripta*, in press quant-ph/9802054.

[34] W. H. Zurek, "Reduction of the Wavepacket: How Long Does it Take?" Los Alamos preprint LAUR 84-2750 (1984); pp. 145-149 in the *Frontiers of Nonequilibrium Statistical Physics: Proceedings of a NATO ASI held June 3-16 in Santa Fe, New Mexico*, G. T. Moore and M. O. Scully, eds. (Plenum, New York, 1986).

[35] H. J. Groenewold, *Int. J. Theor. Phys.* **4**, 327 (1971).

[36] G. Lindblad, *Comm. Math. Phys.* **28**, 245 (1972).

[37] A. S. Holevo, *Problemy Peredachi Informatsii* **9**, 9-11 (1973).

[38] J. R. Anglin, R. Laflamme, W. H. Zurek, and J. P. Paz, *Phys. Rev.* **D52**, 2221-2231 (1995).

[39] R. P. Feynman, "Quantum Mechanical Computers", *Optics News*, reprinted in *Found. Phys.* **16**, 507-531 (1986).

Index

Name Index

Printed in the United States
by Baker & Taylor Publisher Services